Multimodal Surveillance

Sensors, Algorithms, and Systems

For a list of related Artech House titles,
please see the back of this book.

Multimodal Surveillance

Sensors, Algorithms, and Systems

Zhigang Zhu
Thomas S. Huang
Editors

ARTECH
HOUSE

BOSTON | LONDON
artechhouse.com

Library of Congress Cataloging-in-Publication Data
A catalog record for this book is available from the U.S. Library of Congress.

British Library Cataloguing in Publication Data
A catalogue record for this book is available from the British Library.

ISBN-13: 978-1-59693-184-8
ISBN-10: 1-59693-184-1

Cover design by Yekaterina Ratner

© 2007 ARTECH HOUSE, INC.
685 Canton Street
Norwood, MA 02062

10 9 8 7 6 5 4 3 2 1

Contents

CHAPTER 17

SATware: Middleware for Sentient Spaces   387

Foreword

Surveillance is an integral part of human existence—we are naturally endowed with the aptitude to observe, process, and extract information from our surroundings in order to respond to the environment in an appropriate manner. Teaching machines to do the same things is, however, now recognized to be a challenging task that engages several different scientific fields, including machine vision, pattern recognition, sensor networks, distributed computing, information fusion, and signal processing. Rapid advances in sensor design, processor technology, digital archiving, and algorithm development have now made it possible for us to design and install automatic surveillance systems for increased security requirements at critical locations such as airports and street corners. This means questions like: Is this environment hazardous to humans?, What is the identity of the individual at the airline counter?, Is this subway station too crowded?, Has there been a traffic violation or accident?, and Are the people on the street corner behaving suspiciously? may be easily posed to machines equipped with video, audio, or thermal sensors for perceiving their environment. Indeed, the field of automated surveillance has experienced tremendous growth over the last decade due to increased attention on national security and the deployment of surveillance cameras in public spaces. In most situations, the images captured by these cameras are currently monitored by human operators. The goal is to replace the human operators so that objective, consistent, and real-time decisions can be made about events and people.

Multimodal surveillance combines multiple sources of information presented by different sensors in order to generate a more accurate and robust interpretation of the environment. Consolidating information from diverse sources in an effective manner is a challenging task and has broad applications in several different fields besides surveillance. This book introduces the reader to (1) novel sensing mechanisms employed by machines to discern their environment, (2) state-of-the-art algorithms to process and fuse the sensed information in order to interpret the environment; and (3) applications where multimodal surveillance have tremendous impact. Given the importance of this topic, this book will be extremely beneficial to readers who are interested in carrying out research in automated surveillance or who are keen on understanding the progress made in this important field. The editors of this book are distinguished researchers who have made significant contributions in surveillance. They are to be commended for putting together this edited volume that contains chapters written by leading experts on a number of important topics in surveillance.

Anil K. Jain
Michigan State University
July 2007

Preface

Video surveillance has attracted a lot of attention through public media, professional workshops and conferences, and government-funded programs, such as the DARPA Video Surveillance and Monitoring (VASM) program (1995 to 1999). However, although video surveillance is probably still the most popular form of surveillance today, there are many forms of observation or monitoring. In fact, the word *surveillance* has historically been used to describe observation from a distance by various electronic or other technological means, for example, telephone tapping, directional microphones, communications interception, covert listening devices or bugs, subminiature cameras, closed-circuit televisions, GPS tracking, electronic tagging, and reconnaissance aircraft. Therefore, surveillance in nature should include multiple modalities.

In recent years, we witness a rapid growth of research and development in surveillance (including biometrics), using multimodal sensors including video, audio, thermal, vibration, and various other kinds of sensors, in both civilian and military applications. Multimodal sensor fusion, in general, or multimodal biometrics, in particular, has been covered by quite a few other books in recent years. However, this book is the first that covers the current state of the art in multimodal surveillance and tries to address various aspects of multimode surveillance, including sensors, algorithms, and systems, with biometrics as one of its key components.

This edited book is a collection of sample works contributed by leading experts in the emerging field of multimodal surveillance. The main criterion for including a chapter in the book is that it addressed some important issues of multimodal surveillance. We also considered the distribution of the works from different parts of the world and the balance of the contributors from academia, government, and industry. The chapters are based on recent research projects funded by both government and industry, including AFSOR, AFRL, ARL, ARO, DARPA, MURI, NIJ, NSF, and ONR in the United States; the Natural Science Foundation in China (NSFC); BSIN (Dutch) and Royal Commission (UK) in Europe; and some leading research laboratories and companies in the field. A list of the contributors and their affiliations is included at the end of the book.

Here is the organization of the book. After an introductory chapter that offers historical notes on multimodal surveillance and provides an overview of the book, we group the works into three equally important parts: Part I, Multimodal Sensors and Sensing Approaches; Part II, Multimodal Fusion Algorithms; and Part III, Multimodal Systems and Issues. Part I consists of four chapters (Chapters 2–5) on multimodal sensors and sensing approaches. The sensors discussed are either multimodal sensors in nature (Chapters 2 and 3) or are novel sensors that are components of multimodal surveillance systems (Chapters 4 and 5). Some of them are high-end sensors for high fidelity and/or long-range surveillance, whereas others target low-cost solutions. In Part II, we have five chapters (Chapters 6–10) on the multimodal integration of various sensing modalities, using various algorithms. The first two chapters (Chapters 6 and 7) in this part discuss multimodal audiovisual integration for two different tasks—automatic speech recognition and human detection and tracking. Chapters 8, 9, and 10 discuss various approaches of multimodal biometrics, by combining face recognition with three other types of sources: ear, palm, and gait, respectively. Part III contains seven chapters (Chapters 11–17) on various system issues in building and/or supporting multimodal surveillance systems. Those issues include: multimodal sentient computing, information representations, system architectures/frameworks/workbenches, usability issues, real-time performance, 24/7 operations, automatic environment modeling, system evaluations, and distributed infrastructure (middleware). A brief summary of each chapter can be found in Chapter 1.

We realize that the collection in this book of samples consists merely of some representative works in the field and is therefore not complete. However, we hope this collection will stimulate more interest in the research and development of mutlimodal surveillance techniques, including the further development of mutlimodal sensors, multimodal data fusion algorithms, and mutlimodal surveillance systems. With this in mind, the targeted audiences of this book include: researchers in computer vision, multimodal/multimedia data processing, sensor fusion, sensor networks, biometrics and surveillance; professors and graduate students using the book for upper-level graduate courses, such as seminars; and government and industrial personnel (officers, project managers, project directors, and system developers) for applications in multimodal sensors, surveillance, and data integration.

We want to thank everyone who helped in the preparation of this book and related events. First we would like to thank Wayne Yuhasz, executive acquisitions editor at Artech House, who initiated the book project. His enthusiasm and tireless support carried the book project from the starting point to the finish line. We would like to express our deep appreciation to the anonymous reviewers of both the book proposal and the first draft of the book chapters. Their positive responses and constructive suggestions guided us in forming the final shape and contents of the book. We are grateful to developmental editor Barbara Lovenvirth, whose strict formatting guidelines seemed impossible at the beginning to such a diverse group of authors, but they turned out to be a guarantee of the success of the book project. We also wish to acknowledge Kevin Danahy, sales and marketing manager, and Igor Valdman, art director, at Artech House Publishers.

Above all, we would like to thank all the authors of the 16 main chapters in the three parts. They are the ones who have done the real work and have proudly presented to our readers the fruits of many years of their hard work. We treasure the experience of interacting with so many great thinkers and practitioners in multimodal surveillance. We have enjoyed reading every chapter, and we trust you will too.

Zhigang Zhu
CUNY City College and Graduate Center
Thomas S. Huang
University of Illinois at Urbana-Champaign
Editors
July 2007

CHAPTER 1

Multimodal Surveillance: An Introduction

Zhigang Zhu and Thomas S. Huang

1.1 Multimodal Surveillance: A Brief History

In *Merriam-Webster OnLine*, the word *surveillance* literally means "watching over" (in French) [1]. In *Wikipedia*, an online free encyclopedia, surveillance is defined as "the process of monitoring the behavior of people, objects or processes within systems for conformity to expected or desired norms in trusted systems for security or social control" [2].

Surveillance began as the art of watching over the activities of persons or groups from a position of higher authority. Surveillance has been an intrinsic part of human history. Sun Tzu's *The Art of War*, written in Chinese about 2,500 years ago, discusses in one whole chapter (among 13 chapters) how spies should be used against enemies [3]:

> Thus, what enables the wise sovereign and the good general to strike and conquer, and achieve things beyond the reach of ordinary men, is *foreknowledge*. Now this foreknowledge cannot be elicited from spirits; it cannot be obtained inductively from experience, nor by any deductive calculation. Knowledge of the enemy's dispositions *can only be obtained from other men*. [*The Art of War*, Chapter XIII: The Use of Spies; italics for emphasis]

This tells us that "foreknowledge" cannot be obtained from reasoning, experience, or calculation; it can only be obtained from other men (i.e., spies or human sensors). Then, five different classes of other men were discussed in great detail. In fact, in the Old Testament in Numbers, one of the five law books, twelve men, each from one of the twelve Israeli ancestral tribes, were sent out to explore the land of Canaan before Israelites went into the promised land (Numbers 13:17–20, NIV):

> When Moses sent them to explore Canaan, he said, "Go up through the Negev and on into the hill country. See what the *land* is like and whether the *people* who live there are strong or weak, few or many. What *kind of land* do they live in? Is it good or bad? What *kind of towns* do they live in? Are they unwalled or fortified? How is the soil? Is it fertile or poor? Are there *trees* on it or not? Do your best to bring back some of the *fruit* of the land." [italics for emphasis; it was the season for the first ripe grapes]

Interestingly, this gives us a picture of multisensory and multimodal explorations for an important surveillance mission. It was multisensory ("twelve men"). It was multimodal (not only exploring "people" and their attributes, but also "land," facilities, soil, vegetation, and so forth). It was distributed ("go up through" and "on into"). Finally, Moses also needed to resolve differences or even conflicts in the information gathered (e.g., "we worked like grasshoppers to them" versus "we will swallow them up" in the paragraphs following the verses quoted above).

Of course, it is modern electronic and computer technologies that have pushed surveillance toward the science end of watching over. Closed-circuit TV (CCTV)—where the picture is viewed or recorded, but not broadcast—initially developed as a means of security for banks. Today it has developed to the point where it is simple and inexpensive enough to be used in home security systems, and for everyday surveillance. However, although video surveillance is the most popular form of surveillance, there are all forms of observation or monitoring, not just visual observation. In fact, the word *surveillance* has historically been used to describe observation from a distance by means of electronic equipment or other technological means, for example: eavesdropping, telephone tapping, directional microphones, communications interception, covert listening devices (or bugs), subminiature cameras, closed-circuit television, GPS tracking, electronic tagging, reconnaissance aircraft, and so forth. Therefore, surveillance in nature should include multiple modalities.

The greatest impact of computer-enabled surveillance is due to the involvement of a large number of organizations in surveillance operations. State and security services still have the most powerful surveillance systems, because they are enabled under the law. But today's levels of state surveillance have increased in the era of antiterrorism, and by using computers, they are now able to draw together many different information sources with multimodal sensors to produce profiles of persons or groups in society. Many large corporations now use various form of passive surveillance. This is primarily a means of monitoring the activities of staff and for controlling public relations. But some large corporations, such as supermarkets and financial firms, actively use various forms of surveillance to monitor the activities of customers who may impact their operations. In the military, intelligence, surveillance, target acquisition, and reconnaissance (ISTAR) [4] links several battlefield functions together to assist a combat force in employing its sensors and managing the information they gather. Information is collected on the battlefield through systematic observation by deployed soldiers and a variety of electronic sensors. Surveillance, target acquisition, and reconnaissance are methods of obtaining this information. The information is then passed to intelligence personnel for analysis, and then to the commander and the commander's staff for the formulation of battle plans.

In recent years, we have witnessed a rapid growth in research and development of surveillance and biometrics using multimodal sensors, including video, audio, thermal, vibration, and various kinds of other sensors, in both civilian and military applications. This edited book is a collection of sample works contributed by leading researchers in the emerging field of multimodal surveillance. We roughly group the works into three equally important parts:

1. Multimodal sensors and novel sensing approaches;
2. Multimodal data fusion algorithms;
3. Multimodal systems and related issues.

A multimodal surveillance system usually consists of all the three components. Sensors are important since they are the front end of a surveillance system. Algorithms are important since they are the core that enables the system to work. Other system-level issues are also important for building a real multimodal system. In this book we also include works that discuss various tasks of surveillance applications. Most of the chapters address *people and activity issues* in tasks such as human and vehicle tracking, human identification via biometrics, speech detection and recognition, and so forth. A few chapters also deal with *environment modeling and monitoring issues* for surveillance—mainly in Chapter 15 on multimodal 3-D scene modeling, some in Chapter 4 on target selections for remote hearing, and in Chapter 11 and Chapter 17 on sentient computing and sentient spaces.

The criterion for including a chapter in the book was that it addressed some aspects of multimodal surveillance. We also considered the distribution of the works from different parts of the world and the balance of the contributors from academia, government, and industry. In the following, we will give a brief summary of each chapter, in the context of each of the three parts.

1.2   Part I: Multimodal Sensors and Sensing Approaches

In Part I, we include four chapters (Chapters 2–5) on multimodal sensors and sensing approaches. The sensors discussed are either multimodal sensors in nature (e.g., in Chapters 2 and 3) or novel sensors that are components of multimodal surveillance systems (e.g., in Chapters 4 and 5). Some of them are high-end sensors for high fidelity or long-range surveillance, whereas others target low-cost solutions.

1.2.1   The ARL Multimodal Sensor (Chapter 2)

In Chapter 2, Houser and Zong describe the ARL multimodal sensor, which is a research tool for target signature collection, algorithm validation, and emplacement studies at the U.S. Army Research Laboratory (ARL). ARL has a significant program involving the development of unattended ground sensors (UGS) that address a variety of military and government missions. ARL's program involves practically every aspect of sensor development including devices, detection and fusion algorithms, communications, and command and control. One element of the ARL UGS program involves the development of low-cost sensing techniques for the urban environment, and one embodiment of this effort is the multimodal sensor (MMS). The program objectives of this effort were to develop a *networked personnel detection sensor* with the following major criteria: low cost in volume production, support MOUT (military operations in urban terrain) operations, and employ nonimaging sensor diversity techniques. The MMS was an early prototype intended to demonstrate that low-cost sensing techniques were suitable for the urban environment and were a viable alternative to higher cost and fidelity sensors for some applications. The MMS is used today as a demonstration system and a test bed for many facets of urban sensing. This chapter will describe many aspects of the MMS design, including hardware, software, and communications. The detection algorithms will also be described, including collection of target signatures and validation of algorithm performance. Finally, MMS usage in a *force protection* application will be described, including issues encountered when integrating into a larger system.

1.2.2 Design and Deployment of Visible-Thermal Biometric Surveillance Systems (Chapter 3)

In Chapter 3, Socolinsky discusses the smart deployment of visible-thermal biometric surveillance systems at Equinox Corporation. The combined use of visible and thermal infrared imagery has proven valuable for a wide range of surveillance tasks, including facial recognition. However, the added value must be weighed against the increased cost and complexity, as well as the unique challenges that arise from the use of a bimodal system. The primary question is when and why we should specify a fused imaging solution. Whenever such a solution is appropriate, we must strive to provide the simplest system meeting the required performance. This will ensure the most reliable operation, shortest time to deployment, and lowest operational cost. In this chapter, the author explores some of the issues surrounding design and deployment of fused visible-thermal systems for *biometric surveillance*, including the appropriate choice of sensor wavelengths for given scenarios, optical designs best suited to particular tasks, and cost/performance tradeoffs. The problems of cross-modality matching and availability of legacy data is considered in the context of *face recognition*. While human facial recognition is used as the focus of discussion, much of the chapter is applicable to other forms of surveillance.

1.2.3 LDV Sensing and Processing for Remote Hearing in a Multimodal Surveillance System (Chapter 4)

In Chapter 4, Zhu and his research team at the City College of New York discuss a novel laser Doppler vibrometry sensor for *remote hearing* in a multimodal surveillance system for human signature detection. The system consists of three types of sensors: an infrared (IR) camera, a pan/tilt/zoom (PTZ) electro-optical (EO) color camera, and the laser Doppler vibrometer (LDV). In particular, the LDV is explored as a new noncontact remote voice detector. In their research, they found that voice energy vibrates most objects (targets) in the surrounding environment of the speaker, and the vibrations can be detected by an LDV at a large distance. Since signals captured by the LDV are very noisy, algorithms have been designed with Gaussian bandpass filtering, Wiener filtering, and adaptive volume scaling to enhance the LDV voice signals. The enhanced voice signals are intelligible from LDV signal returns via targets without retro-reflective finishes at short or medium distances (less than 100 meters). By using retro-reflective tapes, the distance could be as far as 300 meters. However, the authors point out that manual operations to search and focus the laser beam on a target (with both proper vibration and reflection) are very difficult at medium and large distances. Therefore, IR/EO imaging for *target selection and localization* is also discussed. Future work remains in LDV sensor improvement, automatic LDV targeting, and intelligent refocusing for long-range LDV listening.

1.2.4   Sensor and Data Systems, Audio-Assisted Cameras, and Acoustic Doppler Sensors (Chapter 5)

In Chapter 5, novel low-cost sensor and data systems are presented by Smaragdis, Raj, and Kalgaonkar at Mitsubishi Electric Research Labs in Cambridge, Massachusetts. Specifically, they describe two technologies: *audio-assisted cameras* and acoustic Doppler sensors for *gait recognition*. They make various arguments for why audio analysis can be a useful complement to video cameras and show how Doppler sensing can be applied for gait recognition at a very low cost. Although the two subjects are treated separately, they are by no means meant to be kept as such. Future plans involve the convergence of these two projects that will allow the enhancement of audio and camera systems by Doppler sensing and help improve Doppler estimates by multimodal inputs.

1.3   Part II: Multimodal Fusion Algorithms

In Part II, we include five chapters (Chapters 6–10) on the multimodal integration of various sensing modalities, using various algorithms. The first two chapters (Chapters 6 and 7) in this part discuss multimodal audiovisual integration for two different tasks—automatic speech recognition and human detection and tracking. Chapters 8, 9, and 10 discuss the various approaches of multimodal biometrics by combining face recognition with three other different types of sources: ear, palm, and gait, respectively.

1.3.1   Audiovisual Speech Recognition (Chapter 6)

In Chapter 6, Chu, at the IBM T.J. Watson Research Center, and Huang, at the University of Illinoisal at Urbana-Champaign, present a novel method for automatic audiovisual speech recognition. This chapter considers the fundamental problem of multimodal fusion in the context of pattern recognition tasks in surveillance and human-computer interactions. The authors propose a novel sensory fusion method based on the *coupled hidden Markov models* (CHMM) for audiovisual speech modeling. The CHMM framework allows the fusion of two temporally coupled information sources to take place as an integral part of the statistical modeling process. An important advantage of the CHMM-based fusion method lies in its ability to model asynchronies between the audio and visual channels. They describe two approaches to carrying out inference and learning in CHMM. The first is an exact algorithm derived by extending the forward-backward procedure used in the HMM inference. The second method relies on the model transformation strategy that maps the state space of a CHMM onto the state space of a classic HMM and therefore facilitates the development of sophisticated audiovisual speech recognition systems using existing infrastructures. Audiovisual speech recognition experiments on two different corpora validate the proposed fusion approach. In both evaluations, it is clearly demonstrated that the CHMM-based intermediate integration scheme can utilize the information in the visual channel more effectively than the early integration methods in noisy conditions. The audiovisual systems can achieve higher recognition accuracy than audio-only systems, even when matched acoustic conditions are ensured.

1.3.2 Multimodal Tracking for Smart Videoconferencing and Video Surveillance (Chapter 7)

Many applications such as interactive multimedia, videoconferencing, and surveillance require the ability to track the 3-D motion of the subjects. *Particle filters* represent an attractive solution for the tracking problem. They do not require solution of the inverse problem of obtaining the state from the measurements, and the tracking can naturally integrate multiple modalities. In Chapter 7, Zotkin, Raykar, Duraiswami, and Davis at the University of Maryland present a framework for multimodal tracking using multiple cameras and multiple microphone arrays. In order to calibrate the resulting distributed multisensor system, the authors propose a method to automatically determine the 3-D positions of all microphones in the system using at least five loudspeakers. The method does not require knowledge of the loudspeaker positions but assumes that for each loudspeaker, there exists a microphone very close to it. They derive the *maximum likelihood estimator*, which reduces to the solution of the nonlinear least squares problem. A closed-form approximate solution that can be used as an initial guess is derived. They also derive an approximate expression for the estimator covariance using the implicit function theorem and Taylor series expansion. Using the estimator covariance matrix, the authors analyze the performance of the estimator with respect to the positions of the loudspeakers; in particular, they show that the loudspeakers should be far away from each other, and the microphones should lie within the convex hull formed by the loudspeakers. They verify the correctness and robustness of the multimodal tracker and of the self-calibration algorithm both with Monte-Carlo simulations and on real data from three experimental setups. The authors also present practical details of system implementation.

1.3.3 Multimodal Biometrics Involving the Human Ear (Chapter 8)

Multimodal biometrics is one of the important functionalities in a multimodal surveillance system. In Chapter 8, a multimodal biometrics system involving the human ear is presented by Middendorff and Bowyer at the University of Notre Dame. The authors note that recent years have seen substantial and increasing interest in exploring the use of the human ear as a source for biometrics. In any biometric scenario where one might acquire images of the face, images of the ear could also potentially be acquired. Interestingly, reports in the literature suggest that, using the same *PCA-style recognition algorithm* and face and ear images of similar quality from the same subjects, the recognition rate achieved using the ear as a biometric is similar to that achieved using the face. The combination of the ear and the face then has the potential to achieve higher accuracy. Also, a biometric system that could analyze both ear and face could still produce a result if either one is not able to be imaged for some reason. Therefore, there are both accuracy and coverage advantages to be gained in a multimodal ear + face biometric system.

Researchers have also explored using both 3-D and 2-D data for ear biometrics. Results of several studies suggest that a 3-D shape of the ear allows greater recognition accuracy than a 2-D shape. This is likely due to pose variation between the gallery and probe images being more readily handled in the case of 3-D shape matching. However, 3-D sensors are currently both more expensive than normal cameras and also more restrictive in terms of conditions for image acquisition. For this reason, the authors explore the degradation of 2-D *"eigen-ear"* performance due to pose change between gallery and probe and consider approaches to dealing with this problem.

1.3.4   Fusion of Face and Palmprint for Personal Identification Based on Ordinal Features (Chapter 9)

In Chapter 9, researchers in the Center for Biometrics and Security Research, Chinese Academy of Sciences, present a new multimodal biometric identification system by fusing face and palmprint features to improve identification performance. Effective classifiers based on the so-called *ordinal* features are first constructed from images of faces and palmprints, respectively. Then several strategies, including *sum, product, max and min rules* and *linear discriminant analysis* (LDA), are employed for fusing them on a dataset that consists of the face and palmprint data of 378 subjects, with 20 pairs for each. Experimental results have shown the effectiveness of the proposed system.

1.3.5   Human Identification Using Gait and Face (Chapter 10)

Recognition of humans from arbitrary viewpoints is an important requirement for applications such as intelligent environments, surveillance, and access control. For optimal performance, the system must use as many cues as possible from appropriate vantage points and fuse them in meaningful ways. In Chapter 10, Chellappa and his research team at the University of Maryland discuss issues of human identification by the fusion of face and gait cues from a monocular video. The authors use a view invariant gait recognition algorithm for gait recognition and a face recognition algorithm using particle filters. They employ decision fusion to combine the results of gait and face recognition algorithms and consider two fusion scenarios: *hierarchical* and *holistic*. The first employs the gait recognition algorithm when the person is far away from the camera and passes the top few candidates to the face recognition algorithm. The second approach involves combining the similarity scores obtained individually from the face and gait recognition algorithms. Simple rules like the SUM, MIN, and PRODUCT are used for combining the scores. The results of fusion experiments are demonstrated on the NIST database, which has outdoor gait and face data of 30 subjects.

1.4   Part III: Multimodal Systems and Issues

In Part III, we include seven chapters (Chapters 11–17) on various system issues in building and/or supporting mutlimodal surveillance systems. Those issues include: multimodal sentient computing, information representations, system architectures/frameworks/workbenches, usability issues, real-time performance, 24/7 operations, automatic environment modeling, system evaluations, and distributed infrastructure (middleware).

1.4.1 Sensor Fusion and Environmental Modeling for Multimodal Sentient Computing (Chapter 11)

In Chapter 11, Town in the University of Cambridge Computer Laboratory presents an approach to multisensory and multimodal fusion and environment modeling for *sentient computing*. Visual information obtained from calibrated cameras is integrated with a large-scale indoor surveillance system known as SPIRIT. The SPIRIT system employs an ultrasonic location infrastructure to track people and devices in an office building and model their state in order to provide a sentient computing environment for context-aware applications. The vision techniques used include background and object appearance modeling, face detection, segmentation, and tracking modules. Integration is achieved at the system level through the metaphor of *shared perceptions*, in the sense that the different modalities are guided by and provide updates to a shared world model. This model incorporates aspects of both the static environment (e.g., positions of office walls and doors) and the dynamic environment (e.g., location and appearance of devices and people). Fusion and inference are performed by *Bayesian networks* that model the probabilistic dependencies and reliabilities of different sources of information over time. It is shown that the fusion process significantly enhances the capabilities and robustness of both sensory modalities, thus enabling the system to maintain a richer and more accurate world model.

1.4.2 An End-to-End eChronicling System for Mobile Human Surveillance (Chapter 12)

The second chapter in this part is about a multimodal personal mobile surveillance system. Rapid advances in mobile computing devices and sensor technologies are enabling the capture of unprecedented volumes of data by individuals involved in field operations in a variety of applications. As capture becomes ever more rich and pervasive, the biggest challenge lies in the developments of *information processing and representation tools* that maximize the utility of the captured multisensory data. The right tools hold the promise of converting captured data into actionable intelligence, resulting in improved memory, enhanced situational understanding, and more efficient execution of operations. These tools need to be at least as rich and diverse as the sensors used for capture, and they need to be unified within an effective *system architecture*. In Chapter 12, an initial attempt to an end-to-end e-chronicling system for mobile human surveillance is presented by Pingali and his colleagues at the IBM Thomas J. Watson Research Center.

The system combines several emerging sensor technologies, state-of-the-art analytic engines, and multidimensional navigation tools into an end-to-end electronic chronicling solution.

1.4.3   Systems Issues in Distributed Multimodal Surveillance (Chapter 13)

Surveillance systems inherently have the human tightly coupled in the system loop. In most cases, the human operator is the decision maker who will act upon (or not act upon) the feedback of the system. In the aspect of usability, video surveillance systems are different from other software applications in that they are time-critical and emphasize the accuracy of users' responses. Decisions often need to be made accurately in real time. Furthermore, what is of interest to a particular surveillance system user can vary greatly, and the security forces using the system are not, in general, advanced computer users. Therefore, the usability issue becomes crucial to a multimodal surveillance system. In Chapter 13, such systems issues in distributed multimodal surveillance are discussed by Yu (ObjectVideo, Inc.) and Boult (University of Colorado at Colorado Springs). The authors argue that viable commercial multimodal surveillance systems need to be *reliable* and *robust,* and they must be able to *work at night* (perhaps the most critical time). They must handle small and nondistinctive targets that are as far away as possible. As with other commercial applications, end users of the systems must be able to operate them properly. In this chapter, the authors focus on three significant inherent limitations of current surveillance systems: the effective accuracy at relevant distances, the ability to define and visualize the events on a large scale, and the usability of the system.

1.4.4   Multimodal Workbench for Automatic Surveillance Applications (Chapter 14)

In Chapter 14, a multimodal workbench for automatic surveillance applications is presented by Datcu, Yang, and Rothkrantz of Delft University of Technology in the Netherlands. The framework (i.e., the workbench) is designed to facilitate the communication between different processing components. The authors argue that in modern surveillance applications, satisfactory performance will not be achieved by a single algorithm, but rather by a combination of interconnected algorithms. The proposed framework specification emphasizes only the presence and role of its processing components. In other words, it considers only the description of the processing components along with their interconnections, not the implementation details. This framework is centered on the *shared memory paradigm*, which allows loosely-coupled asynchronous communication between multiple processing components, both in time and space. The shared memory in the current design of the framework takes the form of extended markup language (XML) data spaces. This suggests a more human-centric paradigm to store, retrieve, and process the data. Because the framework implementation itself relies on the philosophy of shared XML data spaces, special attention is given on how to integrate the two underlying technologies—namely, the processing components and XML data management. An example is given of a surveillance application built on top of the framework. The application uses several cameras and microphones to detect unusual behavior in train compartments. The authors note that although the framework has been designed for an automatic surveillance–oriented application, it can be adopted as a basis for any kind of complex multimodal system involving many components and heavy data exchange. The specification fully complies with the requirements of data manipulation in a multidata producer/consumer context in which the availability of data is time-dependent, and some connections might be temporarily interrupted.

1.4.5 Automatic 3-D Modeling of Cities with Multimodal Air and Ground Sensors (Chapter 15)

Three-dimensional models of urban environments are useful in a variety of applications, such as urban planning, training and simulation for disaster scenarios, virtual heritage conservation, and urban surveillance. For surveillance applications, the models provide a lot of useful information about background, context, and occlusions to facilitate moving target detection and tracking. In Chapter 15, Zakhor and Frueh at the University of California at Berkeley present a fast approach to the automated generation of textured 3-D city models with both high details at ground level and complete coverage for a bird's-eye view, using multimodal air and ground-sensing approaches. A close-range facade model is acquired at the ground level by driving a vehicle equipped with *laser scanners* and a *digital camera* under normal traffic conditions on public roads; a far-range digital surface map (DSM) containing complementary roof and terrain shape is created from *airborne laser scans*, then triangulated, and finally texture-mapped with *aerial imagery*. The facade models are registered with respect to the DSM by using Monte Carlo localization and then merged with the DSM by removing redundant parts and filling gaps. The developed algorithms are evaluated on a dataset acquired in downtown Berkeley and elsewhere.

1.4.6 Multimodal Biometrics Systems: Applications and Usage Scenarios (Chapter 16)

To fully assess the utility of multimodal biometric techniques in real-world systems, it is useful to consider usage scenarios and applications in which multimodal techniques are implemented. In Chapter 16, Thieme presents results from an evaluation conducted by International Biometric Group (IBG) and funded by the National Institute of Justice. The evaluation compared the accuracy of various multimodal fusion and normalization techniques based on data generated through commercial *fingerprint*, *face* recognition, and *iris* recognition systems. This evaluation attempted to situate results in the context of typical biometric applications and usage scenarios.

1.4.7 SATware: Middleware for Sentient Spaces (Chapter 17)

In Chapter 17, researchers at the University of California at Irvine present SATware: a stream acquisition and transformation (SAT) middleware they are developing to analyze, query, and transform *multimodal sensor data streams* to facilitate flexible development of *sentient environments*. A sentient space possesses the capabilities to perceive and analyze a situation based on data acquired from disparate sources. A multimodal stream–processing model in SATware and its elementary architectural building blocks are discussed. These include a distributed runtime system that permits injection, execution, and interconnection of stream-processing operators, a declarative language for the composition of such operators, an operator deployment module that optimizes the deployment of stream-processing operators, the concept of *virtual sensors* that encapsulates stream-processing topologies to create semantic sensing abstractions, and an infrastructure directory for storing the availability of resources. The authors describe how this basic architecture provides a suitable foundation for addressing the challenges in building middleware for customizable sentient spaces. SATware is implemented in the context of Responsphere—a pervasive computing, communication, and sensing *infrastructure* deployed at UC Irvine that serves as a unique testbed for research on situation monitoring and awareness in emergency response applications. SATware provides a powerful application development environment in which users (i.e., application builders) can focus on the specifics of the application without having to deal with the technical peculiarities of accessing a large number of diverse sensors via different protocols.

1.5   Concluding Remarks

Multimodal sensing, data processing, and system integration are still in a fast-growing stage of research and development. Much work needs to be done to ensure successful real-world, large-scale surveillance applications. The collection in this book is just a sample of some representative works in the field, and it is therefore by no means complete. However, we hope this collection will stimulate more interest in the research and development of mutlimodal surveillance techniques, including the further development of mutlimodal sensors, multimodal data fusion algorithms, and mutlimodal surveillance systems.

References

[1]   *Merriam-Webster OnLine*, http://www.m-w.com.

[2]   *Wikipedia*, the free encyclopedia, http://en.wikipedia.org/wiki/Surveillance.

[3]   Tzu, Sun, *The Art of War*, translated by Lionel Giles, http://www.chinapage.com/sunzi-e.html.

[4]   *Wikipedia*, ISTAR, http://en.wikipedia.org/wiki/ISTAR.

PART I

Multimodal Sensors and Sensing Approaches

The ARL Multimodal Sensor: A Research Tool for Target Signature Collection, Algorithm Validation, and Emplacement Studies

Jeff Houser and Lei Zong

2.1 Introduction

The U.S. Army Research Laboratory (ARL) has a significant program involving the development of UGS (Unattended Ground Sensors) that addresses a variety of military and government missions. ARL's program involves practically every aspect of sensor development, including devices, detection and fusion algorithms, communications, and command and control. One element of the ARL UGS program involves the development of low-cost sensing techniques for the urban environment, and one embodiment of this effort is the multimodal sensor (MMS).

The program objectives of this effort were to develop a networked personnel detection sensor with the following major criteria: low cost in volume production, support for MOUT (military operations in urban terrain) missions, and employment of nonimaging sensor diversity techniques. The MMS was an early prototype intended to demonstrate that low-cost sensing techniques were suitable for the urban environment and a viable alternative to higher-cost and higher-fidelity sensors for some applications. The MMS is used today as a demonstration system and a test bed for many facets of urban sensing.

This chapter will describe many aspects of the MMS design, including hardware, software, and communications. The detection algorithms will also be described, including the collection of target signatures and validation of algorithm performance. Finally, MMS usage in security and surveillance applications will be described, including usage of the sensors in a larger system.

2.2 Multimodal Sensors

The multimodal sensor was developed as part of an Army technology objective (ATO) program called Disposable Sensors (DS). The DS program was originally a 5-year program to develop very-low-cost sensing technology suitable for MOUT (military operations in urban terrain) applications. The MMS was the first prototype developed under DS and intended to provide early feedback on a number of issues ranging from hardware, algorithms, communications, and operations. The DS program ended prematurely due to budget shortfalls, but ARL has

Figure 2.1 The ARL multimodal sensor.

continued to develop low-cost sensing technologies as part of its internal research and development program.

The multimodal sensor (MMS) is shown in Figure 2.1. This prototype sensor system is based upon the principles of sensor diversity and simultaneously detects multiple physical parameters emitted by the target. This approach is intended to mitigate the known weaknesses exhibited by single-mode sensors under varying operational settings. The physical parameters sensed are seismic, acoustic, and thermal using an accelerometer, a microphone, and a passive infrared (PIR) transducer, respectively.

The objective of multimodal sensing is to increase detection accuracy and reduce false alarm rates, while operating in a complex and cluttered environment. This early prototype is designed with a path toward low-cost, volume production. All technologies and techniques are consistent with small, low-cost battery operation and automated assembly techniques. In this section, we describe the multimodal sensor, focusing on system architecture, major system components, personnel detection algorithms, and principles of operation.

2.2.1 Enclosure

The sensor enclosure is weather-resistant in order to survive reasonable environmental conditions expected during field tests. It is fabricated from black Delrin, a low-cost durable plastic material that is machinable, UV-resistant, and dimensionally stable. Power is provided by an external battery pack that slides into grooves machined into the bottom of the enclosure. The entire enclosure measures 3 inches × 6 inches × 1.5 inches without the battery or antenna.

The top deck of the enclosure contains four colored LEDs used for debugging or training purposes. When used in training mode, the LEDs provide the user with a visual indication of detection capability for individual transducers. The LEDs are disabled under normal operation. A field-of-view indicator is etched into the top surface to aid in pointing the sensor during emplacement.

The front surface contains PIR optics and an orifice that couples sound energy to an internally mounted microphone element. A tapped hole is provided to capture the battery pack's attachment screw. The rear surface contains two toggle switches, an antenna connector, and battery connector. The switches control power and operational modes, while the antenna connector supports an articulating omnidirectional whip antenna. The external battery's power cable mates with the battery connector.

2.2.2   System Description

As shown in Figure 2.2, the MMS electronics section consists of three modules: sensing, processing, and communications. The sensing module contains the transducer components used for detection. The transducers were selected by weighing their expected benefit against such considerations as costs, size, power, signal processing requirements, and consistency with volume production. This parameter evaluation resulted in the following transducer selection: microphone, passive infrared (PIR) and single-axis accelerometer. The microphone detects

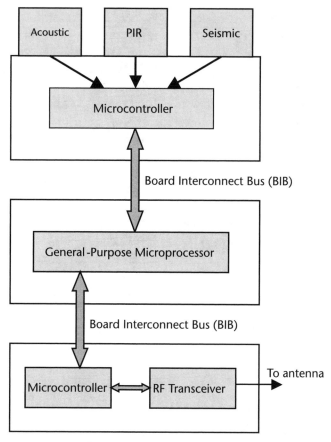

Figure 2.2   Sensor system overview.

acoustic events, such as human voice and transient events, such as door closures, footfalls, and so forth. The microphone is located internally and is acoustically coupled to an orifice located on the front face of the enclosure. The PIR sensor is used to detect human movement by sensing heat signatures. The PIR optics is configured for short range indoor usage and provides an approximate 100-degree azimuthal field of view to increase coverage area. The accelerometer detects physical displacement in the vertical axis and can detect human activity such as walking or running.

There are four analog signal chains in the sensing module: three for transducers and one for battery voltage. Each analog channel is conditioned and digitized before streaming into an 8-bit low-power RISC microcontroller. Sampling frequencies vary for different modalities, with the highest assigned to the microphone, while PIR, accelerometer, and battery are sampled at lower rates. The sampling rates are selected to provide sufficient bandwidth for the signal of interest while minimizing the processing load presented to the microcontroller. The microcontroller packages data bytes into packets and makes the packets available to the processing module through the board interconnect bus (BIB).

The processing module is a single-board embedded computer called the Stargate [1]. The board features an Intel Xscale processor that has performance and power management schemes adequate for embedded sensing applications and is found in handheld devices such as PDAs (personal digital assistants). The Stargate provides the developer with a rich set of standard peripherals, such as PCMCIA, CompactFlash, many types of interface standards including USB, asynchronous serial, and Ethernet. In addition, it interfaces directly to the MICA series mesh networking radios available from the same manufacturer.

The Stargate is shipped with a version of embedded Linux operating system (OS), which greatly reduces the development effort for custom software. The embedded Linux OS offers developers the essential tools needed to execute custom applications, while keeping the environment close to the desktop Linux OS to minimize the learning curve. The embedded OS also offers an optimized floating-point emulator for computationally intensive tasks.

The communications module is a wireless radio frequency (RF) transceiver commercially available from the same manufacturer. The MICA series radios (or motes) [2] feature mesh networking capabilities in a low-power form factor well suited to this application. The mote features an 8-bit low-power microcontroller and a multichannel transceiver. The microcontroller hosts TinyOS [3], a simple but efficient operating system, and network-related firmware that performs networking tasks such as packet routing, collision recovery, network discovery, and so forth. TinyOS has a component-based architecture and is designed for low-power wireless network systems. The operating system is a collection of components that can be selected based upon application needs. The component-based architecture lends flexibility in application design and allows creation of reusable components.

2.2.3 Algorithms

The application software that executes on the processing module controls the system's overall functionality. Key components of the software include modules for detection for each modality and for data fusion. A software interface to the sensing module decodes packets, and extracts and forwards data streams to the appropriate detection module. When detection occurs, the fusion module calls upon a software interface to the RF module to compose and transmit a detection message to the network. Logically, the system operates as shown in Figure 2.3.

The entire personnel detection algorithm [4] consists of four major parts: detection algorithms for each of the three modalities and an overall data fusion algorithm. Each detection algorithm ingests data from the appropriate transducer channel and processes it in real time. The data fusion algorithm observes the results from the individual algorithms and decides whether a detection event is to be declared.

The acoustic detection algorithm looks for changes in energy relative to the background. The targeted frequency for acoustic events is audible human voice band below 2 kHz. To better adapt to the environment, acoustic data samples are also used to update the background levels dynamically. The algorithm examines the data samples and compares energy content in various frequency bands against those of the background. If the differences in energy exceed preset thresholds, the event is flagged for potential acoustic detection.

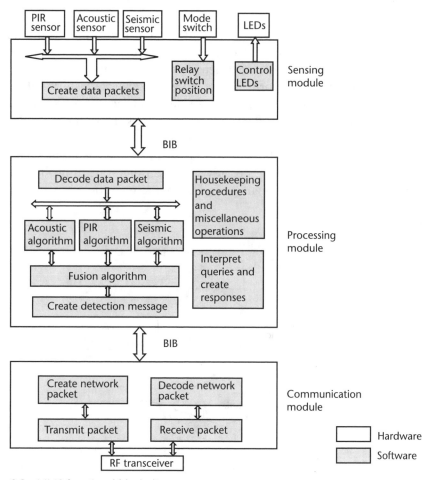

Figure 2.3 MMS functional block diagram.

In addition to the human voice, the algorithm can also detect transient acoustic events such as door closures, dropped or kicked objects, or other short-term noise resulting from human movement. A sharp, short-lived acoustic event translates into a broadband event in frequency, which fits a particular feature that the algorithm looks for. The acoustic algorithm uses results from the voice and transient detectors to determine possible acoustic detections. Figure 2.4 illustrates the logical operation of the acoustic algorithm.

The PIR transducer detects thermal energy emitted from an object relative to the background scene. When a person passes through the field of view, the PIR transducer senses the difference in thermal energy relative to the background. The algorithm compares the sampled data to a preset threshold as a simple, yet effective, detection mechanism. An optimal preset threshold value was selected by assessing the unit-to-unit variance in many transducers.

The MEMS-based accelerometer is used to detect vibration in the vertical axis. When placed upon the floor or other horizontal surface, it can sense the displacement resulting from walking or running. The seismic algorithm is similar to the acoustic algorithm, looking for energy differences between the signal and the background. The targeted frequency is the human gait frequency, which is below 100 Hz. Figure 2.5 illustrates the logical flow for both the PIR and seismic algorithm.

Each second in time, the three detection algorithms read a block of data, process it, and produce a detection result. The data fusion algorithm examines the results to determine whether a detection event should be declared. Results from individual detection algorithms are stored in separate history buffers and used by the fusion algorithm to correlate the results in time and across modalities. Once a detection event is reported, the algorithm passes the result to a software interface to formulate and inject a message into the network.

Other software modules perform basic housekeeping tasks, such as sending out periodic heartbeat messages to advise the sensor's state of health, interpreting and responding to user commands, and changing operational states. These modules,

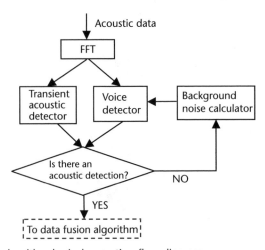

Figure 2.4 Acoustic algorithm logical operation flow diagram.

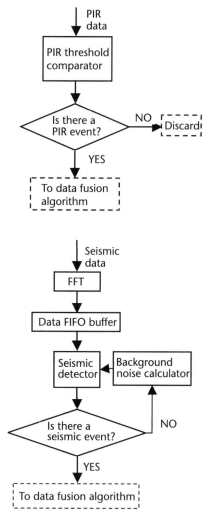

Figure 2.5   Logical operation flow diagrams for PIR and seismic algorithms.

along with the algorithms, must fit into a resource-constrained hardware platform. The processing board consists of a general-purpose fixed-point processor, which is inherently inefficient at FFT computations. Much effort was expended in optimizing the algorithm and computation blocks to meet the real-time requirement.

2.2.4   Communications

Upon the detection of a target, the processing module sends a detection packet to the communication module for transmission to the network. When the communication module receives a message, it decodes the data and passes it to the processing module for action. Some of the possible actions include enabling or disabling a particular modality, setting various thresholds, and status query. More will be discussed under the topic of MMS management in Section 2.4.5.

2.2.5   Principles of Operation

When the sensor is powered on, the OSs in the processing and communication modules boot up separately. The sensing module does not have an OS and becomes operational once the PIR transducer has initialized.

When the processing module boots up, the embedded Linux OS automatically launches the sensor application. The first action the application takes is to collect background data from the acoustic and seismic channels in order to establish a noise floor. The noise floor is continuously and dynamically evaluated to reflect any changes in the environment. The sensor application also tries to establish connection with a gateway by broadcasting messages into the network. If a gateway node exists, the sensor and gateway pair will complete a handshake process to exchange information and add the new sensor to the network. Once the handshake process and the background data collection are complete, the sensor is in the operational state.

During operation, the sensor monitors for activity and sends out detection and housekeeping messages as described above. It also monitors network traffic, looking for commands and housekeeping messages from the gateway. Sensor management functions can be processed while the sensor is in the operational state.

To preserve battery life, the sensor can be put to sleep when not used. The processing module stops executing the detection algorithms and goes into a sleep state. The communication module remains active to relay messages from other sensors to the gateway node and monitor for a wake-up command. More details about the network operation will be discussed in the next section.

2.3   Multimodal Sensor (MMS) System

The previous section described the makeup and behavior of a single MMS, and this section will describe a network of MMS sensors. The network architecture will be discussed along with other network-level entities required for operation. A complete system is shown in Figure 2.6 including a PDA, a gateway, and six sensors. This configuration was used in the C4ISR OTM experiment, as described in Section 2.6.

2.3.1   Multimodal Sensors

The sensors [5] form an ad hoc wireless mesh network as each node establishes a connection path to a gateway node, either directly or indirectly. If a node is not within one hop of the gateway, it uses other nodes as relay points to communicate with the gateway. All sensor messages flow upstream to the gateway.

The MMS sensors are nominally configured to operate in a centralized network. Neighboring sensors do not share information, and all traffic flows upstream to the gateway for any additional processing such as tracking and multisensor fusion. ARL is also involved with distributed tracking [6] techniques that could easily be applied by the MMS sensor network.

Figure 2.6   An MMS system including gateway and PDA.

2.3.2   Multimodal Gateway (mmGW)

The gateway is the central processing point of the network. All sensor information flows to the gateway, which can respond to an individual sensor or all sensors. The gateway performs many functions, but its fundamental task is to act as a communications bridge between the MMS network and higher-level entities.

Upon startup, each sensor sends an initialization packet to the gateway, which responds with a request for sensor status. Upon receiving the sensor status, the gateway synchronizes the sensor clock and allows it to become an active member of the network. When a sensor detects an event, it sends a report to the gateway, including information such as detection time and detecting modalities. Each sensor sends heartbeat messages periodically during periods of inactivity to provide updated status to the gateway. The gateway keeps track of all active sensors and their most recent activities, along with other pertinent information such as operational mode, threshold settings, and battery level. If the gateway has not received messages from an active sensor for an extended duration, the gateway marks the sensor inactive.

The gateway hardware electronics consist of two primary modules and an optional module. The two basic modules are similar to those in a sensor: communication and processing modules. Since the gateway communicates directly to sensors, it has the same communication module as a sensor. Like the sensor, the processing module is a Stargate processor; however, it hosts a gateway-specific application. The gateway software performs all sensor management features mentioned above, keeping track of the current state of the entire network. Unlike the sensor application, which is computationally intensive, the gateway software is more communication-intensive in nature.

Since the gateway tracks the current state of the network, it also acts as the access point for user interfaces. Currently, there are two ways to access and control the network: within and from above. The PDA operates within the MMS network and communicates with the gateway either directly or via the sensor network.

The PDA provides a certain degree of mobility, enabling communication and control anywhere within the network coverage area. The PDA can also provide a graphic representation of the current network status and display detection events in real time. It also provides sensor management features to fine-tune the sensors to a desirable state. More about the PDA will be discussed in the next section.

The alternative means of network access is from above the MMS level through a third-party communication and control application. In this case, an optional communication module mentioned previously might be needed for communication. Through this communication mechanism, the MMS system can become a subsystem within a larger system, providing a piece of the local surveillance picture. Now that the MMS sensors are connected to a higher-level network, command and control can be exercised from any point available on the network. One of the applications discussed in Section 2.6 is an example of the MMS system acting as a subsystem.

2.3.3 PDA

A PDA node, shown in Figure 2.7, is used to view sensor information sent to and from the gateway. The PDA node resides on the same network as multimodal sensors and can move around anywhere within the network, providing a certain degree of mobility. The PDA provides three basic services: sensor survey, emplacement verification, and tactical display.

The sensors do not possess localization capability because they were designed to be low-cost and operate primarily in a GPS-denied environment. The PDA provides this service when a GPS is available using a GPS receiver connected via a CompactFlash slot. In this mode, the user places the PDA in proximity to

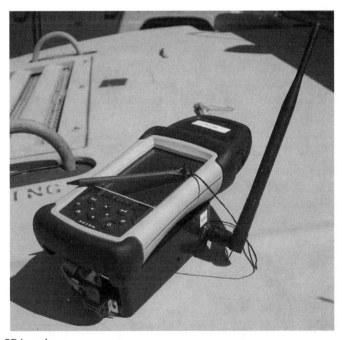

Figure 2.7 A PDA node.

the sensor and waits for notification of a valid GPS solution. The user then enters the sensor ID and sends the position to the gateway for storage. When GPS service is not available, the user can manually locate the sensors on a graphic display.

Emplacement verification ensures that the sensor has joined the network, and that its emplacement will support the desired mission. The user receives notification when the sensor checks into the network that verifies the radio link and basic sensor operation. The user has the ability to verify that the sensor will perform its intended mission by acting as a test target to verify proper detection. For example, the user can monitor detection results on the PDA while moving through the sensor field to verify correct emplacement. When selecting an emplacement site, the PDA's network status indication will indicate whether a network connection is available at the candidate site prior to emplacement.

Sensor management capabilities allow the user to control sensor operation in a number of ways. Individual sensors can be placed in a low-power state in support of premission emplacement or power conservation. Individual sensor performance can be tuned by adjusting transducer detection parameters. In addition, each of the three detection modalities can be individually enabled or disabled by command.

The PDA can also provide tactical display capability, independent of GPS availability. When GPS is available, the PDA automatically displays relative sensor position graphically by scaling the absolute positions to fit the display. When GPS is not available, the user has the option to hand draw a map and drag sensor symbols from a palette to the desired location. In both modes, the symbols blink when detection occurs as an alert to the operator. A separate tabular display is also available along with optional audible alerts.

2.4 Typical Deployment

Technologists can provide suggested usage models that accompany new or experimental systems. These models can be used as guidelines by warfighters in the development of such doctrine. The suggested usage models for the MMS system are missions involving detection of personnel and vehicles. The MMS was designed specifically for urban operations and is effective in detection of personnel and vehicles in urban settings. In addition, testing has shown that MMS systems are also effective against personnel and vehicles in rural settings such as trails, woods, and fields.

The following sections will describe the basic deployment procedures for the MMS system. Particular variations due to change in venue will be discussed when relevant; otherwise, the procedures will be somewhat generic. The procedures are organized into five basic steps: mission planning, emplacement, survey and test, operation, and sensor management.

2.4.1 Mission Planning

In the context of UGS, mission planning involves the selection of emplacement sites for each sensor to achieve the desired mission. Broadly speaking, the expected

Figure 2.8 Predeployment testing of a sensor system.

mission types for MMS sensors are either surveillance or security, or a combination of the two. Figure 2.8 shows a soldier performing a premission test of the sensors. The sensors are laid out upon the ground to facilitate a quick emplacement test.

2.4.2 Sensor Emplacement

There are two main parameters involved in sensor emplacement: relative sensor location and individual sensor placement. The former determines overall area coverage as related to mission tactics and is constrained by network connectivity, while the latter determines individual sensor detection effectiveness. This section will describe issues related to both parameters as they relate to emplacement.

The relative sensor placement is largely driven by cost. The intent of the MMS program was to provide low-cost personnel detection technology to support MOUT (military operations in urban terrain) missions. One particular MOUT operation originally envisioned for this technology was that of securing cleared buildings. This would provide some level of automation and thereby reduce the level of manpower required to stand watch over a cleared structure. This mission drove much of the early engineering decisions and specifically impacted communications choices. In particular, an early assumption was made that the network would be densely populated, as many sensors would be located throughout the building.

The assumption of closely spaced sensors was consistent with low production costs since low-power communications require fewer system resources, which, in turn, helps hold costs down. The fundamental requirement that drives MMS spacing is radio connectivity. The sensor radio transmitters each present approximately 5 mW of power to a low-gain omnidirectional antenna. The antenna configuration was selected to remove any pointing requirements and thereby allow for any type of physical network configuration. In summary, the radio link constrains the relative sensor spacing.

Individual sensor placement is constrained by such parameters as field of view and detection range. The sensors are designed to be placed flat upon the floor or ground and facing the target. For indoor missions involving rooms, the sensors work best when placed in the far corner of the room facing back toward the entryway. The wide angle PIR optics and acoustic directionality provide coverage for typical sized rooms in residential or commercial structures. The sensor is optimized for this type of space and can cover most all of the room to account for movement through any windows, doors, or stairs within the room.

The sensor can also be placed within passageways and near entrances of buildings or structures. In addition, sensors placed in passageways can also act to facilitate communications between sensors located within distant rooms. A bread-crumb-trail analogy is often used to describe this configuration.

When used outdoors, the sensors can be placed along side roads and trails to detect personnel or vehicles. They can also be used to form full or partial perimeters to monitor avenues of approach. Although the optimal sensor orientation is flat and facing the target, other orientations are certainly possible, resulting in varying amounts of degradation for the individual transducers.

For example, the seismic sensor is aligned normal to the bottom surface of the sensor in order to detect vibration in the vertical axis. A slight tilt will marginally reduce the vertical component of the signal presented to the now-tilted transducer. Further, there are other vibration components in the horizontal plane that will begin to couple into the tilted sensor. Other transducers exhibit similar effects, which result in graceful degradation in performance caused by pointing inaccuracies.

2.4.3   Survey and Test

Once the sensor has been emplaced, it must be energized, tested, and surveyed to record its location. Upon application of power, the sensor goes through a startup procedure described in Section 2.2.5. In summary, the sensor application initializes and the radio searches for a network connection.

In order to provide sensor location information to the gateway, the sensor must be surveyed. The PDA contains a GPS receiver and wireless connection to the MMS network. The user places the PDA near the sensor and waits for indication of a valid GPS position. At this point the user commands the PDA to record the current position, and it is transmitted to the gateway for storage. Certain gateway configurations maintain this position in nonvolatile memory in the event of power faults or system resets.

Once the survey is complete, the user is free to test the operational status of the sensor using the PDA. Since the PDA is part of the MMS network, it possesses the infrastructure necessary to monitor and display detection reports. This capability can also be used to build a tactical display, if desired.

2.4.4 Operation

The PDA contains external networking assets that allow it to become a part of the MMS sensor network. This connectivity can be combined with other inherent PDA assets to form the basis of a monitoring system. A custom interface application is used to present sensor data to the user in visual and audible formats. This allows local users access to sensor data if they are within range of the MMS network.

Because the sensors are networked, data can be presented to users at any given level above the gateway. Once data enters the network, there is no fundamental limit concerning its distribution. But practically speaking, the organizational and engineering constraints determine the limits of distribution. Once at or above the level of the MMS gateway, MMS sensor data can be fused with other sensor data to provide higher levels of abstraction.

2.4.5 Sensor Management

Once the system is operational, the user may manage the operation of the sensor network at various levels. For example, the performance of individual sensors can be controlled by adjusting specific parameters and thresholds for each transducer. In addition, sensor and network level modes and parameters can be controlled as well. For example, energy conservation may be achieved at the sensor or network level through control of sleep cycles.

As with operational control, management control can be asserted within and above the MMS network. The specific interface for management may vary and depend on the resources of the host platform.

From this point forward, the sensor deployment is an iterative process: sensors may be added, moved, or deleted as required to achieve the desired level of situational awareness.

2.5 Algorithm Development

The first step in the algorithm development process was to collect personnel signatures in representative urban environments using the MMS waveform capture capabilities. The signatures were then used by developers to identify salient features to be used as a basis for detection algorithms. Candidate algorithms were then implemented and subjected to simulation and operational testing to validate performance. The details of this activity are presented in this section.

2.5.1 Signature Collection

The MMS systems were designed to work in urban environments such as buildings, trailers, caves, underground structures, and so forth. In order to test hardware and algorithm performance in these venues, we collected signature data in two representative locations. The first environment was a vacant basement level of a concrete office building. Although the particular floor was unoccupied, the environment was cluttered by common activity from ventilation, machinery, and elevator noises. As depicted in Figure 2.9, the outside area adjacent to the vacant offices is wooded, resulting in extremely low background noise levels.

The vacant offices contain reinforced concrete floors with tile covering, metal walls, large windows, and some office furniture. Figure 2.10 depicts the interior floor plan of this venue selected to represent an urban industrial, commercial, or residential high rise building.

The second environment chosen was an unoccupied temporary building, very similar in construction to a double-wide trailer, or mobile, home. The trailer is situated near a service road and a busy two-lane public road shown in Figure 2.11. From within the trailer, traffic noise can be heard during the daytime, resulting in high levels of background noise. The building frame and exterior are constructed of metal. The floor consists of plywood and metal sheets under carpet, and the interior walls are constructed of wood paneling over metal framing.

Referring to Figure 2.12, the interior contains two small corner offices on the northern wall and a large enclosed central area. The central area contains closets, plumbing and heating service, and restrooms. The figure also shows the position of interior support columns aligned in an array. This environment was selected to

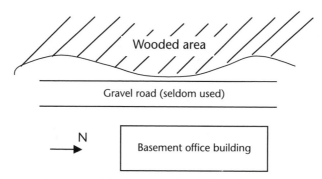

Figure 2.9   Exterior environment of the basement-level offices (figure not to scale).

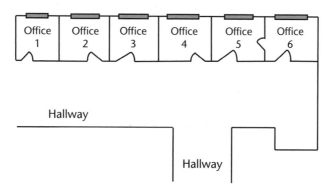

Figure 2.10   Interior environment of the basement level offices (figure not to scale).

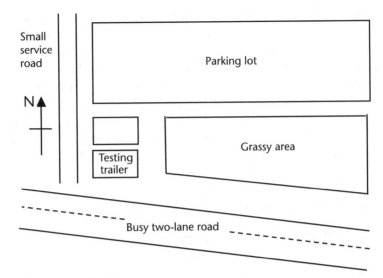

Figure 2.11 Exterior environment of temporary building (figure not to scale).

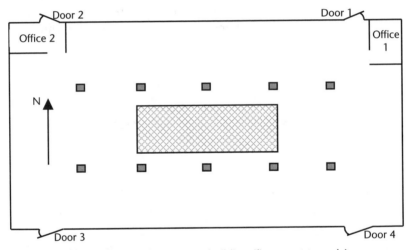

Figure 2.12 Interior environment of temporary building (figure not to scale).

represent suburban residential structures, though more out of convenience than for accuracy.

2.5.1.1 Scenarios

At the beginning of each data collection run, the MMS collects background data for a predetermined amount of time (normally several minutes). Background data is needed to determine the noise floor and thresholds and also to allow the PIR transducer time to initialize. Figure 2.13 shows the test scenario used for the basement office tests. A typical collection run consists of two people walking up and down the hallway at a normal pace. The sensor, now collecting raw data, is located in the far corner of one office.

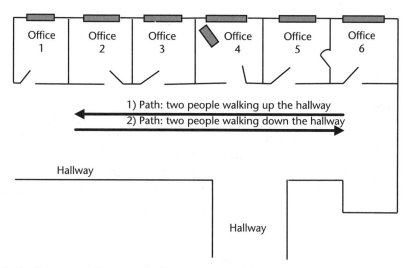

Figure 2.13   Basement office scenario (figure not to scale).

The test targets were instructed to use normal walking speeds and speaking volumes when speech was included. Normal level was subjective in these situations and varied from one person to the next. However, normal walking pace was defined to be the pace a person walks in a continuous motion without getting out of breath, or approximately 3 to 4 mph. The guideline for normal speech volume was a conversational volume, where a person close by (within 3 ft) could hear without difficulty. Several variations of the collection included jogging, running fast, and walking while talking.

The trailer test scenario is depicted in Figure 2.14, where a typical run consisted of two people entering through door 1, walking along the north side of the trailer, passing in front of office 2, and exiting through door 3. After pausing outside of door 3, the subjects then reversed direction and traced their original path back through

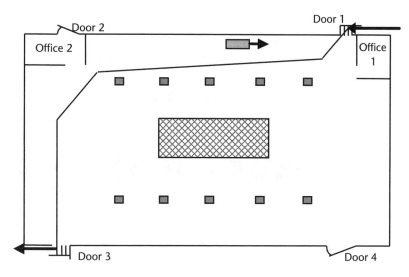

Figure 2.14   Temporary building scenario (figure not to scale).

door 1. Test variations included running out of office 1, and movement without talking. Soft-soled shoes were worn by both personnel performing the tests.

2.5.1.2 Target Features

Figure 2.15 shows the raw data collected in the basement office environment. The data is plotted in panels from top to bottom for acoustic, PIR, and seismic detection modes, respectively. The sensor was placed in a far corner from, and pointed toward, the doorway, as shown in Figure 2.13. A time-frequency series, or dynamic spectrum, is shown for the acoustic and seismic data in Figure 2.16. Figures 2.17 and 2.18 are similar plots of data taken in the trailer environment. The sensor was placed close to, and pointed parallel with, the north wall as shown in Figure 2.14.

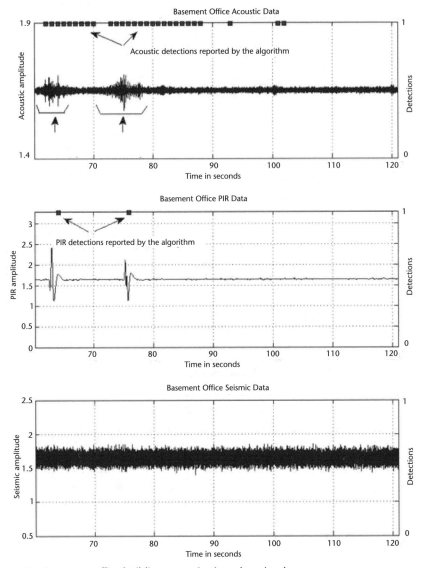

Figure 2.15 Basement office building scenario time domain plot.

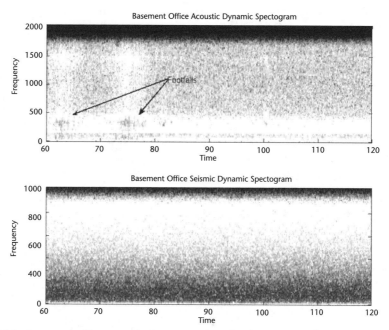

Figure 2.16 Basement office scenario frequency domain plot.

Observation of the time-domain plots will show temporal symmetry caused by the initial and subsequent return paths of the test subjects near the sensors. Also shown on the time-domain plots are the performance of the algorithm as a function of time, as described earlier in this section. The frequency content of the acoustic and seismic data is shown in Figures 2.16 and 2.18, respectively, to provide another perspective. PIR processing is performed so efficiently in the time domain that Fourier analysis is not justified for this transducer.

Two personnel were observed in both scenarios, though they were talking in the trailer but not in the building. Observation of the acoustic spectrogram in Figure 2.18 shows the presence of voice and footfall features, whereas the basement data in Figure 2.16 only shows footfalls since the targets were not speaking.

One of the reasons for the multimodal sensor approach is readily seen by comparing the frequency domain plots for the seismic sensors. Note that the trailer data clearly shows periodic footfalls but the basement data does not. This is because the basement construction is extremely solid and the floor is not displaced to the same extent as in the temporary building. Further, the basement sensor is considerably farther away from the target's closest approach than the trailer sensor. Sensor performance is a strong function of the environment and the target.

In spite of the inconsistent performance of the seismic detector, the PIR and acoustic detectors clearly recognize the target in both cases.

2.5.2 Validation

Candidate algorithm performance was validated using simulation and in-target testing techniques. Algorithms were developed and simulated using a high-level

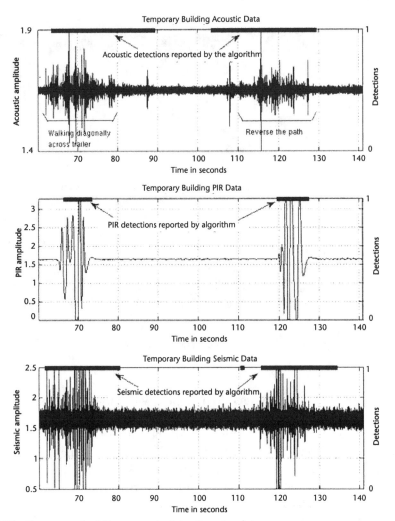

Figure 2.17   Temporary building scenario time-domain plot.

language in order to rapidly evaluate performance during early stages of development. Mature algorithms were then implemented on the target processor (i.e., the actual sensor platform) for system level testing. Finally, the final sensor algorithms were validated by repeating the original scenarios used to collect the signature data.

2.5.2.1 Simulation

In order to rapidly evaluate the performance of early candidate algorithms, they were first implemented in a high-level language on a PC workstation. The signature files collected by the MMS were then used as test vectors to evaluate performance. This environment provided an efficient means to tune, optimize, and graphically evaluate performance without the need to implement each candidate on the sensor platform. Simulation results were then overlaid on the time-domain plots as indicated by the small squares at the top of Figure 2.15 and Figure 2.17. The small

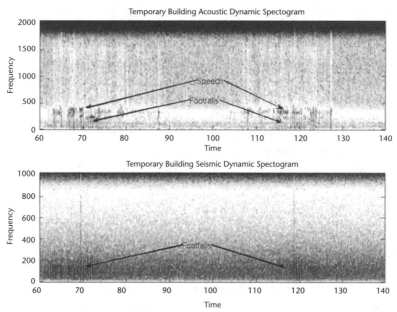

Figure 2.18   Temporary building frequency-domain plot.

squares at the top of the respective waveform indicate a section of data that contained a detection event. This allowed the developer to graphically evaluate the performance by visually observing target features and detection results on the same figure. From this vantage point, developers could quickly tune and optimize the high-level language version while observing results.

Once this testing infrastructure was in place, evaluating the performance of new and different scenarios was a simple process. Many scenarios could be collected using MMS data capture and the stored files evaluated in the high level environment. Once an algorithm was vetted using this process, it would then be subjected to in-target testing.

2.5.2.2   In-Target Testing/Real Time

Algorithms that successfully passed the simulation process were then implemented in C programming language and compiled to run in a target Linux environment. Once implemented, or ported, to the target system, the algorithm must be validated to verify that it is a true adaptation of the high-level version. This process is complicated by the fact that the algorithm is now operating in real time and interacting with target hardware and software components. In this type of testing, implementation issues are encountered that relate to the physical constraints of the hardware.

Some of these constraints are memory and processor limitations, integration with other software modules, and limited visibility into the system during evaluation. This process required a considerable level of creative technical effort in order to fit memory and processor intensive tasks such as FFT-based algorithms onto an integer-point processor with a real-time, low-latency requirement. Once the implementation issues have been resolved and the detection algorithms are stable, in-target evaluation can begin.

In-target evaluation involves the recreation of the original scenario. However, now the sensor is being used as a detector rather than a collection asset. Real-time detection performance is assessed along with performance in a wireless networked environment.

Real-time validation involves a gateway, an external display device, and a sensor loaded with the detection algorithms. A network of a single sensor is set up and the scenarios are performed. Various scenarios and environments are used as well during this validation. The real-time result can be observed on the display device in order to assess the important issues such as system latency.

2.6   MMS Applications

The MMS is currently used as a research platform for low-cost sensing technologies and has been adapted for several particular applications since its creation. This section will describe the operation of the MMS in three separate applications, as well as some of the modifications and enhancements required in each case. The three applications of MMS are (1) cave and urban assault ACTD, (2) NATO LG-6 demonstration, and (3) C4ISR OTM experiment.

2.6.1   Cave and Urban Assault (CUA)

The DS program participated in the cave and urban assault ACTD (advanced concepts technology demonstration) in the fall of 2005. This was the first operational test of the prototypes in a relevant environment: a MOUT training site used to train soldiers for urban warfare. In addition, the CUA ACTD provided soldiers to operate the system and provide feedback to the developers on many aspects of the system.

2.6.1.1   Equipment Configuration

The sensors were packaged into kits containing 12 MMSs, 4 imaging sensors, a communications gateway, and a portable handheld display. MMS sensors were provided by ARL, and other system components were provided by the respective DS program participants. The radios within each kit were configured to operate on one of three frequency channels, so that three sensor fields could be deployed and operate independent of one another.

The original MMS configuration included a 2.4-GHz radio (Crossbow MICAz) to maintain compatibility with the imaging sensor. The program selected this radio because its higher data rates would accommodate the imaging sensor's increased payload. The MMS would later migrate to a 900-MHz radio (MICA2), made by the same vendor, in order to avoid the congested 2.4-GHz band.

2.6.1.2 Test Environment

This particular MOUT site is essentially a mock village containing residential, commercial, religious, and local government structures. The overall area of the site is approximately 1 km^2 with roadways constructed of hard packed dirt, with and without gravel cover.

Most buildings were constructed of concrete block with a typical number of doors and window openings. Commercial and government buildings typically contained flat roofs, while the residential buildings had wooden pitched roofs. Most doorways contained doors of hollow wood construction, and window openings were either covered with Plexiglas or wooden shutters that could be opened or closed. All buildings were sparsely furnished with furniture appropriate to the structure, and all floors were fabricated using poured concrete.

Testing was conducted at the system level using both imagers and nonimaging MMSs in several building types. Other tests were conducted to separately evaluate imagers and nonimaging sensors, but this section will cover only operational testing specific to MMSs.

Stand-alone MMS testing was conducted along a street containing six detached (i.e., single family) residential dwellings shown in Figure 2.19. All six were constructed identically with a single main level containing four rooms and a half-basement containing one room. Each building had front, side, and rear entrances and a small covered front porch. Each building also had a backyard with planted grass lined by a low concrete block wall.

2.6.1.3 Test Scenarios

Three basic tests were conducted in this portion of the MOUT site: detection of (1) dismounted personnel moving through the six residential buildings, (2) dismounted personnel outside the buildings, and (3) mounted personnel

Figure 2.19 Residential structures.

driving into the neighborhood. This section will briefly describe the test scenario and performance of the sensors in this setting.

All six houses were instrumented identically for the first personnel detection test, with one MMS placed in each of the four rooms on the main level. Sensors were typically placed on the floor in the rear corner of each room, as described in Section 2.4, though variations in furnishings resulted in modified sensor placement. In these cases, sensors were placed on window sills and horizontal furniture surfaces as needed.

A nonscale schematic of the neighborhood is shown in Figure 2.20. The six houses are located at the southern edge of the figure and facing north. The round black symbols indicate the MMS positions within the house, and the black arrows indicate the positions and headings of the MMS located outside the houses. The diamond indicates the position of the gateway that fed sensor data to the remote handheld display located across the street at a position indicated by the pentagon.

This first test consisted of a single dismount moving sequentially through each room of a house and then proceeding to the next house in turn until all rooms (main level only) of all homes had been occupied. The test subject was quiet for this portion of the test, but there were times when the subject was directed to stop and talk on a handheld radio to test the acoustic performance of the sensors. In each case, every sensor correctly detected the subject in every room, and the operator was able to track movement through the homes on the remote display.

The second test was designed to detect dismounted personnel attempting to move along the east-west road that passed in front of the six homes. Multiple test runs were conducted in order to vary the approach pattern of the personnel. In this testing, the sensors were placed outside the houses and facing the two-lane–width road. Additional sensors were placed across the street in order to provide better

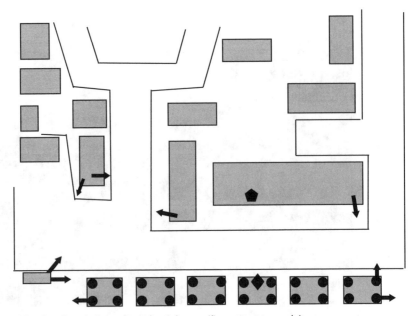

Figure 2.20 Overhead view of residential area (figure not to scale).

coverage. Two test subjects were directed to walk or run into the neighborhood using every possible entrance onto the main road. In each case the personnel were detected, and the operator was able to follow the subject's movement on each run.

In this test, each sensor did not necessarily detect the dismounts due to target range from the sensor. For example, if the targets moved along the southern side of the street, then all sensors on that side detected movement while sensors located on the northern side did not always detect the targets. Performance was determined by road width and the individual setback distance of the sensor from the edge of the road. Sensors were set back from the road for maximum concealment.

The final test consisted of two personnel driving a large SUV (sport utility vehicle) along the same road that passed in front of the houses using the same sensor location as the previous test. Each possible road approach was taken, and each run resulted in positive detection of the vehicle. In this case, all sensors detected the vehicle on each pass.

2.6.2   NATO LG-6

The MMSs were demonstrated at the NATO (North Atlantic Treaty Organization) LG-6 (Land Group Six) experiment conducted in the spring of 2006. The MMS participation in this experiment was limited to a few scenarios conducted in a temporary MOUT site.

The DS program had concluded by this time, and the ARL sensor team began an internal collaboration to develop a display capability for the MMS data. This also required the development of a new communications gateway. The significance of the NATO test is that it marked the first use of the MMS system with ARL-only hardware.

The communications gateway consisted of a 900-MHz radio and processor mounted together in a metal enclosure. The gateway contained a serial port to transfer data between the gateway and a portable computer used to display sensor data. The display software also contained modest sensor management functionality used primarily to survey the sensor positions, as the PDA had not yet been developed.

2.6.2.1   Environment

The test venue at APG consisted of a compound with a perimeter defined by an 8-ft-high wall of concrete construction with a gated entrance. The internal structures consisted of modular freight shipping containers that can be transferred by truck, rail, and ship. The containers were constructed of corrugated metal with wooden floorboards installed. The containers were placed within the compound to simulate single and two-story buildings when stacked appropriately. The containers, approximately the size of railroad boxcars, were modified with the addition of internal stairs used to access the upper level.

A number of personnel and vehicle detection scenarios were conducted using the MMS network. The tests and results will not be enumerated here, but then fell into two broad classes: personnel and vehicle detection. Personnel were detected within the compound from within and outside of the structures. Sensors were placed along the asphalt street just outside the perimeter wall to detect foot traffic outside the compound. Military and civilian vehicles were detected driving along the street outside the compound as well as entering and exiting the compound.

The system was monitored by ARL operators from across the street from within a similar container building. The operators were able to observe the scene and the display monitor concurrently to assess sensor performance. The performance of the MMS sensors in this setting was consistent with the previous CUA demonstration. The sensors detected all personnel and vehicles within their effective range with at least one of the sensing modalities.

2.6.3 C4ISR OTM

The MMS was also used at the C4ISR OTM (Command, Control, Computers, Communication, Intelligence, Surveillance, and Reconnaissance On-The-Move) Experiment in mid-2006. The sensors were used to provide security for a reconnaissance platoon conducting a number of surveillance operations over the course of the two-week experiment. A significant integration activity was required to integrate the MMS system into the framework defined by the experiment designers.

2.6.3.1 Equipment Configuration

In addition to the MMS system, low-cost imaging sensors were developed and employed in this experiment along with an emplacement aid, or PDA, to provide sensor location for the users. Finally, a new communications gateway was developed to allow medium-haul communications from the sensor field back to the vehicle. The sensors were configured into three kits, each containing six MMSs, one gateway, one imager, and one PDA. Each kit was configured to operate independently on one of three available radio frequencies during the experiment.

Functional descriptions of the MMS, PDA, and generalized communications gateway were presented in Section 2.3. This particular gateway implementation and imaging sensor will be discussed only to the extent that they relate to the MMS.

The MMS gateway for this experiment contains a longer range data radio developed at ARL called the Blue radio. The Blue radio was developed to provide a medium-haul wireless network for UGS applications. Placing the Blue radio into the MMS gateway allowed the sensor field to be moved beyond the range of the lower powered MMS radios and thereby increased the area coverage of the system. Sensor data flowed from the MMS sensor network into the gateway, where it then entered the Blue radio network. Blue radios were used in the vehicles and the imaging sensors.

The imaging sensor contained a low-cost daylight (i.e., CMOS) imager of the type commonly used in cellular telephones. A Blue radio was mounted within the sensor to transmit acquired images through the Blue radio network with nodes at (1) the MMS gateway, (2) the imager, and (3) the vehicle. Upon reaching

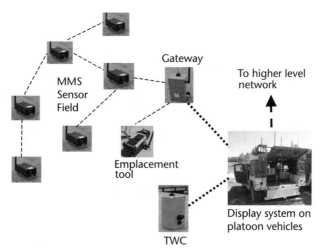

Figure 2.21 ARL sensor network for C4ISR OTM.

the vehicle, sensor data was distributed to higher network levels in a manner prescribed by the experiment designers. Figure 2.21 provides a schematic view of the network configuration.

2.6.3.2 Environment

The experiment venue was a very large wooded area during mid-2006. Multiple experiment runs were conducted in open, MOUT, and wooded areas during the course of the experiment. The soil conditions were generally sandy, and the weather was seasonable though dry.

2.6.3.3 Test Scenario

The basic mission for the sensor system was platoon security during the conduct of their various reconnaissance missions assigned throughout the experiment. The sensor systems were typically deployed to provide perimeter security for elements of the platoon, which contained three vehicles, and each was assigned a specific sensor system. The data from each sensor field traversed its MMS and Blue radio networks to arrive at its respective vehicle. Upon reaching the vehicle, the sensor data entered the next higher networking level and was able to flow to any asset at that level as allowed by network distribution doctrine.

Most of the sensor security missions were designed to detect personnel and vehicular traffic along controlled avenues of approach, such as woodland trails and roads. Although most of the missions using MMSs were for platoon security, some scenarios involved surveillance. In this application, the sensors were used to detect the advance of opposition forces into an MOUT site.

In summary, the MMSs were used for both security and surveillance missions involving the detection of personnel and vehicles. These missions were conducted in rural (woodland and open area) and MOUT settings.

2.7 Summary

The multimodal sensor is the result of ARL experimentation with small, low-cost sensor technology suitable for urban military operations. Since its initial design, it has been adapted for use in three relevant exercises in addition to considerable laboratory testing. During this time, the MMS has provided ARL researchers with valuable insight into the development of low-cost sensors. In addition to the sensor performance perspective, the MMS has proven to be a vehicle to advance knowledge in many aspects of sensing.

The MMS has provided feedback relative not only to individual sensor performance but also to the performance of networks of sensors. MMSs have been tested individually, in small isolated networks, and as part of much larger networks with theater-wide scope. These tests have resulted in much better insight into data distribution and usage concepts. This insight bridges many of the expected disciplines such as electronics, algorithms, signal processing, and communications. However, equally important feedback has deepened our understanding of how soldiers interact with small sensors.

The concept of sensor diversity has spawned a considerable development effort of multimodal signature collection and archival at ARL. In addition to the waveform and storage capability of the MMS, ARL has developed a hyper-modal data collector (HMDC) tool that currently collects collocated, coregistered datasets from a wide array of transducer types. The combined signature collection capability of the MMS and HMDC is currently supporting further algorithm development among ARL and collaborators in other application areas such as multisensor fusion and distributed fusion and tracking.

References

[1]   "Stargate datasheet," http://www.xbow.com/Products/Product_pdf_files/Wireless_pdf/ Stargate_Datasheet.pdf.

[2]   "MICA2 datasheet," http://www.xbow.com/Products/Product_pdf_files/Wireless_pdf/ MICA2_Datasheet.pdf.

[3]   "TinyOS FAQ," http://www.tinyos.net/faq.html.

[4]   Damarla, T., et al., "Personnel Detection Algorithm for Disposable Sensors Using Multi Sensor Data," *Proc. of MSS BAMS: 2006 Meeting of the MSS Specialty Group on Battlespace Acoustic and Seismic Sensing, Magnetic and Electric Field Sensors*, Laurel, MD, August 2005.

[5]   Zong, L., J. Houser, and T. Damarla, "Multi-Modal Unattended Ground Sensors," *Proc. of Defense Security Symposium, International Society of Optical Engineering (SPIE)*, Orlando, FL, April 2006.

[6]   Pham, T., and H. Papadopoulos, "Distributed Tracking in Ad-Hoc Sensor Networks," *2005 IEEE/SP 13th Workshop on Statistical Signal Processing*, July 17–20, 2005, pp. 1226–1231.

CHAPTER 3

Design and Deployment of Visible-Thermal Biometric Surveillance Systems

Diego A. Socolinsky

Automatic video surveillance in uncontrolled outdoor settings is a very challenging computer vision task. Nearly infinite variability of environmental factors and the open-ended goals of many surveillance problems conspire to create situations where even the most advanced detection, tracking, and recognition algorithms falter. While the common academic response to such challenges is to develop new, more powerful algorithms capable of handling a broader range of conditions with acceptable performance, this course of action is sometimes not appropriate from the industrial-commercial point of view. Sometimes systems must be deployed sooner than would allow for the development cycle of complex new algorithms and must be more robust than most such algorithms can be expected to be on short notice. Under those circumstances, one may look toward better data quality as one means of improving performance while remaining close to the existing state of the art algorithmic technology. This is often the motivation for deployment of multimodal surveillance systems in the real world.

Data quality can be quantified in many ways but intuitively refers to the amount of relevant information conveyed by the data. As far as video surveillance is concerned, numerical gauges such as signal-to-noise ratio (SNR), pixel resolution, sensitivity, and frame rate are common indicators of image quality that can be used as predictors of algorithmic performance. The joint use of multiple imaging modalities is one means of improving some of the quality measures of the input data, in hopes of ultimately improving overall system performance. For example, if the relevant task is detection of moving objects throughout the full diurnal cycle, then the addition of a low-light capable sensor to an existing standard visible surveillance camera will increase the system SNR during nighttime, and thus presumably improve overall detection performance. However, the complexity of the system is also increased by adding a second sensor, not to mention its cost. Is the addition of a low-light camera necessary, or could we instead replace the visible camera outright and use the low-light sensor alone? Would a different sensor technology be better altogether? What is the added performance versus the added cost for different sensor configurations? Will the choice of sensors have an impact if the relevant task changes to include object recognition in addition to detection? The purpose of this chapter is to address some of these questions from a practical point of view.

Since surveillance often requires operation in daytime and nighttime conditions alike, we will focus specifically on bimodal systems in which at least one sensor is capable of imaging in the dark. This will lead us naturally to the combination of thermal infrared imagery with a purely reflective modality, such as standard visible imagery. The following examples are based on the author's experience designing, deploying, and exploiting multimodal imaging systems. As such, they should not be taken as prescriptive, but rather as one practitioner's point of view.

3.1 A Quick Tour Through the (Relevant) Electromagnetic Spectrum

Before discussing the relative benefits of different sensor designs and configurations, we should briefly review the nature of the imagery in each band of the spectrum. Figure 3.1 shows wavelengths from just below 0.4 µm up to 14 µm. The human eye is sensitive to radiation roughly in the range between 0.4 µm and 0.7 µm, depending on individual variation. Blue colors are perceived toward the low end of that range, while reds are near the top. Imagery captured in this range is purely reflective, meaning that the photons sensed by the focal plane array originate at a light source and bounce off the target object and into the camera. This is also true of imaging in the next two slices of the spectrum: near-infrared (NIR), which ranges from 0.7 µm to about 1 µm; and shortwave infrared (SWIR), which comprises the range from 1 µm to 2.5 µm. Moisture suspended in the atmosphere is mostly responsible for the absorption bands 2.5 µm to 3 µm, and 5 µm to 8 µm, which is why we normally do not image in those wavelengths. Between 5 µm and 8 µm lies the midwave infrared (MWIR) spectrum. This is an interesting modality, as it has both reflective and emissive properties—that is, photons impinging on the focal plane array fall into two categories: reflected ones, much as in the lower wavelengths; and emitted ones, which are radiated by the target object by virtue of its temperature and are independent of external illuminants. Finally, the range between 8 µm and 12 µm is known as longwave infrared (LWIR) and consists primarily of emitted radiation. Note that regardless of wavelength, smooth objects such as a mirror will reflect radiation, and thus no imaging modality is completely invariant to illumination effects.

Figure 3.2 shows a face simultaneously imaged in the visible, SWIR, MWIR, and LWIR spectra. Even though both visible and SWIR imagery are strictly reflective, it is interesting to note the fairly pronounced difference in the appearance of the subject between the two modalities. For most people, hair has a much higher albedo in the SWIR, resulting in light-hair-colored images.

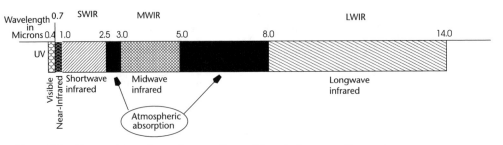

Figure 3.1 Nomenclature for various portions of the electromagnetic spectrum.

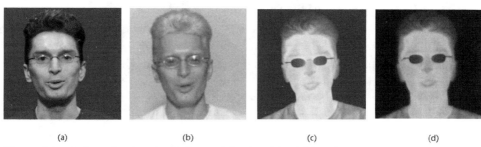

(a)	(b)	(c)	(d)

Figure 3.2   A face simultaneously imaged in the (a) visible spectrum: 0.4 μm to 0.7 μm; (b) shortwave infrared: 0.9 μm to 1.7 μm; (c) midwave infrared: 3.0 μm to 5.0 μm; and (d) longwave infrared: 8.0 μm to 14.0 μm.

Likewise, most clothing dyes have higher albedo in the NIR and SWIR than in the visible spectrum, thus clothing often appears much brighter in those modalities. The change in appearance when we move to longer wavelengths is quite dramatic, as we start to see the emissive component taking over. The images in Figure 3.2 were acquired indoors, and thus the reflective aspects of MWIR are not emphasized, resulting in rather similar MWIR and LWIR images. An interesting fact, well known to anyone familiar with thermal imagery, is that glass is completely opaque in the MWIR and LWIR, as can be seen by looking at the eyeglasses in Figure 3.2(c) and Figure 3.2(d). Glass also has very low emissivity in the MWIR and LWIR (compared to skin), which, combined with the fact that room temperature is normally lower than skin temperature, explains why it appears darker than skin. This has obvious consequences for any biometric application that exploits the appearance of the human face.

In this chapter we will focus primarily on visible, NIR, and LWIR imagery for surveillance and biometric applications. These modalities give us the most obvious complementarity, and using them in combination is normally the most cost-effective path to high performance.

3.2   Why and When to Use a Fused Visible-Thermal System

Thermal infrared imagery has a long history of development and application in the military realm. This is primarily due to the fact that, as mentioned in Section 3.1, thermal imagery is somewhat unaffected by level of illumination, relying on emitted rather than reflected radiation. In other words, it works in the dark. This is a critical advantage for military operations, since it allows for nighttime imaging without the need for external lighting. In addition to making operation in the dark a possibility, thermal imagery (most notably LWIR) provides more consistent target appearances throughout the diurnal cycle, thus simplifying the task of recognition. All in all, LWIR appears to be the optimal imaging modality: it works night and day, and objects appear consistently uniform regardless of illumination.

On the other hand, thermal imaging has its disadvantages. First, there is the issue of cost. Thermal cameras range from very expensive to outrageously expensive, depending on wavelength, resolution, and sensitivity. On a pixel-per-pixel basis, thermal cameras are anywhere from a few dozen to a thousand times more expensive than their visible counterparts. This point alone makes the use of thermal imagery, let alone fused imagery, hard to justify for a large number of applications. However, there are those situations where performance is more important than price, and thermal sensors are justified. Many, but not all, military and national security applications fall within this category. Beyond cost, thermal cameras have considerably lower resolution than visible ones, with the typical resolution still hovering around 320×240 pixels. Higher resolution sensors exist (up to about a megapixel in rare cases) but are almost exclusively oddities available only in minimal quantities and with exorbitant prices. Lower resolution naturally leads to the need for higher magnification optics, which leads us to mention that, not surprisingly, infrared lenses are far more expensive than standard visible ones, often by a factor of 10 or more. Add to this the fact that the selection is much more limited, and you quickly realize that your next thermal imaging project may very well require custom optics.

Two more related issues can be considered as disadvantages for thermal imaging. Most writing on signs and markings on vehicles are not discernible under most circumstances. This is a clear problem for applications such as driver vision enhancement, since a driver whose sole visual input comes from a thermal sensor would be unable to read most traffic signs. Also, any identifying markings on military vehicles would be invisible, potentially complicating friend-versus-foe identification. Along similar lines, untrained operators have a harder time orienting themselves and correctly identifying objects in thermal imagery than in visible imagery. This is not surprising, since visible imagery is, by definition, our native modality. The unfortunate consequence is that situational awareness is sometimes decreased by the use of thermal imaging alone.

Visible imaging technology has complementary strengths and weaknesses with respect to thermal imaging. Very high resolution cameras are available for affordable prices. In fact, the price per pixel has been steadily decreasing at a fast pace for years. A wide variety of lenses is available off the shelf to suit almost any application. Given the widespread use of visible cameras, a large amount of legacy data is available covering most targets of interest, both humans and vehicles. Setting aside for a moment the operational feasibility of the identification task, we have enough visible facial imagery of many well-known terrorists to train a biometric surveillance system to find them. The same cannot be said of thermal imagery. On the other hand, visible imagery is heavily affected by ambient illumination—so much so that the appearance of a target can vary dramatically from one illumination condition to another, making matching and recognition extremely challenging. So, even if we had a properly trained face recognition system to spot wanted individuals, our accuracy would be hampered by lighting variations. Of course, visible imagery is of limited use in darkness, so nighttime operation is severely restricted. On the plus side, visible imagery is useful for situational awareness, since our brains are very good at interpreting it.

Imaging in the NIR and SWIR can in some sense be seen as an extension of visible imaging, with some advantages and some drawbacks. Through the use of image intensification, the NIR spectrum is ideal for night vision applications. In fact, that is the most commonly used modality in the area. While digital intensified night vision technology is relatively new, it is certainly mature enough to be deployed. The cost is considerably higher than for visible imagery, but not so much so as to make it prohibitive. Available resolution exceeds a megapixel, so it is comparable with high-quality visible video cameras, but image noise is quite a bit higher. As for SWIR, technology is still fairly new, and thus the cost is comparable to some thermal systems. However, SWIR has proven to be extremely promising for outdoor night vision applications, given the SWIR radiance of the nighttime sky. Even though SWIR imagery is of a reflective nature, most objects have different enough albedo in the SWIR and visible ranges as to make crossmodality matching more of a challenge (Jeff Paul, program manager for DARPA MANTIS programs, personal communication, 2005).

After this review of pros and cons for the different relevant modalities, we begin to see why a system combining reflective and thermal imaging might be useful. Assuming that the cost of the thermal component can be justified, a combined system has some distinct advantages. At any time in the diurnal cycle when both sensors are capable of operating, we get the best of both worlds. The thermal component provides consistent imagery regardless of illumination, albeit at lower resolution. The (nominally) visible component provides high resolution data, although degraded by illumination effects. In combination, these two streams of data are more robust than either one alone. As an example, multiple studies by different groups have shown that for the task of cooperative face recognition under favorable lighting conditions, the joint use of visible and LWIR imagery provides increased performance over either modality alone [1–10]. Figure 3.3 shows an example set of receiver-operator-characteristic curves for the same face recognition

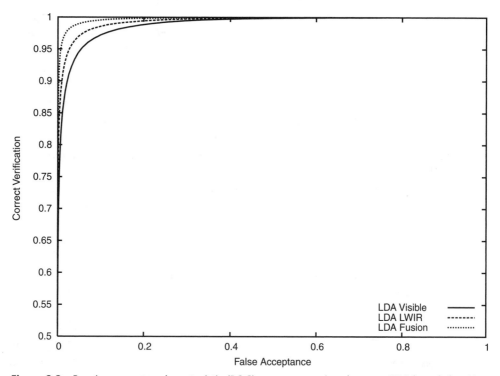

Figure 3.3 Receiver operator characteristic (ROC) curve comparing the same LDA-based algorithm for visible, LWIR, and fused visible/LWIR modalities.

Figure 3.4   Screenshot from a biometric surveillance system using a combination of LWIR and NIR imagery.

algorithm applied to visible and LWIR imagery, and a score-based fusion of the two. The equal error rate for the fused result is about half that of either modality alone. Under unfavorable illumination conditions, the advantage can be even higher.

Tracking, is another application where visible/NIR and thermal modalities are often synergistic, especially as targets move through shadows and strong illumination variations or where complementary phenomenology can be used to filter out noise. Figure 3.4 shows an application where thermal imaging is used for pedestrian detection and tracking, which in turn controls a pan-tilt mount with a low-light NIR sensor and high magnification optics. In this case, the thermal sensor provides good quality low-noise imagery on which to perform background modeling and people-tracking throughout the diurnal cycle using a wide-field-of-view lens. The slaved NIR sensor is fitted with a long focal length lens that can capture facial images for biometric surveillance at distances of up to 200 ft. Note that a wide field visible sensor would not allow nighttime operation, while an intensified NIR (I2) sensor would be quite noisy and would make background modeling more difficult. At the same time, a narrow field thermal sensor would require very expensive optics, as price generally goes up with focal length, and would not allow for use of legacy visible biometric data. Altogether, this system shows how visible and thermal imagery can be used in a combination that exploits their respective strengths within a simple design.

An example where the two modalities are more tightly coupled is shown in Figure 3.5. This is a custom-fused visible-LWIR system for tracking of maritime objects in a harbor environment. During operation, the system is mounted in a floating buoy and provides surveillance and security for harbor operations. One of the main challenges for detection and tracking of maritime objects from a

Figure 3.5 Fused visible-thermal system developed for harbor surveillance operations. When deployed, the system is mounted in a floating buoy.

surface-mounted sensor is the constantly moving water surface, including waves and whitecaps. While this presents difficulties on both modalities, it does so in different ways and at different times. In order to best exploit the complementarity between modalities at all depths of field, this system is configured using a dichroic splitter, which allows very accurate pixel-to-pixel coregistration of channels. This should be contrasted with the previously described system, where the visible and thermal channels were only coarsely coregistered via absolute camera calibration.

3.3   Optical Design

One critical factor in the design of visible-thermal systems is the optical configuration of the sensors. The choice of configuration influences the packaging requirements and overall size of the system, has a direct impact on cost, and dictates what kinds of processing are possible using the resulting imagery. Perhaps the most common configuration, and certainly the simplest, is to use boresighted sensors mounted rigidly with parallel optical axes (Figure 3.6). Simplicity is the main advantage of this arrangement, which requires nothing beyond a rigid mounting platform or rail and can be executed very inexpensively. As long as accurate coregisration between sensors is not required, or the objects of interest are far away from the sensor rig, this should be the first method used by anyone experimenting with fused imagery. The obvious disadvantage is that we have essentially created a bimodal stereo rig, and a depth-dependent disparity between corresponding features in the two modalities is inevitable. Sometimes this is not a problem. If the objects of interest are far away from the cameras, then the disparity after calibration will be minimally dependent on relief of the objects, and thus the

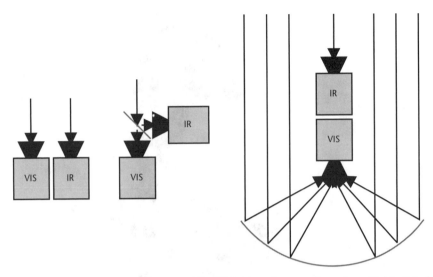

Figure 3.6 Left: Boresighted cameras. Middle: Dichroic splitter configuration. Right: Catadioptric configuration.

images may be considered coregistered. Alternatively, exact coregistration may not be required at all, as in the bimodal surveillance system in Section 3.2. In that case, one modality was used to guide the other, but pixel-level correspondences were not used and therefore not needed.

Other tasks, however, require pixel-level coregistration accuracy that cannot be attained with boresighted sensors. An example in biometrics stems from the fact noted above that eyeglasses are opaque in the thermal infrared (see Figure 3.2). Most face recognition systems, regardless of imaging modality, make use of eye locations as a means to geometrically normalize imagery prior to further processing and classification. This is simply because the eyes are easily detectable features that are rigidly encased in the skull and do not change relative position with facial expressions. Unfortunately, if a subject is wearing eyeglasses, the eyes are occluded in the thermal infrared and cannot be accurately located.[1] A fused visible-thermal system with pixel-level coregistration between channels can be used to overcome this hurdle, as long as conditions are favorable for the visible sensor, by simply transferring the eye locations from the visible image over to the registered thermal image.

Another example comes from background modeling algorithms commonly used for detection of moving objects. In this case, a statistical model is built for each image pixel based on some local feature such as color. In a visible-thermal system, this can be done successfully only if both channels are coregistered to sufficient accuracy so that local statistics are computed over corresponding scene locations on both channels.

Lastly, we draw another example from the emerging field of immersive visualization for perception enhancement. Over the last few years, a number of

---

1.   Eye locations for an eyeglass-wearing subject in the thermal infrared can be estimated from a segmentation of the glasses, but due to variability in eyeglass design, this estimate is normally not accurate enough for robust normalization.

wearable visible-thermal systems have been developed that give the user a fused visualization of imagery captured from a head-worn camera pair in real time. By combining visible or intensified NIR imagery with thermal infrared, users are afforded enhanced situational awareness and increased ability to navigate treacherous terrain and detect humans and vehicles in adverse conditions. In order for the fused imagery to accurately combine features from both modalities, the sensors must be coregistered with high accuracy. Otherwise, the fused output may contain artifacts obscuring real scene features, or creating nonexistent ones.

There are a few optical designs we can use to obtain pixel-level coregistration, and as can be expected, accuracy and compact packaging are directly related to cost. The simplest design uses a pair of cameras with aft optics and a dichroic splitter, as seen in Figure 3.5 and Figure 3.6. In this arrangement, cameras are mounted with perpendicular optical axes, each fitted with its own lens. A dichroic splitter is placed in the optical path, tilted at a 45-degree angle to each optical axis. Using appropriate mechanical mounts for the cameras and mirror, as well as magnification matched lenses, we can reduce the baseline between sensors to near zero through careful calibration. We should note that while it is theoretically possible to perfectly align the sensors with this design, it is in practice a difficult and time-consuming process to do so. Most systems with this optical design have a residual amount of depth-dependent misregistration, though it is much smaller than with boresighted sensors. A critical cost consideration in this case is the size of the dichroich splitter, since the materials and coatings involved can be quite expensive. This in turn suggests that using smaller cameras mounted closer to the splitter may be a cost-saving design consideration. Since the cameras are mounted with perpendicular optical axes, this arrangement may also result in a larger overall package than if we used boresighted cameras.

If we consider the path of incoming light to each of the sensors in a bimodal system using a dichroic splitter, we quickly realize that we essentially have a pair of boresighted sensors, with the key difference that we can reduce the baseline below the physical size of the cameras. That is, in a simply boresighted arrangement, we cannot get the optical axes any closer than allowed by the size of the cameras without the respective shells coming into contact. By using a dichroic splitter, we are able to divert one of the optical paths to the side, where we have more space to maneuver the cameras without encroaching on each other. In this fashion, we can decrease the baseline to practically zero. An alternative means of reducing the baseline is to use a catadioptric design, as shown in Figure 3.6. Such optics are most commonly seen in telescopes, where a small mirror is placed at the focus of the larger light-gathering mirror and used to produce an image on the eyepiece. The presence of the small mirror in the optical path simply diminishes the overall amount of light impinging upon the larger mirror and decreases the t-number of the telescope. We can replace the small mirror with a focal plane array to produce a catadioptric camera. Note that we can insert a second small refractive camera in the obscuration created by the first FPA. The resulting bimodal system can be calibrated to have minimal baseline by appropriately locating the mirrors and cameras. The main drawback of this configuration is that it is limited to very narrow fields of view (or high magnifications), usually around five degrees or less.

This is well-suited for telescopic applications, but not to more general surveillance tasks. Successful fused visible-thermal weapon sights have been built based on this optical design. That application is a particularly good fit, since a narrow field of view is standard and the catadioptric arrangement results in a package that can be made small in the two dimensions perpendicular to the optical axes, but is larger along the common axis. As weapon sights are roughly cylindrical in shape, the usability constraints are well matched to the feasibility constraints.

The previous two designs allow for very close coregistration of the sensors. Their principal drawback rests on the fact that they require close alignment of two or more physical objects (cameras, lenses, mirrors) of nonnegligible mass. Even when carefully calibrated into coregistration, these systems are vulnerable to misalignment due to vibrations or shock. In some operating conditions these cannot be avoided, so maintaining the utmost rigidity between components becomes an important design parameter. This is sometimes at odds with allowing sufficient degrees of freedom during calibration in order to bring the channels into coregistration. The ideal design provides complete positional flexibility during calibration while becoming completely rigid for day-to-day operation. Needless to say, this can be challenging. The only general advice in that respect is that smaller sensors, by virtue of their lower mass, make a rigid system easier to achieve.

The ultimate level of coregistration between channels can be achieved only by using a hybrid focal plane array. A few companies and research laboratories have produced arrays containing a combination of visible and LWIR pixels (other hybrid combinations also exist). While a discussion of the technology is beyond the scope of this chapter, suffice it to say that this level of alignment can hardly be improved. On the other hand, all hybrid arrays are experimental at this point and do not have the individual performance of their single-modality brethren. We cannot even establish a cost comparison, since hybrid arrays are a true rarity. We can safely expect them to be expensive once they reach production. An interesting point, usually overlooked when discussing hybrid visible-LWIR focal plane arrays, is the need for a lens that can simultaneously focus visible and LWIR light on the chip. Recall that a system with aft optics utilizes separate lenses for each modality. When both modalities are combined onto the same chip, we are forced to use a single lens. A simple solution is to use a catadioptric lens—in essence a curved mirror that reflects light at all wavelengths onto its focusing point. As we remarked above, however, such a lens is only feasible for very narrow fields of view. If we require a wider field of view, as almost any surveillance system would, we need to use a refractive lens. At the moment, only one company (the author's employer) manufactures a refractive lens capable of simultaneously focusing visible and LWIR light with a 20-degree field of view and has designs reaching 40 degrees. Readers familiar with the concept of chromatic aberration in visible lenses will realize that the problem here is much larger, since one must simultaneously focus wavelengths almost 14 μm away, rather than a mere 0.5 μm. Obviously, the combination of a hybrid visible-LWIR focal plane array with a refractive hyperachromatic lens has the potential to yield the most compact and highest performing package, although certainly not at the lowest cost. A few such systems have been built.

3.4   Choice of Sensors

The previous section addressed optical design considerations for a fused visible-thermal system. In this section, we concentrate on the choice of wavelengths and sensors. From a phenomenological viewpoint, it is advantageous to combine a modality with primarily reflective characteristics with one that is primarily emissive. This is not a technically accurate distinction, since some objects will be reflective in practically all wavelengths of interest. Smooth surfaces have this property and can be seen to be highly reflective throughout the spectrum from visible through LWIR. Nonetheless, the distinction is practically useful and can guide our design.

At temperatures encountered in everyday scenes, objects essentially do not radiate in wavelengths up to and including SWIR. This means that all light reaching a sensor originated in an external source, bounced off the object in question, and then reached the sensor. Consequently, any change on the nature, location, or intensity of the illuminant will affect the resulting image. At those same temperatures, objects naturally radiate in the MWIR and LWIR wavelengths (and above). Up to scaling by the emissivity of the surface, the emitted power is dictated by Planck's law and tends to grow from the MWIR to the LWIR.[2] As a result, a larger proportion of the light measured from an object in the LWIR tends to be emitted, resulting in lower variability in that modality with respect to sources of illumination. For some materials, illumination invariance is not so much a consequence of Planck's law, but rather of their surface emissivity. An important example is human skin. A considerable body of research now exists on the subject of thermal infrared face recognition [1–9]. The original impetus for this research was the fact that thermal imagery of the face shows little dependence on illumination. This fact can be easily verified in the LWIR but is equally easily seen to be false in the MWIR. The reason is that human skin has emissivity near unity in the LWIR but only about 75% in the MWIR [11]. In combination with the difference in emission from the face at skin temperature (approximately 33°C), this results in moderate illumination dependence for MWIR face images. While indoor data collections may not reveal this, data acquired in sunlight conditions will easily show a much stronger variability in the MWIR than in the LWIR. For biometric surveillance purposes, LWIR will provide a more invariant signature.

Longwave infrared sensors also have the most attractive price-point of all thermal modalities, primarily thanks to the fact that LWIR focal planes based on microbolometer technology can operate without cooling. Uncooled cameras consume much less power, are smaller, and cost a fraction of the price of their cooled counterparts, making them attractive for portable applications. In turn, cooled sensors have much higher sensitivity and lower noise. If measuring small variations in temperature or emissivity is the goal, then a cooled MWIR camera is probably the right choice. Researchers have used cooled MWIR sensors to

2.   This statement is true for temperatures encountered in everyday scenes, say, from freezing to boiling water. The statement is not true in general, as Planck's curve does not monotonically increase with respect to wavelength.

map the patterns of blood vessels, opening up new avenues for short-range biometrics [12].

These considerations indicate that when designing a system with reflective and thermal sensors, LWIR prevails in terms of cost and signature robustness, while MWIR has the advantage in terms of sensitivity (and thus often probability of detection). The selection of reflective sensor is driven primarily by the availability of ambient light. If the system is to be used in daylight, or with ample artificial illumination, there is little reason to use anything other than a standard visible camera. The choice between color and grayscale depends largely on the application. Background modeling and foreground detection benefit from color, as does object tracking. Recognition and classification in a surveillance scenario are less dependent on color[3] and may benefit more from the higher resolution available in grayscale cameras. Sensitivity and dynamic range are also usually better in monochrome sensors and are both critical when operating in outdoor scenarios.

If operation throughout the diurnal cycle is a requirement, a reflective modality other than visible will be necessary (although it can be supplemented with visible imagery whenever there is sufficient light). The option is between one of the various intensified NIR technologies and SWIR. While SWIR has recently been shown to have very promising performance for outdoor nighttime operation (Jeff Paul, program manager for DARPA MANTIS program, personal communication, 2005), it remains expensive and somewhat challenged in terms of resolution compared to NIR. As SWIR technology improves, it may very well become the default modality for reflective night-vision use, but it is still too early to tell. Intensified NIR is the more established modality in this arena. Traditionally used with the help of direct-view devices, it has recently transitioned to the digital domain through the use of a few competing technologies. Fiber-optically coupled CCDs (ICCD), electron-bombarded active pixel sensors (EBAPS) and MCP CMOS are the most competitive instantiations of digital intensified NIR currently in the market. They are similar in terms of light sensitivity, noise, and MTF, with a small advantage going to each one in different comparisons. In terms of price, they rival the cost of uncooled LWIR microbolometers, so a system incorporating both should be reserved for situations where around-the-clock performance is an absolute necessity. We should keep in mind that I2 imagery is not just "visible that works at night." In fact, there are considerable trade-offs to be made when moving to I2 imagery, primarily in terms of noise and MTF. While modern digital I2 cameras have resolutions exceeding one megapixel, the MTF (sharpness, in plain terms) of the imagery is far from that of a comparable visible camera, so the effective resolution (in terms of resolving power) is actually lower. Also, as an inevitable consequence of intensification and photomultiplication, I2 sensors have considerably more noise than their visible counterparts. At low light levels, the signal-to-noise ratio can be severely diminished, to the point that automated processing of the imagery can be very challenging. Some digital I2 technologies, mainly EBAPS, have daytime performance approaching that of a visible camera, but this is the exception rather than the rule. When specifying an I2 sensor as a

---

3. Except perhaps in the situation where we are interested in locating, say, red cars.

visible replacement, we must be prepared to give up some daytime performance in the interest of a larger increase at night. Alternatively, a three-camera system can be built, where the thermal camera is used in combination with either a visible or intensified sensor, depending on lighting conditions. Needless to say, complications mount quickly as we add more sensors, so every system designer would do well to heed Occam's razor.

3.5   Biometrically Enabled Visible-Thermal Surveillance

When designing a multimodal surveillance system meant to detect and track humans, we should keep in mind that a requirement for recognition capabilities cannot be far behind. As automated surveillance systems become more widespread, customers will inevitably want to go beyond the ability to detect and track, requesting the ability to identify the detected individuals. Some such systems have already been developed, and some have even been publicly tested, much to the chagrin of their corporate parents. Regardless of currently achievable performances, we can be assured that biometrically enabled surveillance systems will be deployed in various contexts within the next decade. This section addresses the intrinsic issues of a visible-thermal system as they pertain to biometric identification, and in particular face recognition.

One of the first questions asked by someone unfamiliar with visible-thermal face recognition is, "Can you recognize someone from a driver's license picture, given their thermal image?" We should quickly review the nature of visible-thermal face recognition to see why this is a great question. When we speak of face recognition using fused visible-thermal imagery, we normally refer to a process whereby, either at enrollment or testing time, image pairs are collected using any one of the sensor configurations described above. For all existing publicly described systems, low-level matching is performed in a somewhat independent manner on each channel. That is, no system compares the visible image in one pair to the thermal image in the other. Some systems combine both modalities into a single image and then proceed with matching, but that has essentially the same effect. As of the time of writing, there are no face recognition systems capable of accurately matching images across the visible-thermal divide. This is not surprising, given the great disparity in appearance between those modalities. On the other hand, it is also not inconceivable that such a system could be built, since both images are generated by the same underlying geometric object by a process that can be physically modeled. A simple, though low-performance, version of a cross-modality recognizer could use relative facial feature locations as its underlying biometric signature. Such features can be automatically located in both modalities, and face recognition based on facial feature locations (without appearance) has been done before [13]. Unfortunately, the accuracy of that matching method is very low and does not lend itself to real-world applications.

So, the current state of the art in visible-thermal face recognition matches each channel essentially separately[4] and combines the results afterward. The

---

4.   This is not strictly true, since information gleaned from one channel, such as eye locations, is brought to bear on the other. However, the low-level comparison occurs separately for each channel.

unfortunate consequence of any such method is that images in both modalities are required both at enrollment and testing time. Without a matching thermal image, having someone's driver's license picture only allows us to do face recognition in the visible domain, even if we can capture a visible-thermal pair at the time of identification. This leads immediately and abruptly to the fact that there is virtually no useful legacy data to prime a visible-thermal biometric surveillance system. We can contrast this situation with that for visible biometric surveillance. In that case, there is normally a large quantity of face imagery acquired previously for a variety of reasons: driver's licenses, passports, visa applications, work IDs, and so forth. Setting aside the issue of performance, we see that these can be used as gallery data for a visible biometric surveillance system. The same is not possible for a visible-thermal system.

The above objection seems to rule out the use of visible-thermal imagery for face recognition in a realistic scenario, since it is unlikely (though not impossible) that thermal images will become standard for driver's licenses or passports. There is, of course, a valid application in access control, where the subject is voluntarily enrolled into the system, and for which visible-thermal face recognition has been shown to yield superior performance [4, 10]. We will not dwell on that scenario, as this chapter is focused on surveillance. Instead, we will propose an alternative deployment where fused visible-thermal imagery is not only applicable but also improves performance. The key functionality in the proposed scenario is reacquisition. This is quite common in surveillance problems, as exemplified in the following two applications. First, consider a security system monitoring the perimeter of a secure installation, say an embassy. Aside from detecting approaching pedestrians, and perhaps testing their identity against a watch list, we may be interested in detecting individuals who are spotted often during a given period of time but are not on our watch list. For relatively short periods of time, this would detect loitering individuals. For longer periods of time, it would detect individuals who come by the facility often. Both of these types of individuals may deserve further investigation by a human operator as a potential security risk. In such a system, enrollment happens on the fly, using the same sensor used for recognition. Therefore, if a fused visible-thermal system is deployed, it can be used without regard for lack of legacy data. As long as one of the available modalities coincides with that of the watch list images (likely visible), there should be no performance degradation for the watch list task, while performance for loitering detection (reacquisition) is improved by the addition of a thermal modality. This is especially helpful in order to extend the diurnal coverage of the surveillance system.

A related application of reacquisition can be drawn from casino surveillance, although other environments would do just as well. In casino surveillance, just as in loitering detection, enrollment of individuals into a watch list is performed on the fly, using the same sensor for detection and recognition. These are typically installed in clear domes throughout the casino floor and are used to monitor players and detect unwanted individuals. If some or all such sensors were replaced with visible-thermal systems, performance would improve for all individuals enrolled using the fused system, while it would remain equal for those enrolled using the legacy visible-only system. As casino floors often have uneven illumination and are sometimes rather dim, the addition of a thermal sensor would add considerable robustness to the recognition process.

Other biometric surveillance applications have important aspects that can be cast as reacquisition tasks and would thus directly benefit from a multimodal approach without incurring the penalty imposed by lack of legacy enrollment data. The key to a successful deployment of a multimodal system in such an environment is to incorporate the legacy modality in the new system, so as to preserve performance for existing enrollment data and improve it for subsequent enrollments.

3.6 Conclusions

The use of visible (or NIR) and thermal sensors in combination opens up new possibilities for surveillance applications. It allows us to transfer tried and true techniques that work in favorable illumination conditions to realistic outdoor scenarios. By leveraging the relative strengths of reflective and emmisive modalities, fused sensor systems are often capable of operating throughout the diurnal cycle. Their main advantage is the robustness to environmental variation, which ultimately results in higher performance. The success of any such deployment ultimately depends not only on performance, but also on reliability and cost. These three factors are intimately linked to the design of the sensor platform itself, which in turn dictates the possible algorithmic avenues. Proper design is critical to customer acceptance, and in the end has a major influence on the likelihood of deployment.

Real-world deployment is often dependent on measurably acceptable performance for a given task. It is very important to realize that properly defining the task has great influence in perceived and measured performance, and ultimately may mean the difference between an academic exercise and a commercially deployed system. For the case of visible-thermal biometric surveillance, we showed that despite the fact that visible-thermal face recognition can be academically shown to outperform visible alone, if the task is ill-suited to the technology, acceptable performance cannot be achieved. On the other hand, a properly selected task will greatly benefit from the added performance afforded by a visible-thermal system. This reasoning is applicable to many current aspects of biometric surveillance, both unimodal or multimodal. Unless the application realm is properly delimited, performance cannot be expected to be adequate, regardless of algorithmic or hardware progress. It will be of great benefit to the field to concentrate on achieving excellent results in feasible yet relevant scenarios, rather than mediocre ones in infeasible ones.

This chapter constitutes a brief foray into design issues for visible-thermal systems, with emphasis on biometric surveillance. We have purposely kept the exposition at a relatively high level, so as not to be mired by details. In this fashion, it will hopefully prove a useful introduction for the uninitiated and a subject of fruitful discussion for the more seasoned practitioner.

References

[1]   Socolinsky, D., et al., "Illumination Invariant Face Recognition Using Thermal Infrared Imagery," *Proc. CVPR*, Kauai, HI, December 2001.

[2]   Prokoski, F., "History, Current Status, and Future of Infrared Identification," *Proc. of IEEE Workshop on Computer Vision Beyond the Visible Spectrum: Methods and Applications*, Hilton Head, SC, 2000.

[3]   Socolinsky, D., and A. Selinger, "A Comparative Analysis of Face Recognition Performance with Visible and Thermal Infrared Imagery," *Proc. ICPR*, Quebec, Canada, August 2002.

[4]   Socolinsky, D., and A. Selinger, "Face Recognition with Visible and Thermal Infrared Imagery," *Computer Vision and Image Understanding*, July–August 2003.

[5]   Wolff, L., D. Socolinsky, and C. Eveland, "Face Recognition in the Thermal Infrared," in *Computer Vision Beyond the Visible Spectrum*, I. Pavlidis and B. Bhanu, (eds.), New York: Springer Verlag, 2004.

[6]   Chen, X., P. Flynn, and K. Bowyer, "Visible-Light and Infrared Face Recognition," *Proc. of the Workshop on Multimodal User Authentication*, Santa Barbara, CA, December 2003.

[7]   Chen, X., P. Flynn, and K. Bowyer, "PCA-Based Face Recognition in Infrared Imagery: Baseline and Comparative Studies," *International Workshop on Analysis and Modeling of Faces and Gestures*, Nice, France, October 2003.

[8]   Cutler, R., "Face Recognition Using Infrared Images and Eigenfaces," Technical report, University of Maryland, April 1996, http://www.cs.umd.edu/rgc/pub/ir_eigenface.pdf.

[9]   Wilder, J., et al., "Comparison of Visible and Infra-Red Imagery for Face Recognition," *Proc. of 2nd International Conference on Automatic Face and Gesture Recognition*, Killington, VT, 1996, pp. 182–187.

[10] Socolinsky, D., and A. Selinger, "Thermal Face Recognition in an Operational Scenario," *Proc. of CVPR 2004*, Washington, D.C., 2004.

[11] Wolff, L., D. Socolinsky, and C. Eveland, "Quantitative Measurement of Illumination Invariance for Face Recognition Using Thermal Infrared Imagery," *Proc. of CVBVS*, Kauai, HI, December 2001.

[12] Buddharaju, P., I. Pavlidis, and P. Tsiamyrtzis, "Pose-Invariant Physiological Face Recognition in the Thermal Infrared Spectrum," *Biometrics '06*, 2006, p. 53.

[13] Sakai, T., M. Nagao, and T. Kanade, "Computer Analysis and Classification of Photographs of Human Faces," *Proc. First USA-JAPAN Computer Conference*, 1972, pp. 55–62.

CHAPTER 4

LDV Sensing and Processing for Remote Hearing in a Multimodal Surveillance System

Zhigang Zhu, Weihong Li, Edgardo Molina, and George Wolberg

4.1 Introduction

Recent improvements in laser vibrometry [1–6] and day/night infrared (IR) and electro-optical (EO) imaging technology [7, 8] have created the opportunity to create a long-range multimodal surveillance system. Such a system would have day and night operation. The IR and EO video system would provide the video surveillance while allowing the operator to select the best targeting point for picking up audio detectable by the laser vibrometer. This multimodal capability would greatly improve security force performance through clandestine listening of subjects that are probing or penetrating a perimeter defense. The subjects may be aware that they are observed but most likely would not infer that they could be heard. This system could also provide the feeds for advanced face and voice recognition systems.

Laser Doppler vibrometers (LDV) such as those manufactured by Polytec [2] and B&K Ometron [3] can effectively detect vibration within 200m with a sensitivity on the order of 1 μm/s. These instruments are designed for use in laboratories (0m to 5m working distance) and field work (5m to 200m) [2–6, 9]. For example, these instruments have been used to measure the vibrations of civil structures like high-rise buildings, bridges, towers, and so forth, at distances of up to 200m. However, for distances above 200m, it will be necessary to treat the target surface with retro-reflective tape or paint to ensure sufficient retro-reflectivity. At distances beyond 200m and under field conditions, the outgoing and reflected beam will pass through the medium with different temperatures and thus different reflective coefficients. Another difficulty is that such an instrument uses a front lens to focus the laser beam on the target surface in order to minimize the size of the measuring point. At a distance greater than 200m, the speckle pattern of the laser beam induces noise, and signal dropout will be substantial [10]. Finally, the visible laser beam is good for a human to select a target, but it is not desirable for a clandestine surveillance application.

The overall goal of this work is to develop an advanced multimodal interface and related processing tools for perimeter surveillance using the state-of-the-art sensing technologies. In the foreseeable future, human involvement in all three stages—sensors, alarm, and response—is still indispensable for a successful surveillance system. However, the studies of the capabilities of these sensors— infrared (IR) cameras, visible (EO) cameras, and the laser vibrometers (LDVs)— are critical to surveillance tasks. IR and EO cameras have been widely studied and used in human and vehicle detection in traffic and surveillance applications [8]. However, literature on remote acoustic detection using the emerging LDVs is rare. Therefore, we mainly focus on the experimental study of LDV-based voice detection in the context of a multimodal surveillance system. We have also set up a system with the three types of sensors for performing integration of multimodal sensors in human signature detection. Hence, this chapter also briefly discusses how we can use IR/EO imaging for subject detection, and target selection for LDV listening.

We first give an overall picture of our technical approach: the integration of laser Doppler vibrometry and IR/color imaging for multimodal surveillance. One of the important issues is how to use IR and/or color imaging to help the laser Doppler vibrometer to select the appropriate reflective surfaces in the environment. Then we discuss various aspects of LDVs for voice detection: basic principles and problems, signal enhancement algorithms, and experimental designs.

Speech enhancement algorithms are applied to improve the performance of recognizing noisy voice signals detected by the LDV system. Many speech enhancement algorithms have been proposed [11–14], but they have been mainly used for improving the performance of speech communication systems with noisy channels. Acoustic signals captured by laser vibrometers need special treatment since the detected speech signals may be corrupted by more than one noise source, such as laser photon noises, target movements, and background acoustic noises (wind, engine sound, and so forth).

The quality of the LDV signals also strongly depends on the reflectance properties of the surfaces of the target. Important issues like target surface properties, size and shape, distance from the sensors, sensor installation, and calibration strategy are studied through several sets of indoor and outdoor experiments. Through this study, we have gained a better understanding of the LDV performance, which could guide us for improving the LDV sensors and acquiring better signals via multimodal sensing. We provide a brief discussion on some future work in LDV sensor improvements and multimodal human signature detection.

4.2 Multimodal Sensors for Remote Hearing

There are two main components in our approach of multimodal human signature detection (Figure 4.1): the IR/EO imaging video component and the LDV audio component. Both the IR/EO and LDV sensing components can support day and night operation even though it would be better to use an IR camera to perform the surveillance task during the nighttime. The overall approach is the integration of the IR/EO imaging and LDV audio detection for a long-range surveillance task, which has the following three steps:

- *Step 1. Subject detection and tracking via the IR/EO imaging module.* The subjects could be humans or vehicles (driven by humans). This could be achieved by motion detection and human/vehicle identification, either manually or automatically.

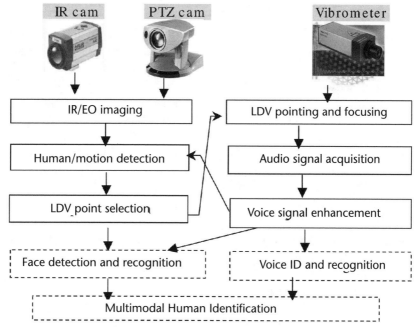

Figure 4.1   System diagram.

- *Step 2. Audio targeting and detection by the LDV audio module.* The audio signals could be human voices or vehicle engine sounds. We mainly consider the human voice detection. The main issue is to select LDV targeting points, not necessarily on the subjects to be monitored, provided by the IR/EO imaging module, in order to detect the vibration caused by the subjects' voices.
- *Step 3. Face/vehicle shots of best views triggered by the feedback from audio detection.* By using the audio feedback, the IR/EO imaging module can verify the existence of humans and capture the best face shots for face recognition. Together with the voice recognition module, the surveillance system could further perform human identification and event understanding.

Figure 4.1 shows the system diagram. In our current experiments, the first step was performed manually. But a large body of automated algorithms exists in literature [e.g., 8] that can be readily applied here. In the text of the following sections, we will point out where IR/EO modules come into play for LDV remote hearing. The third step is also well-studied, including face detection and recognition, audio/video speech recognition [15–17], and multimodal biometrics [18]. The focus of the chapter will be on the second step, including LDV target detection, LDV pointing, LDV voice detection, and signal enhancements.

For enabling the study of the multimodal sensor integration for human signature detection, we have acquired the following sensors: a laser Doppler vibrometer (LDV) OFV-505 from Polytec, a ThermoVision A40M infrared camera from FLIR, and a Canon color/near IR pan/tilt/zoom (PTZ) camera.

4.2.1 The LDV Sensor

The laser Doppler vibrometer from Polytec [2] includes a controller OFV-5000 with a digital velocity decode card VD-6 and a sensor head OFV-505 (Figure 4.2). We also acquired a telescope VIB-A-P05 for accurate targeting at large distances.

The sensor head uses a particular helium neon red laser with wavelength of 632.8 nm and is equipped with a super long-range (SLR) lens. It sends the interferometry signals to the controller, which is connected to the computer via an RS-232 port. The controller box includes a velocity decoder VD-06, which processes signals received from the sensor head. There are a number of output signal formats from the controller, including an S/P-DIF output and digital and analogue velocity signal outputs.

To receive and to process the signal from the controller, we use a low-cost Audigy2 ZS audio card with a built-in S/P-DIF I/O interface on the console of the host computer. This audio card can receive the digital signals from the controller and play them back through the audio outputs on the console machine. It can also save the received signals as audio files, for example, in mp3 or wav format. The main features of the LDV sensor and the accessories are listed as follows:

- *Sensor head OFV-505*. HeNe (Helium-Neon) laser, $\lambda = 632.8$ nm, power < 1 mW; OFV-SLR lens ($f = 30$ mm) for range 1.8m – 200 + m, automatic focus;
- *Controller OFV-5000*. Lowpass (5, 20, 100 kHz), highpass (100 Hz), tracking filters; RS-232 interface for computer control;
- *Velocity decoder VD-06*. Ranges: 1, 2, 10, and 50 mm/s/V selectable; Resolution 0.02 µm/s under 1 mm/s/V range (2 mv/20V); 350-kHz bandwidth analog output; 24 bit, 96 kHz max; Digital output on S/P-DIF interface.

We also developed a software system [19, 20] to configure the controller and process the received LDV digital signals. This system communicates with the controller via the RS-232 interface by sending commands to the controller to change the device parameters and to monitor the status of the device. This system also has integrated some LDV signal processing and enhancement components, which will be described in Section 4.3.2.

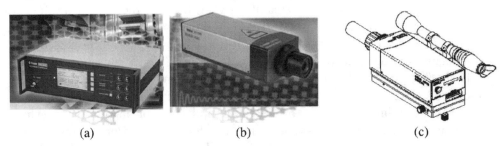

| (a) | (b) | (c) |

Figure 4.2 The Polytec LDV (a) controller OFV-5000, (b) sensor head OFV-505, and (c) telescope VIB-A-P05.

4.2.2   The Infrared Camera

The FLIR ThermoVision A40M IR camera has the following features that make it suitable for human and vehicle detection: *temperature range* of −20°C to 500°C, *accuracy (% of reading)* ±2°C or ±2%; 320×240 focal plane array with uncooled microbolometer detector, *spectral range* 7.5 to 13 μm; 24° FOV lens, *spatial resolution* 1.3 mrad and with built-in focus motor; *Firewire output*—IEEE-1394 8/16-bit monochrome and 8-bit color; *video output*—RS170 EIA/NTSC or CCIR/PAL composite video for monitoring on a TV screen; and ThermoVision Systems developers kit (C++) for software development. Figure 4.3 shows an example where a person sitting in a dark room can be easily detected clearly by the FILR ThermoVision far-infrared camera. Furthermore, the accurate temperature measurements also provide important information for discriminating human bodies from other hot/warm objects. After the successful detection of humans, surrounding objects, such as the doors or walls in this example, can be searched in the environment whose vibration with audiowaves could reveal what the people might be saying.

Note that the FILR ThermoVision IR camera is a far-infrared thermal camera. It does not need to have active IR illumination, and it is suitable for detecting humans and vehicles at a distance (Figure 4.4).

Figure 4.3   A person sitting in a dark room can be seen clearly in the IR image, and the temperature can be accurately measured. The reading of the temperature at the cross (Sp1) on the face is 33.1°C.

Figure 4.4   Two IR images before and after a person standing at about 200 ft. The temperature reading at Sp1 changes from 11°C to 27°C. The corresponding images with the person in the scene are shown in Figure 4.5.

Figure 4.5 Two images of a person at about 200 ft, captured by changing the zoom factors of the PTZ camera.

4.2.3 The PTZ Camera

The state-of-the-art, computer-controllable pan/tilt/zoom (PTZ) camera is also ideal for human and other target detection at a great distance. The Canon PTZ camera we acquired has the following properties: 26X optical zoom lens and 12X digital zoom; 1/4″ 340,000 pixel CCD; pan: +/−100°; tilt: +90/−30°; minimum subject illumination 1 Lux (1/30-second shutter speed); and RS-232 computer control interface. Figure 4.5 shows two images of the same scene with two different camera zoom factors. In most of our current experiments, we use this camera and other similar zoomable cameras to assist the LDV target detection and laser pointing/focusing.

4.3 LDV Hearing: Sensing and Processing

4.3.1 Principle and Research Issues

Laser Doppler vibrometers (LDVs) work according to the principles of laser interferometry. Measurements are made at the point where the laser beam strikes the structure under vibration. In the heterodyning interferometer (Figure 4.6),

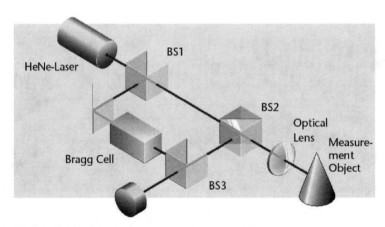

Figure 4.6 The modules of the laser Doppler vibrometer (LDV).

a coherent laser beam is divided into object and reference beams by a beam splitter BS1. The object beam strikes a point on the moving (vibrating) object, and light reflected from that point travels back to beam splitter BS2 and mixes (interferes) with the reference beam at beam splitter BS3. If the object is moving (vibrating), this mixing process produces an intensity fluctuation in the light. Whenever the object has moved by half the wavelength, λ/2, which is 0.3169 μm (or 12.46 microinches) in the case of HeNe laser, the intensity has gone through a complete dark-bright-dark cycle. A detector converts this signal to a voltage fluctuation. The Doppler frequency f_D of this sinusoidal cycle is proportional to the velocity v of the object according to the formula

$$f_D = 2v/\lambda \tag{4.1}$$

Instead of detecting the Doppler frequency, the velocity is directly obtained by a digital quadrature demodulation method [1, 2]. The Bragg cell, which is an acousto-optic modulator to shift the light frequency by 40 MHz, is used for identifying the sign of the velocity.

Interestingly, most objects vibrate while wave energy (including voice waves) is applied on them. Though the vibration caused by the voice energy is very small compared with other vibration, this tiny vibration can be detected by the LDV. Voice frequency f ranges from about 300 Hz to 3,000 Hz. Velocity demodulation is better for detecting vibration with higher frequencies because of the following relation of velocity, frequency, and magnitude of the vibration:

$$v = 2\pi f m \tag{4.2}$$

Note that the velocity v will be large with a large frequency f, even under a small magnitude m. The Polytec LDV sensor OFV-505 and the controller OFV-5000 can be configured to detect vibrations under several different velocity ranges: 1 mm/s/V, 2 mm/s/V, 10 mm/s/V, and 50 mm/s/V, where V stands for velocity. For voice vibration, we usually use the 1 mm/s/V range. The best resolution is 0.02 μm/s under 1 mm/s/V range, according to the manufacturer's specification (with retro-tape treatment). Without retro-tape treatment, the LDV still has a sensitivity on the order of 1.0 μm/s. This indicates that the LDV can detect vibration (due to voice waves) at a magnitude as low as $m = v/2\pi f = 1.0/(2*3.14*300) = 0.5$ nanometer (nm). Note that voice waves are in a relative low-frequency range. The Polytec OFV-505 LDV sensor that we have is capable of detection vibration with a much higher frequency (up to 350 KHz).

There are two important issues to consider in order to use an LDV to detect the vibration of a target caused by human voices. First, the target vibrates with the voices. Second, points on the surface of the target where the laser beam hits reflect the laser beam back to the LDV. We call such points *LDV targeting points*, or simply *LDV points*. Therefore, the LDV points selected for audio detection could be the following three types of targets (Figure 4.7):

1. *Points on a human body.* For example, the throat of a human will be one of the most obvious parts where the vibration with the speech could be

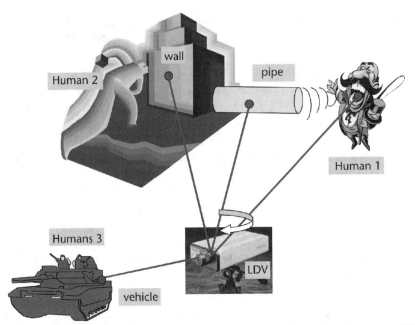

Figure 4.7 Target selection and multimodal display. The laser Doppler vibrometer (LDV) can measure audio signals from tiny vibrations of the LDV points (solid dots in the figure) that couple with the audio sources.

detected by the LDV. However, we have found that it is very challenging, since it is uncooperative: (a) it cannot be easily targeted, especially when the human is moving; (b) it does not have a good reflective surface for the laser beam, and therefore a retro-reflective tape has to be used; and (c) the vibration of the throat only includes the low-frequency parts of the voice. For these reasons, our experiments will mainly focus on the remaining two types of targets.

2. *Points on a vehicle with humans inside.* Human voice signals vibrate the body of a vehicle, which could be readily detected by the LDV. Even if the engine is on and the volume of the speech is low (e.g., in cases of whispering), we could still extract the human voice by signal decomposition, since the human voice and engine noise have different frequency ranges. However, even if the vehicle is stationary, we have found that the body of the vehicle basically does not reflect the HeNe laser suitably for our purposes without applying retro-reflective tape; it is even more challenging to detect a voice when the vehicle is moving. With retro-tape, the signal returns with LDV are excellent when the targets (cars) park at various distances (10m to 50m in our experiments) and also with a large range of incident angles of the laser beam.

3. *Points in the environment.* For perimeter surveillance, we can use existing facilities or install special facilities for human audio signal detection. Facilities such as walls, pillars, lamp posts, large bulletin boards, and traffic signs vibrate very well with human voices, particularly during the relative silence of night. Note that the LDV has a sensitivity on the order of 1 μm/s

and can therefore pick up very small vibrations. We have found that most objects vibrate with voices, and many types of surfaces reflect the LDV laser beam within some distance (about 10m) without retro-reflective treatment. Response is significantly improved if we can paint or paste certain points of the facilities with retro-reflective tapes or paints; operating distances then can increase to 300m (about 1,000 ft) or more.

4.3.2   LDV Audio Signal Enhancement

For the human voice, the frequency range is about 300 Hz to 3 kHz. However, the frequency response range of the LDV is much wider than that. Even if we have used the on-board digital filters, we still get signals that include troublesome, large, slowly varying components corresponding to the slow but significant background vibrations of the targets. The magnitudes of the meaningful acoustic signals are relatively small, adding on top of the low frequency vibration signals. This prevents the intelligibility of the acoustic signals by human ears. On the other hand, the inherent speckle-pattern problem on a normal rough surface and the possible occlusion of the LDV laser beam (e.g., by passing by objects) introduce noises with large and high-frequency components into the LDV measurements (Figure 4.8). This creates very high and loud noise when we listen directly to the acoustic signal. Therefore, we have applied Gaussian bandpass filtering [19] and Wiener filtering [21] to process the vibration signals captured by the LDV. In addition, the volumes of the voice signals may change dramatically with the changes in the vibration magnitudes of the target due to the changes in speech loudness (shouting, normal speaking, whispering) and the distances of the human speakers to the target. Therefore, we have also designed an adaptive volume function [19] to cope with this problem. Figure 4.8 shows two real examples of these two types of problems. Most of the audio clips mentioned in the text can be played in the corresponding online html version of our technical report [20].

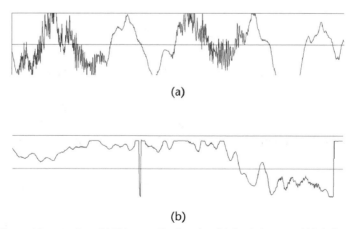

(a)

(b)

Figure 4.8   Two real examples of LDV acoustic signals with both low- and high-frequency noises. (a) "Hello ... Hello" and (b) "I am whispering ... (high frequency noise) ... OK ... Hello (high frequency noise)."

4.3.2.1 The Gaussian Bandpass Filter

We can produce the Gaussian bandpass transfer function by expressing it as the difference of two Gaussians of different widths, as has been widely used in image processing [22], in other words:

$$H(s) = Be^{-s^2/2\alpha_2^2} - Ae^{-s^2/2\alpha_1^2}, \quad B \geqslant A, \alpha_2 > \alpha_1 \tag{4.3}$$

The impulse response of this filter is given by

$$h(t) = \frac{B}{\sqrt{2\pi\sigma_2^2}}e^{-t^2/2\sigma_2^2} - \frac{A}{\sqrt{2\pi\sigma_1^2}}e^{-t^2/2\sigma_1^2}, \quad \sigma_i = \frac{1}{2\pi\alpha_i} \quad (i=1,2) \tag{4.4}$$

Notice that the broader Gaussian in the frequency domain creates a narrower Gaussian in the time domain, and vice versa. We want to reduce the signal magnitude outside the frequency range of human voices, in other words, below $s_1 = 300$ Hz and above $s_2 = 3K$ Hz. The high frequency reduction is mainly controlled by the width of the first (the broader) Gaussian function in (4.3), or α_2, and the low frequency reduction is mainly controlled by the width of the second Gaussian function, or α_1. Since the Gaussian function drops significantly when $|s_i| > 2\alpha_i$, $(i=1, 2)$, respectively, we obtain the widths of the two Gaussian functions in the frequency domain as

$$\alpha_i = s_i/2 \text{ (Hz)}, \quad i = 1,2 \tag{4.5}$$

In practice, we process the waveform directly in the time domain, in other words, by convolving the waveform with the impulse response in (4.4). This leads to a real-time algorithm for LDV voice signal enhancement. For doing this, we need to calculate the variances of the two Gaussian functions in the time domain. Combining (4.4) and (4.5), we have

$$\sigma_i = \frac{1}{\pi s_i} \text{ (sec)}, \quad i = 1,2 \tag{4.6}$$

For digital signals, we need to determine the size of the convolution kernel. Since the narrower Gaussian (with width α_1) in the frequency domain creates a broader Gaussian (with width σ_1) in the time domain, we use σ_1 to estimate the appropriate window size of the convolution. Again, we truncate the impulse function when we have $t > 2\sigma_1$. Therefore, the size of the Gaussian bandpass filter is calculated as

$$W_1 = 2m(2\sigma_1) + 1 = \frac{4m}{\pi s_1} + 1 \tag{4.7}$$

where m is the sampling rate of the digital signal. Typically, we use $m = 48$ Ksamples/second with the S/P-DIF format. Therefore, the size of the window will be $W_1 = 210$ when $S_1 = 300$ Hz. An example is shown in Figure 4.9.

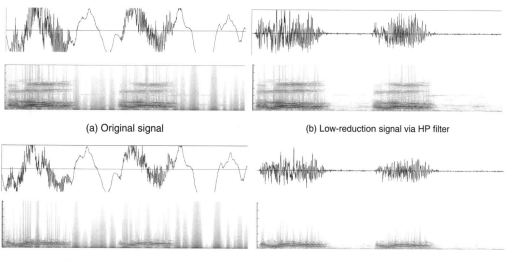

(a) Original signal   (b) Low-reduction signal via HP filter

(c) High-reduction signal via LP filter   (d) Band-pass signal

Figure 4.9 The waveforms and the corresponding spectrograms of (a) original signal (no filter), (b) Gaussian highpass (i.e., low-reduction) filter, (c) Gaussian lowpass (i.e., high-reduction) filter, and (d) Gaussian bandpass filter. In each picture, both the waveform and the corresponding spectrogram are shown. The spectrograms are generated when the FFT window is 1,024 samples, or 21.3 ms as the sampling rate is 48 Ksamples/second. We use $s_1 = 300$ Hz for low-reduction and $s_2 = 3,000$ Hz for high-reduction in this example.

4.3.2.2   Wiener Filtering

After applying the bandpass filter, the noise outside of the voice frequency range is attenuated, but the noise energy falling inside the voice frequency range still exists. This problem is handled using Wiener filtering, which is one of the most effective speech enhancement approaches in literature [12].

Wiener filtering is an optimal filtering technique under the minimum mean squared error (MMSE) criteria. The degraded speech can be modeled as the summation of a clean speech signal and additive noise, which is illustrated as

$$y[n] = s[n] + d[n] \qquad (4.8)$$

where $d[n]$ is additive noise signal, and the clean signal $s[n]$ is independent of the noise $d[n]$. With this assumption, their autocorrelation signals satisfies

$$R_{yy}[n] = R_{ss}[n] + R_{dd}[n] \qquad (4.9)$$

since the power density functions are the Fourier transforms of the corresponding autocorrelation signals, so

$$S_{yy}(\omega) = S_{ss}(\omega) + S_{dd}(\omega) \qquad (4.10)$$

We know that $S_{yy}(\omega) = |Y(\omega)|^2, S_{ss}(\omega) = |S(\omega)|^2, S_{dd}(\omega) = |D(\omega)|^2$, where $Y(\omega)$, $S(\omega)$, and $D(\omega)$ are the Fourier transforms of $y[n]$, $s[n]$, and $d[n]$, respectively. This leads to

$$|Y(\omega)|^2 = |S(\omega)|^2 + |D(\omega)|^2 \tag{4.11}$$

Under the MMSE criteria, the optimal linear filter for speech enhancement is the Wiener filter, and its frequency response is

$$H(\omega) = \frac{S_{ss}(\omega)}{S_{ss}(\omega) + S_{dd}(\omega)} \tag{4.12}$$

After we have already estimated the $\hat{S}_{dd}(\omega)$, the frequency response of the optimal filter is

$$H(\omega) = 1 - \frac{\hat{S}_{dd}(\omega)}{S_{yy}(\omega)} \tag{4.13}$$

The estimate frequency response could have negative values, so in practice zeros are used to replace the negative values. In Wiener filtering, the output signal has the following formula:

$$\hat{s}[n] = IFFT\left\{ \frac{|Y(\omega)|^2 - |\hat{D}(\omega)|^2}{|Y(\omega)|^2} \cdot |Y(\omega)| e^{j\theta_{sn}(\omega)} \right\} \tag{4.14}$$

where $\theta_{sn}(\omega)$ is the phase information of the degraded signal $y[n]$. In practice, the noise is only stationary in a short period of time. In order to update the optimal filter, adaptive noise spectrum estimation is conducted using a moving window, which contains one second of signal. The noise part is assumed to be the low-energy part (20%) of this signal. Then the noise power spectrum ($\hat{D}(\omega)$) is estimated using the noise part of this signal. The optimal filter is therefore updated based on this new noise estimation.

4.3.2.3 Volume Selection and Adaptation

The useful original signal obtained from the S/P-DIF output of the controller is a velocity signal. When treated as the voice signal, the volume is too small to be heard by human ears. When volumes of the voice signals also can change dramatically within an audio clip, a fixed volume increase cannot lead to clearly audible playback. Therefore, we have designed an adaptive volume algorithm. For each frame, for example of 1,024 samples, the volumes are scaled by the following equation:

$$v = \frac{C_{max}}{|Max(x_1, x_2 \ldots x_n)|} \tag{4.15}$$

where v is the scale factor of the volume, C_{max} is the maximum constant value of the volume (defined as the largest short integer, i.e., 32,767), and $x_1, x_2, \ldots, x_n$ are sample data in one speech frame (e.g., $n = 1{,}024$ samples). The scaled sample data stream, $vx_1, vx_2, \ldots, vx_n$, will then be played via a speaker so that a suitable level of voice will be heard. Figure 4.10 shows an example.

4.4   Experiment Designs and Analysis

In order to use an LDV to detect audio signals from a target, the target needs to meet two conditions: reflection to HeNe laser and vibration with voices. Due to the difficulty in detecting voice vibration directly from the body of a human speaker, we mainly focus on the use of targets in the environments nearby the human speaker. We have found that the vibration of most objects in man-made environments caused by waves of voices can be readily detected by the LDV. However, the LDV must get signal returns from the laser reflection. The degrees of signal returns depend on the following conditions:

1. Surface normal versus laser beam direction to ensure that laser could be reflected back to the sensor;
2. Color of the surface with spectral response to 632.8 nm, for example, pure green traffic posts do not reflect red laser light;

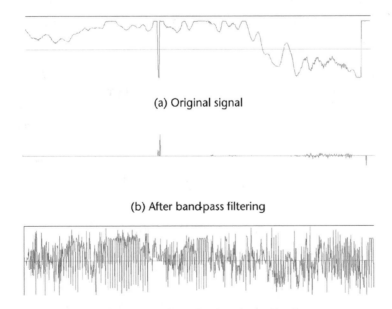

(a) Original signal

(b) After band-pass filtering

(c) Adaptive scaling after band-pass filtering

Figure 4.10   (a–c) The waveform of the original signal and the results of fixed scaling and adaptive scaling, after using suitable filtering. The short audio clip reads, "I am whispering… [noise]… OK… Hello [noise]," which was captured by the LDV OFV-505 from a metal cake box carried by a person at a distance of about 30m from the LDV. The surface of the target was treated by a piece of retro-tape.

3. Roughness of the surfaces—the inherent speckle-pattern problem of the laser reflection greatly affects the signal-to-noise ratio (SNR);

4. The distance from the sensor head to the target, which governs the focus of the laser beam as well as the energy returned.

Without surface finish of retro-reflective tapes or paints, factors such as surface reflectivity, colors, orientations, and distances are all critical in intelligible LDV acoustic signal acquisition. Therefore, IR and color cameras and 3-D computer vision techniques could be very helpful in finding the most suitable LDV targets for remote hearing. Whenever possible, retro-reflective traffic tapes or paints are a perfect solution to the above reflection problems *if* the targets are cooperative—that is to say, if the surfaces of targets can be treated by such tapes or paints. The traffic retro-reflective tapes (retro-tapes) are capable of diffuse reflection in that they reflect the laser beam back in all directions within a rather large angular range. In the following, we will present the experimental results first in real environments and then in somewhat controlled environments for performance analysis.

4.4.1 Real Data Collections

We have performed experiments with the following settings: types of surface, surface directions, long-range listening, through-wall listening, and talking inside of cars. In all experiments, the LDV velocity range is 1 mm/s/V, and a person's speech describes the experiment configurations. A walkie-talkie is used for remote communication only. With each example, we provide links in [20] for both the original LDV audio clip and the processed audio clip. The same configurations (bandpass 300 Hz to 3 kHz, adaptive volume) are used in processing the data for all the experiments, unless indicated. Each audio clip should convey most of the information for the experiment if it is intelligible. Typically, the duration of each clip is about 1 to 3 minutes. Sometimes the signals start with a period of noises (no speaking), but all of them should include large portions of meaningful voices. Note that we only include original data and the corresponding processed data with one fixed configuration of Gaussian filtering. The original clips have very low volume, so usually the listener cannot hear anything meaningful; on the other hand, the processed audio clips are not optimal at all for intelligibility.

4.4.1.1 Experiments on Long-Range LDV Listening

We tested the long-range LDV listening in an open space (Figure 4.11), with various distances from about 30m to 300m (or 100 ft to 1,000 ft). A small metal cake box with retro-tape finish was fixed in front of the speaker's belly. The signal return of the LDV is insensitive to the incident angles of the laser beam, thanks to the retro-tape finish. Both normal speech volumes and whispers have been successfully detected. The size of the laser spot changed from less than 1 mm to about 5 mm to 10 mm when the range changed from 30m to 300m. The noise levels also increased from 2 mV to 10 mV out of a total range of 20-V analog LDV signal. The movements of the body during speaking may cause the laser beam to

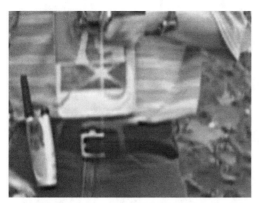

Figure 4.11 Long-range LDV listening experiment. A metal cake box (left) is used, with a piece of 3M traffic retro-tape pasted on. The laser spot can be seen clearly.

briefly stray off the tape, so large noises could be heard in some places. The changes in volume of the speaker's speech could be read by listening to the speech. With longer ranges, the laser is more difficult to localize and focus, and the signal return becomes weaker. Therefore, the noise levels become larger. Within 120m, the LDV voice is obviously intelligible; at a distance of 260m, many parts of the speech could be identified, even with some difficulty. For all the distances, the signal processing plays a significant role in making the speech intelligible. Without processing, the audio signal is buried in the low-frequency large-magnitude vibration and high-frequency speckle noises. We also want to emphasize that automatic targeting and intelligent refocusing is one of the important technical issues that deserves attention for long-range LDV listening, since it is extremely difficult to aim the laser beam at the target and keep it focused. We believe that a multimodal integration approach provides a feasible way to achieve this goal.

We have also tested long-range LDV listening in a corridor of a building at City College (refer to Figure 4.5). We collected two sets of audio clips when a person talked to the same cake box nearby, with and *without* retro-tape. The vibration target (the cake box) was at a distance of about 100m (about 340 ft) from the LDV sensor head. With the retro-tape, the speech is clearly intelligible. The significance of this experiment is that *without* the retro-tape, the speech is still intelligible from LDV detection of targets at a distance as far as 100m. However, it is very noisy. Therefore, with the state-of-the-art sensor technology, we realize that signal enhancement techniques need to be developed that are more advanced than simple bandpass filtering, Wiener filtering, and adaptive volume scaling.

4.4.1.2 Experiments on Listening Through Walls/Doors/Windows

In the second set of experiments, we tested the LDV listening through walls and windows (Figures 4.12 and 4.13). In the experiments of LDV listening through walls, we tried several different cases: laser pointing to walls and pointing to a door from a distance of about 30 ft away, with and without retro-tapes, vibrated by normal speech and whispering when the speaker was walking around in a

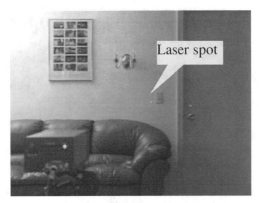 

Figure 4.12 Listening through walls—a person was speaking in a room behind the door (right), while the LDV was listening in the other room through the wall (left).

No Tape   Retro-Tape

Figure 4.13 Listening through windows—a person was speaking outside the house, close to a window, while the LDV was listening inside the room via the window frame. Left: without retro-tape; right: with retro-tape. The signal return strengths can be seen from the back of the sensor.

15×15-ft^2 room, and facing toward or away from the targets that the LDV points to. We have found that the speech is quite intelligible via the walls or the door if retro-tape treatment is used, no matter whether the speaker talks in a normal voice or a whisper, is facing the target or not, or is close or some distance away. Without retro-tape, it is very hard to identify the speech, but we can definitely tell the audio clips include a person's speech.

In the experiments of LDV listening through windows, we used the window frames as vibration targets while a person was speaking outside a house (Figure 4.13). The LDV was inside the second floor of the house, several meters away from the window. The person spoke outside the house, close to the window. Since the window frames are treated with paint, the reflection is good, even though the signal return strength is less than half of that with tape (see the bars in the back of the LDV sensor in Figure 4.13). We have also tested listening via the window frame when the distance between the sensor head and the target was more than 20m (or 64 ft) away. The LDV voice detection almost has the same performance as this short-range example.

4.4.1.3 Experiments on Talking Inside Cars

It is very interesting to consider whether LDV can be used to detect voices from inside a vehicle in which intruders of a perimeter may hide, making plans that they feel nobody will discover. In such cases, the car engine or music may be on, and the persons may whisper quietly inside the car. We simulated such situations by using several cars and minivans at different distances (Figure 4.14). We have tried to detect the vibration caused by human speech inside the car via various parts of the car, (e.g., driver-side door, back of truck, an object inside the car that can be seen through the windows of the car). Contrary to our initial assumption, an automobile body does not offer good retro-reflection of the LDV laser beam for effective listening. Therefore, we have treated the targeting points of the vehicles with retro-reflective tapes. We made the following observations:

1. The speech vibration can be detected via any part of the car (front, side, back, or inside).
2. The speech can be distinguished with the engine and/or music on.
3. Whispers can also cause car-body vibration that is readily detectable by the LDV.

15m, driver side door, with and without retro-tape

50m, back truck, without retro-tape

15m, balloon inside, with retro-tape

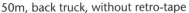

Figure 4.14 Talking inside cars.

4. With retro-tape treatment, the speech is clearly intelligible; without retro-tape, human speech can be detected in short ranges, within about 15m to 20m, but the SNR will be very low when the target is far, for example, greater than 50m. However, the signals can always be identified as human voices at such distances without retro-tape.

4.4.1.4 Experiments on Types of Surfaces

We have tested many different surfaces, both indoors and outdoors, natural and man-made. In addition to car bodies, window frames, walls, doors, and the metal cake box we have shown above, we have also tested LDV measurements via traffic signs, building pillars, and lampposts, with or without using retro-tapes (Figure 4.15). For example, a black building pillar in Figure 4.15 provides excellent LDV voice-listening capability without any retro-tape treatment. Traffic signs and a wooden shed seem also to offer good media for carrying the voice signal from speakers to the LDV sensors.

We have found that the LDV could obtain signal returns for voice reading from most objects, without retro-tape treatment, if the distance between the sensor head and the target is short: within 10m for most objects, and within 100m for quite a few objects with good vibration and reflection, such as a metal cake box with paint finish, traffic signs, and so forth. With retro-tape, the LDV works very well for almost all objects at large distances. The largest distance we have tested is about 300m.

4.4.1.5 Experiments on Surface Directions

In real applications, it is also important that the LDV can get a signal return when the laser beam aims at a target in various directions. In theory, the laser beam could be reflected back from the surface of an object by fine-adjusting the targeting location in such a way that the surface normal on the scale of the wavelength

Building Pillar, 10m, without retro-tape   Traffic Sign, about 40m, laser on white paint

Shed, 40m, w/ retro tape   Shed 40m, w/o retro   Lamp Post, 15m with retro-tape

Figure 4.15 Targets in the environment: building pillar, traffic sign, wooden shed, and lamp post.

Figure 4.16 Experiments on surface directions.

of the laser beam is in the direction of the laser beam. This can be achieved by maximizing the signal return of the LDV sensor when performing localization and focusing. However, in practice this could be very problematic, since small physical movements of either the sensor or the targets may change the speckle patterns on rough surfaces [1]. This leads to a very noisy signal, typically as white noise with large magnitudes. A very practical solution is retro-reflective tape or paint surface finishing, if allowed. Such retro-tapes (retro-paints) have the properties of reflecting lights from a wide range of directions. Figure 4.16 shows an experiment of LDV voice detection through a normal mailbox with retro-tape treatment. The voice signals are still intelligible, even when the angle between the surface normal and the laser beam is as large as 80 degrees.

4.4.2 LDV Performance Analysis

In order to get a more quantitative understanding of the LDV performance, we performed three sets of experiments in the lab with controlled conditions, by controlling the voice sources and measuring several parameters of the LDV signal returns.

In the first set of experiments, we measured the signal return strengths, vibration magnitudes, and signal-to-noise ratio (SNR) from various vibration targets, with and without retro-tapes. We used a standard computer speaker to play a 6-second audio clip from a presentation, and the sound vibrated objects in the lab. The objects we measured and compared are the small metal cake box we have used, a whiteboard, a wooden door, and a concrete wall. The LDV laser beam was pointed to each target, and an autofocus was performed before each measurement. The distance of the targets to the LDV was within several meters (about 10 ft).

The Polytec LDV provides a function to get *signal return strengths* in the range of 0 to 512. Typically, the LDV audio signal will not be intelligible if the strength is lower than 10. Meanwhile, a signal *magnitude* is measured as the average signal magnitude of the 6-second audio clip. This tells us the vibration magnitudes of the object with the source sound. In order to get an SNR, we also obtained LDV measurements under the same conditions, but without playing the audio source. This means that the vibration of the targets was caused only by background acoustic noises (e.g., air conditioning machine), physical noises (e.g., from

Table 4.1 Signal Return Strengths of the Objects from 10 Feet (the Full Strength is 512)

	Cake Box	Wooden Door	Concrete Wall	Whiteboard
Retro-tape	512	512	512	512
No tape	200	230	120	250

movements of the sensor head and the targets), and optical/electronic noises. Then, a *pseudo* SNR could be obtained by taking the ratio between the average magnitudes of the LDV signals with and without playing the source voice clip. Table 4.1 lists the LDV signal return strengths when the LDV audio clips were collected. It shows that the signal return strengths were full for all four objects we tested when the retro-tape was used. This in fact indicates the excellent reflection capability of the retro-tape. Without retro-tape, the signal return strengths indicate the reflection properties of the objects. The metal cake box (with paints), wooden door (with white paint), and the plastic whiteboard all have almost half full signal return. The signal return from the concrete wall (with white paint) is almost a quarter of the full strength. Note that all the measurements were performed at a distance of about 10 ft. Within that short distance, the voice signals without retro-tape treatment almost have the same quality as those with retro-tape treatment.

Table 4.2 lists the average signal magnitudes from the four objects, with and without retro-tape, and with and without playing original sound. In each case we performed the measurements twice, so two average numbers appear in each table entry. Clearly, the signal magnitudes with and without the meaningful sound source are comparable. This indicates that the background noises cannot easily be removed if they share the same frequency ranges. However, if the magnitudes of the background noises mainly come from the low frequency part (below 300 Hz) and the high frequency part (above 3,000 Hz), the bandpass filter will reduce them.

In the second set of experiments, we tested the LDV focusing and read signal strengths on various objects without retro-finish (Figure 4.17, Table 4.3). The scene image [Figure 4.17(a)] was the intensity image of a 640 × 480 color image from a digital camera using ambient fluorescent lighting. The LDV sensor head was placed 8 ft from the wooden yardstick in the scene, the wall being more than 9 ft from the sensor. The focusing and signal strength data for each point was collected manually using the Java LDV application program we have developed. For each LDV point, the signal strength was obtained from a different amount of readings, the minimum being 44. Only the first 44 readings were used in analyzing each point's average, median, and standard deviation. In addition, the focus parameter (range 0 to 3,000) was also recorded for each point, and the approximate value of the red component (range of 0 to 255) was also read from the red channel of the scene [Figure 4.17(b)]; the LDV points were also labeled on the image.

Table 4.2 Magnitudes of LDV Signals and Background Noise (Under 1 mm/s/V)

	Retro	Cake Box	Wooden Door	Concrete Wall	Whiteboard
Signal	Yes	200/149	175/177	78/75	86/75
	No	200/202	155/131	105/95	82/113
Noise	Yes	174/163	159/186	71/72	72/85
	No	190/175	157/146	99/104	64/75

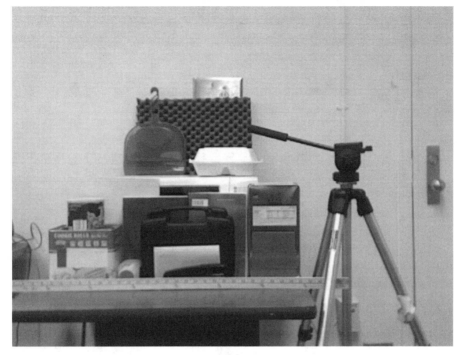

(a)

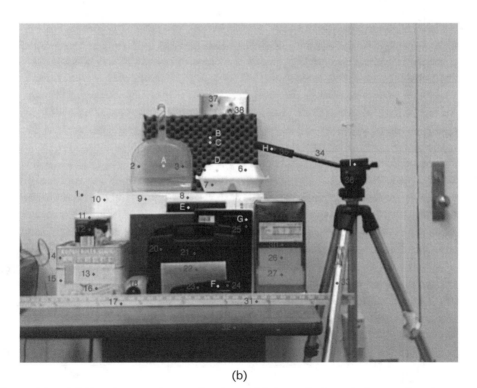

(b)

Figure 4.17 LDV focusing experiments. Top image (a) shows the intensity image of a color scene. Bottom image (b) shows the labeled "red channel" of the color scene in an intensity image.

Table 4.3 Laser Beam Focusing on Various Objects

	Point Description	Red	Focused?	fp	sa	sm	sd
1	Wall (left of scene)	210	yes	1244	112.0	114	35.0
2	Translucent red shovel (left, in front of wall)	191	yes	1137	16.7	16	5.9
3	Translucent red shovel (right, in front of green foam)	130	yes	1155	17.0	13	9.2
4	Green foam (on light side slant of tips)	93	yes	1221	19.1	19	12.0
5	White styrofoam (on light edge, top left)	248	yes	900	19.7	20	5.4
6	White styrofoam (flat, top right)	255	yes	925	1.9	1	2.3
7	White styrofoam (flat, bottom left)	186	yes	891	1.1	1	0.4
8	Microwave door (top white)	238	yes	869	178.0	178	3.4
9	Pink sheet (microwave behind)	228	yes	850	239.0	239	5.1
10	Pink sheet (air behind)	245	yes	855	111.0	109	6.7
11	Yellow lid	197	yes	689	211.0	210	10.0
12	Blue container	22	yes	689	147.0	144	27.0
13	White retro-tape	195	yes	580	457.0	455	4.5
14	Red cookie roll can	172	yes	662	102.0	102	82.0
15	White cookie roll can	195	yes	663	262.0	262	2.7
16	Cookie figure, cookie roll can	172	yes	667	151.0	154	26.0
17	Wooden yardstick (left side)	210	yes	646	141.0	148	21.0
18	Teal plastic lid	7	yes	669	112.0	112	13.0
19	Purple book	48	yes	801	19.3	20	5.1
20	Matte black case (left)	9	yes	706	41.2	38	26.0
21	Matte black case (middle)	10	yes	703	55.7	55	17.0
22	Blue book	104	yes	691	241.0	244	7.5
23	Black stapler (left-middle)	10	yes	686	1.3	1	0.7
24	Black stapler (right)	8	yes	686	16.0	18	5.9
25	Green book (slightly below top right)	5	yes	839	12.3	12	2.0
26	Red retro-tape (top)	135	yes	746	307.0	310	47.0
27	Red retro-tape (bottom)	137	yes	855	512.0	512	0.0
28	Metal cake box (top left)	47	yes	839	5.4	5	2.1
29	Metal cake box (top right)	77	yes	844	38.4	40	4.0
30	Metal cake box (middle)	28	yes	853	13.9	14	3.0
31	Wooden yardstick (right)	179	yes	710	126.0	126	5.5

Table 4.3 (Continued)

	Point Description	Red	Focused?	fp	sa	sm	sd
32	Brown table edge	31	yes	444	24.9	24	17.0
33	Tripod steel leg (front)	255	yes	704	139.0	139	8.5
34	Tripod black metal handle	15	yes	884	148.0	151	20.0
35	Tripod black foam on handle (once only)	34	yes	898	1.0	1	0.0
36	Tripod black plastic (only on light edges)	22	yes	789	55.3	57	9.1
37	Outlet metal plate	182	yes	1285	100.0	105	42.0
38	Outlet white plastic	176	yes	1287	105.0	105	21.0
A	Translucent red shovel (center, in front of green foam)	129	no				
B	Green foam (inside crater)	20	no				
C	Green foam (on tip)	44	no				
D	Styrofoam (on flat surfaces)	255	no				
E	Microwave door (top black)	11	no				
F	Black stapler (middle right)	5	no				
G	Green book (top right)	0	no				
H	Tripod black foam handle, same/other spots will not focus	22	no				
I	Tripod black plastic (in shadow, or darker regions)	15	no				

Red: red channel reading; fp: focusing parameter; sa: signal strength average; sm: signal strength median; sd: signal strength standard deviation. Signal strength values below 10 are in bold italic where returned signals are not very meaningful.

Table 4.4 Signal Return Strengths of Various Distances (the Full Strength is 512)

	25 ft	50 ft	75 ft	100 ft
Retro-tape	512	512	460	400
No tape	50–100	14–30	10–30	10

During the data collection, it was noticed that a points angle of reflectance is more important to signal strength than the color of the point, although in the red channel image, slight changes in color were noticed that coincide with the signal strength. An example is the green book (25) in the scene; at point 25 the red component was 5 out of 255, but in the general area above it, where the sensor did not focus, the red component was 0. An unusual observation was that the sensor was able to focus on black/dark materials but generally had low signal strength. Also noted was that fully white materials, such as the white Styrofoam, point D, were unable to be focused. Translucent materials like the dust pan, even when red, seem to be a problem for focusing. During the data collection for point 2, it appeared that the sensor had focused behind the dust pan. This might be the reason why the sensor could not focus at point A. Most materials that failed to focus had low red component, under 44, except for the two cases mentioned previously. On the 16 points where the sensor did focus and the value of the red component was low, under 100 out of 255, the signal strength was not over 100 out of 512, except in 3 cases.

In the third set of experiments, we measured the above properties of the LDV with the cake box we have used in the above experiments at various distances in the corridor outside our lab (refer to Figure 4.5). The distances from the lab to one end of the corridor is 100 ft (about 30m), and to the other end of the corridor is 340 ft (about 100m). In the first subset of the distance experiments, we had measurements both with and without the retro-tape on the cake box. In the second subset of the distance experiments, we only obtained measurements with retro-tapes.

In the first subset of experiments, a human speaker read "LDV project experiment." Table 4.4 lists the LDV signal return strengths in several distance configurations. Note that without retro-tape, the signal return strengths drop dramatically when the distance is above 50 ft, and at 100 ft the signal strength is below 10, which is the threshold for detecting meaningful vibration signals by the LDV. Note that this does not mean we cannot obtain a meaningful LDV voice signal from a target above 100 ft without retro-tape treatment. We have successfully obtained voice signals from the cake box at 300 ft. However, as we have noted, a fine and tedious targeting and focusing of the laser beam needs to be performed so that the signal return strength is above the threshold. As a comparison the signal return strengths with retro-tape are also shown, which do

Table 4.5 Magnitudes of Voices and Noises at Various Distances (1 mm/s/v)

	Retro-	25 ft	50 ft	75 ft	100 ft
Signal	yes	70/76	92/89	157/172	162/175
	no	74/68	148/87	166/136	1,335/1,828
Noise	yes	50/57	83/81	160/137	109/141
	no	48/46	58/78	121/125	1,604/1,778

Table 4.6   Results of Longer Distance Experiments (LDV Velocity Range 1 mm/s/v)

Distance (ft)	Spot Size (mm)	Signal Strength (of 512) (with Orig. Audio)	Noise Mags	Voice Mags (with Proc. Audio)
50	1	458	174	185
100	5–8	227	159	156
150	5–8	248	119	165
200	>10	45	866	670
250	<10	290	118	138
300	10	277	180	258
340	10–15	65	172	249

not have significant drop over distances. The average magnitudes with and without the voice source are shown in Table 4.5. Again, the magnitudes of the signals and the noises are comparable. Note that at 100 ft the signal magnitudes are one order larger than other measurements, which indicates that the signals captured by the LDV are merely noises and therefore meaningless.

In the second subset of distance experiments, we examined longer ranges from 50 ft to 340 ft. We also roughly measured the LDV laser spot size on the target. In all distances, retro-tape was used. Table 4.6 lists the measurements. We have the following observations:

1. The spot size becomes significantly bigger at distances above 100 ft and does not increase significantly within 350 ft, thanks to the super long lens and the auto focus function of the LDV sensor. We should note that even if the measurements of the sizes are made roughly by human eyes, the size correlates with the intelligibility of the audio signals. For example, since we have better focus of the laser beam at 250 ft than at 200 ft, the voice at 250 ft is actually better.

2. The signal strength is almost full when the distance is below 100 ft, and is roughly half full at the distance from 100 to 350 ft. However, the strength number is much lower (around 50) when the laser beam is not perpendicularly directed to the surface of the target and/or not focused well.

3. The SNR is still about 1:1, but the audio signals are intelligible at distances up to 340 ft.

4.4.3   Enhancement Evaluation and Analysis

To evaluate the performance of the speech enhancement by the proposed techniques, both subjective and objective evaluation can be conducted. Here we compare the spectrograms of LDV audio, its enhanced speech signals, and a corresponding clean signal captured at the same time using a wireless microphone. The LDV audio signal in Figure 4.18 was captured 100 ft away by aiming the laser beam at a metal cake box (without retro-reflective finish), and the clean signal was captured using the wireless microphone connected to a laptop placed next to the target (i.e., the metal box) and the speaker. The clean signal [Figure 4.18(f)] was aligned with the LDV signal.

As can be seen in the spectrograms, most of high-frequency energy in the clean signal [Figure 4.18(f)] disappeared in the LDV audio signal [Figure 4.18(a)]. This is

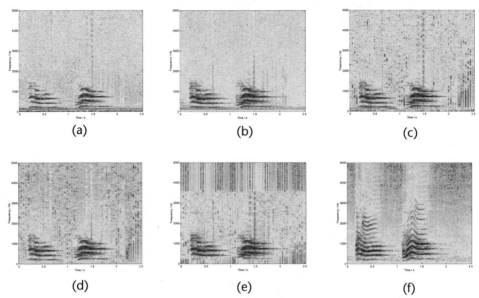

Figure 4.18 The spectrograms of the (a) original LDV signal, (b) Gaussian bandpass filtered signal, (c) Wiener filtered signal, (d) Wiener filtered + Gaussian bandpass filtered signal, (e) Wiener filtered + Hann bandpass filtered signal, and (f) clean signal. All correspond to the speech of "Hello, hello."

because the measured object (the metal cake box) vibrated by voice energy could not vibrate in such high frequency as air or the eardrum does. In Figure 4.18(a), we can also see strong noise in the low frequency part (at the bottom), as well as relative weak noise in high frequency (at the top). Figure 4.18(b) shows the enhanced signal spectrogram by Gaussian bandpass filter, which reduces both the low- and high-frequency noise. Figure 4.18(c) shows the enhanced signal by the Wiener filter only. Figure 4.18(d) shows that the noise is largely attenuated by the combined approach, which is the application of the Wiener filter followed by the Gaussian bandpass filter. As a matter of fact, we have found that the order of the combination does not matter much. Figure 4.18(e) shows the Hann bandpass filter, which almost removes noise with frequency lower than 300 Hz or higher than 3 kHz, but introduces a lot of artificial noise in the middle of the speech frequency. In comparison, the Gaussian plus Wiener filtering gives the best result [Figure 4.18(d)], which is the closest to the clean signal [Figure 4.18(f)].

Several audio clips before and after processing are provided in [21]. Three different types of reflecting surfaces are tested: the small empty metal cake box (with retro-tape), the metal box surface itself (without retro-tape), and the wood hose box surface. They are all 100 ft away from the sensor head. The signal captured from the retro-tape surface contains strong low-frequency noise, but the voice is much clearer to perceive than the other two signals captured on the surfaces that produce strong high-frequency noise. From the results, we can see that the combined approach largely improves the original signals and also outperforms the methods using only one of the filters. For the case with obvious low-frequency noise (reflected from the retro-tape), Gaussian bandpass filtering is more significant than Wiener filtering. Therefore, for this case, the bandpass filter followed by the Wiener filter outperforms the other one. The reason is that the Wiener filtering has the best effect after the high-energy low-frequency part has been removed by the Gaussian bandpass filtering.

4.5   Discussions on Sensor Improvements and Multimodal Integration

4.5.1   Further Research Issues in LDV Acoustic Detection

For the capability of a vibrometer to measure vibration at large distances, Scruby and Drain [1] gave a very good description in the case of ultrasonic measurements. The discussion is also valid for voice detection via an LDV. In application of laser interferometry, we are frequently concerned with surfaces that are not optically finished or are irregular on the scale of the wavelength of the light. Light scattered from different parts of the surface is not phase-related and is consequently diffused over a wide angle. In a given direction of scattering, the resultant may be considered as made up of contributions from a large number of independent sources having the same frequency but a random phase relation. In a slightly different direction, the relative phases change and the resultant is different. This gives rise to an irregular angular distribution of the intensity of the scattered light, a speckle pattern characteristic of monochromatic illumination. The size of the speckle is inversely proportional to the dimensions of the area illuminated. Assuming that there are no problems from atmosphere absorption or refractive index gradients, the sensitivity of reference-beam interferometry is limited by the SNR [1]

$$\frac{<W_s^2>}{<W_N^2>} \propto \frac{W_{s0}\lambda^2}{r_0^2} \tag{4.16}$$

assuming that the photon noise from the reference beam is the dominant source of noise. In the above equation, W_{s0} is the mean light power per unit solid angle scattered from the whole illuminated spot in the direction of the detector (i.e., average over the speckles), r_0 is the radius of the spot size, and λ is the wavelength of the laser. Therefore, the LDV's sensitivity depends on the laser power, the type of rough surface, and the focused spot size. Note that (4.16) only gives a theoretical bound for the LDV noise. However, this may lead to a few improvements for the LDV performance for remote hearing.

- *Reflectance.* As we have found, almost all natural objects vibrate with normal sound waves. However, surface reflectance is a big issue. The relatively poor performance with a rough surface at a large distance is basically due to the fact that only a small fraction of the scattered light (approximately one speckle) can be used because of the coherence consideration. This problem could be minimized by the use of a highly convergent illuminating beam, thus reducing the spot size (the radius r_0) and consequently increasing the speckle size. A more practical choice is to use surface treatment with the state-of-the-art LDV sensors. A useful surface finish for obtaining good signal in LDV is retro-reflecting surface. With this, the light is returned to the sources independently of the orientation of the surface, thus maximizing W_{s0}. Retro-reflecting surface finishes are available as adhesive tapes and paints. The thickness of the finish is not a big issue for the low-frequency acoustic signals. However, in many military applications, particularly in overseas airbase protection, applying retro-tape is not practical. Spray painting might be a better approach. Shooting small retro-vibration bullets is another possible choice. With some smart way to augment surface finishes, LDV remote noncontact listening will be a more attractive alternative to microphone bugs, since LDV obviates the need for a remote power supply and wired communication channels.

- *Wavelengths.* The change of the red laser to invisible such as IR and ultraviolet has several advantages. First, it satisfies the clandestine listening requirement. Second, some surfaces (e.g., windows) may have better reflection to wavelengths other than red. Third, changing the wavelength may increase the sensitivity [1, 16]. From (4.1), decreasing the wavelength will increase the Doppler frequency shift, thus increasing the resolution of the LDV. However, from (4.16) this could decrease the SNR, thus decreasing the sensitivity. Therefore, further research needs to be performed to find the best wavelength for LDV voice detection applications.

- *Power.* Current laser power is limited by eye-safety issues as dictated by laser industry standards. The Polytec OFV-505 uses a laser beam with power less than 1 mW. One may increase the laser power to the upper limits of the standard up to 2 mW. This increase (from 1 mW or 1.5 mW to 2 mW) will only slightly increase the signal return strength and the sensitivity of the LDV (4.16), but these increases will not be significant unless breaking of the eye-safe limit is allowed. The wavelength and capability of longer-range focus (above 1,000 ft) will bring more benefit in the LDV sensitivity than only increasing the power.

4.5.2 Multimodal Integration and Intelligent Targeting and Focusing

When using an LDV for voice detection, we need to find and localize the target that vibrates with voice waves and reflects the laser beam of the LDV, and then aim the laser beam of the LDV at the target. Multimodal integration of IR/EO imaging and LDV listening provides a solution to this problem. Ultimately this will lead to a fully automated system for clandestine listening for perimeter protection. Even when the LDV is used by a soldier in the field, automatic target detection, localization, and LDV focusing will help the soldier find and aim the LDV at the target for voice detection.

We have found that it is extremely difficult for a human operator to aim the laser beam of the LDV at a distant target and keep it focused. In the current experiments, the human operator turns the LDV sensor head in order to aim the laser beam at the target. The laser beam needs to be refocused when the distance of the target is changed. Otherwise, the laser spot is out of focus. As a consequence, it is very hard for the human to see the laser spot at a distance above 10m, and it is impossible to detect vibration when the laser spot is out of focus. In a typical distance experiment we conducted, the target (or an "assistant" target), preferably with a retro-reflective tape pasted, needs to be held by hand and to be moved gradually from the sensor head to the designated distance, while trying to keep the laser spot on it. At about several meters, the LDV needs to be refocused to make the laser spot clearly visible to human eyes. Even with the automatic focus function of the Polytec OFV-505 sensor head, it usually takes more than 10 seconds for the LDV to search the full range of the focus parameter (0 to 3,000) in order to bring the laser spot into focus. Therefore, automatic targeting and intelligent refocusing is one of the important technical challenges that deserves attention for long-range LDV listening. Future research issues include the following three aspects:

- *Target detection and localization via IR/EO imaging.* Techniques for detection of humans and their surroundings need to be developed for finding vibration targets for LDV listening. We have set up an IR/EO imaging system with an IR camera and a PTZ camera for this purpose. In our current experiments, this was performed manually. In the future, we will look into automated approaches for achieving a more intelligent system.
- *Registration between the IR/EO imaging system and the LDV system.* If the two types of sensors are precisely aligned, we can easily point the laser beam of the LDV to the target that the IR/EO imaging system has detected. The geometrical calibration and registration of multimodal sensors also enable distance measurements using triangulation (e.g., between LDV laser beam and color images) for automatic LDV focusing.
- *Automated targeting and focusing.* Our current LDV system has real-time signal return strength measurements as well as the real-time vibration signals. The search range of the focus function can also be controlled by program. Algorithms could be developed to perform real-time laser focus updating by using the feedback of the LDV signal return strengths and the readings of actual vibration signals. We could also make micro movement of the LDV sensor head to track the target to get the best signal return while performing automatic refocusing.

4.6 Conclusions

The LDV is a noncontact, remote, and high-resolution (both spatially and temporally) voice detector. In this work, we have mainly focused on the experimental study on LDV-based voice detection. We also briefly discuss how we can use IR/EO imaging for target selection and localization for LDV listening. We have set up a system with three types of sensors (the IR camera, PTZ color camera, and LDV) for performing integration of multimodal sensors in human signature detection. The basic idea is to provide an advanced multimodal interface of the site (e.g., an air base) to give the operator the best cognitive understanding of the environment, the sensors, and the events.

We mainly discuss various aspects of LDVs for voice detection—basic principles and problems, signal enhancement algorithms, and experimental designs. We focus on the study of LDV *signal acquisition and enhancement for remote hearing.* We investigate the possibility and quality of voice captured by

LDV devices that point to the objects nearby the voice sources. We have found that the vibration of the objects caused by the voice energy reflects the voice itself. After the enhancement with Gaussian bandpass filtering, Wiener filtering, and adaptive volume scaling, the LDV voice signals are mostly intelligible from targets without retro-reflective tape at short distances (< 300 ft, or 100m). By using retro-reflective tape, the distance could be as far as 1,000 ft (300m).

However, without retro-reflective tape treatment, the LDV voice signals are very noisy from targets at medium and large distances. Therefore, further LDV sensor improvement is required. With current state-of-the-art sensor technology, we realize that signal enhancement techniques need to be developed that are more advanced than the simple bandpass and/or Wiener filtering and adaptive volume scaling. For example, model-based voice signal enhancement could be a solution in that background noises might be captured and analyzed, and models could be developed from the resulting data.

We also want to emphasize that automatic targeting and intelligent refocusing is one of the important technical issues that deserves attention for long-range LDV listening, since it is extremely difficult to aim a laser beam at a distant target and keep it focused. We believe that LDV voice detection techniques combined with the IR/EO video processing techniques can provide a more useful and powerful surveillance technology for both military and civilian applications.

Acknowledgments

This material is based on the research sponsored by the Air Force Research Laboratory under agreement number F33615-03-1-6383. Additional funding is provided by NSF under Grant No. CNS-0551598, by CUNY through a CUNY Equipment Grant Competition Program. The U.S. government is authorized to reproduce and distribute reprints for governmental purposes notwithstanding any copyright notation thereon. However, the views and conclusions contained herein are those of the authors and should not be interpreted as necessarily representing the official policies or endorsements, either expressed or implied, of the Air Force Research Laboratory or the U.S. government.

We are grateful to Lt. Jonathan Lee and Mr. Robert Lee at Air Force Research Laboratory (AFRL) for their valuable discussions during the course of this work. We would also like to thank Professor Thomas Huang and his team at UIUC for their collaborations in developing Wiener filtering and exploring multimodal remote hearing. Professor Ning Xiang at Rensselaer Polytechnic Institute (RPI) has provided his consulting services on laser Doppler vibrometers that have led us to a better understanding of this new type of sensor. Professor Esther Levin, with her expertise in speech technology, has provided valuable discussions on speech signal processing.

References

[1]   Scruby, C., and L. Drain, *Laser Ultrasonics Technologies and Applications*, Bristol/ Philadelphia/New York: Adam Hilger, 1990.

[2]   Polytec laser vibrometer, http://www.polytec.com.

[3]   Ometron vibration measurement systems, http://www.imageautomation.com.

[4]   MetroLaser laser vibrometer, http://www.metrolaserinc.com/vibrometer.htm.

[5]   Halkon, B., S. Frizzel, and S. Rothberg, "Vibration Measurements Using Continuous Scanning Laser Vibrometry: Velocity Sensitivity Model Experimental Validation," *Measurement Science and Technology,* Vol. 14, No. 6, 2003, pp. 773–783.

[6]   Laser radar remote sensing vibrometer, http://sbir.gsfc.nasa.gov/SBIR/successes/ss/4-006text. html.

[7]   FLIR systems security ThermoVision cameras, http://www.flir.com/security/index.htm/lang=EN.

[8]   Hammoud, R., *Proc. of Joint IEEE International Workshops on Object Tracking and Classification Beyond the Visible Spectrum (OTCBVS),* 2004, 2005, and 2006.

[9]   Costley, D., J. Sabatier, and N. Xiang, "Forward-Looking Acoustic Mine Detection System," *SPIE 15th Conference on Detection and Remediation Technologies for Mines and Minelike Targets IV,* 2001, pp. 617–626.

[10]   Goodman, J., "Laser Speckle and Related Phenomena," in *Topics on Applied Physics,* Vol. 9, J. C. Dainty, (ed.), Berlin/New York: Springer-Verlag, 1984.

[11]   Cohen, I., "On Speech Enhancement Under Signal Presence Uncertainty," *ICASSP-2001,* May 2001, pp. 167–170.

[12]   Ephraim, Y., H. Lev-Ari, and W. Roberts, "A Brief Survey of Speech Enhancement," in *The Electronic Handbook,* Boca Raton, FL: CRC Press, 2003.

[13]   Hu, Y., and P. Loizou, "A Subspace Approach for Enhancing Speech Corrupted by Colored Noise," *ICASSP-2002,* May 2002, pp. 573–576.

[14]   Vetter, R., "Single Channel Speech Enhancement Using MDL-Based Subspace Approach in Bark Domain," *ICASSP-2002,* May 2002, pp. 641–644.

[15]   Potamianos, G., et al., "Audio-Visual Automatic Speech Recognition: An Overview," in *Issues in Visual and Audio-Visual Speech Processing,* G. Bailly, E. Vatikiotis-Bateson, and P. Perrier, (eds.), Cambridge, MA: MIT Press, 2004.

[16]   Chu, S., and T. Huang, "Audio-Visual Speech Modeling Using Coupled Hidden Markov Models," *Proc. of IEEE International Conference on Acoustics, Speech, and Signal Processing,* 2002.

[17]   Dupont, S., and J. Luettin, "Audio-Visual Speech Modeling for Continuous Speech Recognition," *IEEE Trans. on Multimedia,* Vol. 2, No. 3, 2000, pp. 141–151.

[18]   Ross, A., K. Nandakumar, and A. Jain, *Handbook of Multibiometrics,* International Series on Biometrics, Vol. 6, New York: Springer, 2006.

[19]   Zhu, Z., W. Li, and G. Wolberg, "Integrating LDV Audio and IR Video for Remote Multimodal Surveillance," *2nd Joint IEEE International Workshop on Object Tracking and Classification in and Beyond the Visible Spectrum (OTCBVS'05),* San Diego, CA, June 20, 2005, http://doi.ieeecomputersociety.org/10.1109/CVPR.2005.478.

[20]   Zhu, Z., and W. Li, "Integration of Laser Vibrometer and Infrared Video for Multimedia Surveillance Display," TR-2005006, Computer Science Department, CUNY Graduate Center, AFRL Technical Report, 2006, http://www-cs.ccny.cuny.edu/~zhu/LDV/FinalReportsHTML/CCNY-LDV-Tech-Report-html.htm, April 2005.

[21]   Li, W., et al., "LDV Remote Voice Acquisition and Enhancement," *International Conference on Pattern Recognition (ICPR'06),* Hong Kong, China, August 2006; supplemental material: http://www-cs.ccny.cuny.edu/~zhu/LDV/ICPR06-LDV-Suppl.rar.

[22]   Castleman, K., *Digital Image Processing,* Upper Saddle River, NJ: Prentice-Hall, 1979.

CHAPTER 5

Sensor and Data Systems, Audio-Assisted Cameras, and Acoustic Doppler Sensors

Paris Smaragdis, Bhiksha Raj, and Kaustubh Kalgaonkar

5.1 Introduction

In this chapter we present some of the work underway at Mitsubishi Electric Research Labs on sensor technology. Specifically, we describe two technologies: audio-assisted cameras and acoustic Doppler sensors for gait recognition. The chapter is laid out in two distinct and independent sections. One section describes our work on audio-assisted cameras. The other section describes our work on acoustic Doppler sensors.

5.2 Audio-Assisted Cameras

Nowadays cameras are a part of our daily lives. They are used to monitor a wide variety of settings ranging from retail shops and secure areas, such as military installations, to elevators and blind spots in cars. A primary reason of their widespread use is that they can enable remote and aggregate monitoring of multiple areas by a single human operator. Instead of dispatching multiple specially trained operators on site to manually perform monitoring, one can now install a networked array of cameras to achieve a comparably powerful monitoring ability. This approach, however, has serious drawbacks. The redirection of multiple visual inputs to a significantly smaller number of observers makes it hard to perform careful monitoring and take proper action instantly. Thus, instead of a real-time surveillance, we have reduced the use of a camera as a tool to record and store information to be analyzed later if needed. The results are nonreal-time responses, operator fatigue or errors, and massive amounts of unexplored video data.

The solution to the above problems has been that of smart sensors. We increasingly see new scientific publications and commercial products that employ computer vision algorithms to automatically detect specific situations, thereby removing human input from the process and providing real-time response and tagging of data. Coupled with new commercially available camera specifications such as high resolutions, pan-tilt-zoom (PTZ) abilities, and exotic lens configurations, this has led to an exciting field of research that holds plenty of promise.

In this section we describe our work on audio-assisted cameras; that is, cameras that employ audio cues to enhance their performance. This is an idea that has been motivated by a variety of applications where automatic content analysis is desired but is nearly impossible using today's computer vision technology.

The use of audio sensing to assist visual processing is not a novel concept. In fact, this is something we all perform subconsciously in our daily lives.

Our listening abilities are constantly used to guide our gaze and even make sense of what we see. This same idea transfers in a very straightforward manner to the computer vision world. Our objective is to use as many audio cues as we can to assist the operation of a camera and maximize its utility as a monitoring device.

There are multiple problems that we wish to address with audio assistance of a camera. The most straightforward problem is that of pointing a camera in the right direction. The use of PTZ cameras assures that we can sweep over a wide range of viewpoints, but it doesn't assure that the camera will point to a region of interest. Extending this idea to the human operator's gaze, we often have banks of multiple monitors conveying images from multiple areas, but there is again no guarantee that the human operator will be looking at the right monitor at the right time. There are also temporal problems. Cameras can record continuously, but the useful sections in recordings are often quite short segments. Continuous recording results in increased storage and bandwidth requirements and amounts to a lot of data, which then necessitates careful off-line analysis. A mechanism to automatically record only the "interesting" parts would greatly reduce the need for storage and networking resources, and also provide easy means to access information fast.

Audio is a very valuable mode for the above problems. Spatial audio analysis can be used to help in steering cameras, whereas audio recognition algorithms can help annotate a camera input, thereby not only assisting steering, but also prioritizing and highlighting the appropriate content. Performing some of these tasks using visual processing is not as straightforward, especially considering how poor lighting and visual conditions can complicate visual processing (let alone the complexities of detecting events such as car collisions, and so forth).

In the following sections we will present how we can use audio analysis to help resolve the problems we have posed. We will describe the prototype design and the type of processing that we employ, and then we will describe some applications where we have found audio processing to be a valuable complement to cameras.

5.2.1 Prototype Setup

In this section we present a prototype audio-assisted camera and a description of its computational abilities. The system consists of a regular PTZ camera coupled with an array of microphones. Data is streamed from the camera and the microphones in real time to a computer that performs a series of processing steps as described in the following paragraphs. The objective of this prototype system is to be able to analyze the activities in its deployed site so that it would steer the camera toward the right direction but also provide annotation on the type of events that transpired around it.

The prototype system and its deployment site are shown in Figure 5.1. The camera is complemented by eight microphones (four omnidirectional and four directional). Only the four omnidirectional microphones are used for camera-related actions, the directional ones being employed for further audio processing to separate incoming sounds. The site we deployed the camera in is rich in both visual and audio activity. As shown in Figure 5.1, the camera is centrally mounted on a ceiling of abuilding lobby and is surrounded by multiple doors. There is virtually nonstop movement from people walking between any combination of doors, as

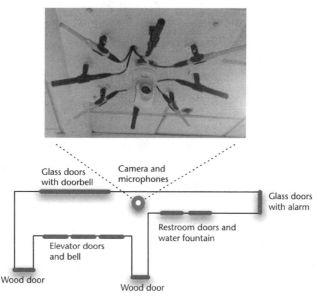

Figure 5.1 The audio-assisted camera prototype (top) and the environment it was set in (bottom).

well as people congregating and participating in conversations. There are also alarms and doorbells that are installed on two of the doors. Our objective was to be able to track the people that are passing by from all directions, classify any activity, and detect any special situations. In terms of audio processing, there are various layers of analysis; in this chapter we will focus on sound recognition, location recognition, and generalized localization.

As one might suspect from the preceding site description, the sounds that are present in the camera's environment are rather limited and quite distinct. The most common occurring sounds are human speech, the sliding of elevator doors, elevator arrival bells, glass doors closing and opening, wood doors closing and opening, a water fountain, a doorbell, and an alarm. Most of these sounds are strongly correlated to a particular position. Because of that we can provide some rudimentary steering ability to the camera using just sound recognition.

5.2.2 Sound Recognition

Generalized sound recognition can be a complicated process when involving multiple sounds, however, even a simple classification solution can perform adequately for our purposes. In order to implement a sound recognition system for the camera, we used a learning-by-example process. We gathered representative samples of all common occurring sounds and divided them into classes. For this particular site we used nine classes: wooden doors, glass doors, elevator doors, restroom doors, speech, water fountain, alarm, doorbell, and elevator bell. The recognition system was based on a Gaussian mixture model (GMM) for incoming spectra. Let us denote each of the nine classes by C_i; each C_i contains a series of training waveforms $x_{i,j}(t)$ so that $C_i \{x_{i,1}(t),\ldots, x_{i,N}(t)\}$. To learn to recognize each sound, we will extract the magnitude of the short time spectra of each example

$x_{i,j}(t)$ using the short time Fourier transform (STFT). To do so we reformat $x_{i,j}(t)$ as $\mathbf{x}_{i,j}(t) = [x_{i,j}(t), ..., x_{i,j}(t + N - 1)]^T$ and obtain $\mathbf{a}_{i,j}(\tau) = \|\mathbf{F}_N \cdot \mathbf{x}_{i,j}(\tau)\|$ where $\mathbf{F}_N$ is the N point Fourier transform matrix and τ is arbitrarily sampled (usually $\tau = Nt$). The dimensionality of these vectors, however, is relatively high, ranging in the few thousand discrete frequencies, and would require significant computational resources to process. In addition to that issue, the entire spectral range of the incoming sounds is largely redundant since there are often just a couple of key spectral regions that are key to recognition. These particular problems can be easily solved using dimensionality reduction techniques that can reduce the computational complexity of the problem in addition to pointing out the most important areas of the spectra that should be used. The most straightforward dimensionality reduction technique to use is that of principal components analysis (PCA) [1], which will keep and consolidate the frequency ranges that exhibit the most energy while removing frequencies that appear to have no significant information content. This step is in essence a linear transformation and is performed simultaneously on the spectra of all the available inputs $(a_{i,j}(\tau) \forall i,j,\tau)$. PCA analysis results into a matrix $\mathbf{W}$, which represents the dimensionality reduction transform. To transform our input features from the magnitude spectral space to the reduced dimensionality space, we simply perform $\mathbf{r}_{i,j}(\tau) = \mathbf{W} \cdot (\mathbf{a}_{i,j}(\tau) - \bar{\mathbf{a}})$, where $\bar{\mathbf{a}}$ is the mean of all $\mathbf{a}_{i,j}(\tau)$ and $\mathbf{r}$ denotes the reduced dimension features. The dimensionality of $\mathbf{r}$ can be selected so as to satisfy computational constraints while maintaining enough data to facilitate recognition. Once the features of all training examples have been obtained, we train a separate GMM for each class of sounds. Doing so will result into a set of Gaussian mixture models G_i for each sound class C_i, which will have adapted to that class's statistics. Assuming our training set is adequately representing each sound class, we can safely assume that for a new input $\mathbf{r}$, the best fitting G_i (i.e., the one that results into a higher likelihood) will reveal the class that $\mathbf{r}$ belongs to.

Deployment of this system yields a very high recognition rate well into 90% accuracy. It is straightforward, then, to associate sound classes with positions of origin to steer the camera. Upon detection of classes like the doorbell class, water fountain, or restroom doors, we can safely assume that there is visual content at their associated position and steer the camera appropriately.

5.2.3 Location Recognition

Although in some cases sound recognition is adequate for pointing the camera toward the proper direction, it is definitely not going to help in the case where we have the same type of sound potentially originating from multiple positions (such as the case of wooden and glass doors, which can occur from either of many positions). Locating the direction of interest in their case is a more complicated task. Just as before, we will employ a learning-based approach to determine the look direction. In the previous case we trained on the timbre of the arriving sounds and disregarded the spatial information that the microphone array had provided us. We will now incorporate this information in order to make a better estimate of the look direction. This time we will use cross microphone features. Let us denote the signal from microphone i as $x_i(t)$ and its corresponding spectra obtained from an STFT as $\mathbf{f}_i(\tau)$.

Assume for simplicity that we have two microphones. The two inputs $\mathbf{f}_1(\tau)$ and $\mathbf{f}_2(\tau)$ will contain certain differences that relate to the spatial attributes of the occurring sound. We obtain two types of features from $\mathbf{f}_i(\tau)$, a cross-mic attenuation feature and a cross-mic phase feature. We define them as: $\mathbf{e}(\tau) = \|\mathbf{f}_1(\tau)/\mathbf{f}_2(\tau)\|$ and $\mathbf{p}(\tau) = \angle \mathbf{f}_1(\tau)/\mathbf{f}_2(\tau)$, where the vector division is element-wise. The vector $\mathbf{e}$ will describe how different the energy of the captured sound between the two microphones is, and $\mathbf{p}$ will describe the accompanying phase difference. It is straightforward to use a Gaussian model to learn $\mathbf{e}$ for each sound class. It isn't straightforward, however, to do the same thing for $\mathbf{p}$. Vector $\mathbf{p}$, being a phase measure, exhibits a wrapping at 0 and π, which results into distributions that a Gaussian cannot model reliably. Therefore, in this case, we need to employ a more sophisticated model that can deal with phase wrapping. One such model is described in [2] and allows modeling of phase variables in terms of Gaussians without being affected by wrapping effects. Using that model for $\mathbf{p}$ and a Gaussian model for $\mathbf{e}$, we can learn the spatial characteristics of each sound class in addition to the sound recognition described before. The combination of these models results into a very reliable estimator that can point to the correct direction and also provide the type of sound that is being heard. An advantage of this approach is that it is immune to traditional localization problems such as reverberation and spurious echoes. The classification of the sound class helps narrow down the candidate positions to just a few (thereby minimizing the potential for error), and the learning-based approach makes use of the reverberation to learn a position, as opposed to traditional localization techniques, which can perform unreliably in the presence of echoes.

5.2.4 Everything Else

The combination of the above classifiers can be very successful for the task at hand, but it is limited by its knowledge. Since we have employed a training methodology, unknown sounds and positions cannot be tracked accurately. A good example of where things might break is that of speech. Speakers can be present anywhere around the camera, and the only prior that we can assume after detection of speech is that the source is a few feet above the floor. In this case we need to have a fallback mechanism that can actually localize sources. Thankfully, source localization using microphone arrays is a mature subject rich in approaches to suit various constraints. For our implementation we employ an algorithm very similar to the one presented by [3] to localize incoming sounds when the trained models do not yield satisfying likelihoods. This approach allows us to deal with multiple sources at the same time and to easily impose priors on the positions if the site geometry is known. Of lesser concern is the case in which we have an unknown sound coming in. The poor fit of this new sound to any existing sound class model will make the camera fall back on the sound localization estimate and flag it as an unknown sound class.

5.2.5 Applications

The model described so far has found a multitude of applications in surveillance. In this section we present a few of them and provide some more evidence that audio assistance of cameras offers a lot of benefits.

One of the earliest incarnations of this work was on traffic monitoring. The task was that of monitoring an intersection using cameras. The problem with this setup was that the cameras were recording continuously, and storage requirements, in addition to the human effort required to scan through the recordings, was costly. In addition to that, there was no guarantee that a serious incident would trigger an immediate response from the command center since most of the time data was recorded and not monitored by a human operator in real time. A solution to these problems would have been to use some smart sensing ability. Unfortunately, approaches based on computer vision solutions were not feasible since the task of detecting a traffic incident is extremely difficult to do visually. In addition to that, poor lighting and weather conditions, which themselves are causes of accidents, often impede visual analysis. By employing an audio-assisted camera we were able to overcome all these problems. The detection of incident-related sounds such as collisions, tire squeals, car horns, screaming, and so forth, is very straightforward using our sound recognition framework. Using a modest training set from data provided by Japan's national police archives, we were able to design classifiers that accurately recognized sounds that relate to traffic incidents and use them to record only a few seconds before the incident and a few seconds after. This approach dramatically decreased the storage requirements and also provided useful information to intersection designers who are now able to easily obtain a glimpse of the most common mishaps without having to sift through large amounts of data.

Another similar application is that of elevator surveillance. We often see surveillance cameras installed in elevators, but an automated visual recognition of an incident is really hard to obtain. The poor lighting conditions and the nonformulaic nature of an elevator crime make it a nearly impossible computer vision task. Just as we have done before, we can turn on the camera when suspicious sounds such as screaming, impacts, and stressed speech are detected. This annotation can be used as an alert to a human operator who is most likely observing multiple video streams at the same time and has few chances of actually spotting an incident as it happens. Just as before, we were able to get a very high recognition rate of staged incidents with relatively simple processing and modest computational requirements.

Building usage is another application that we are using this technology for. Just as in the traffic intersection example, it is often desirable to have statistics on how a building is being utilized. Using audio assistance, we can point cameras to trouble spots, assist surveillance by human operators, and gather more conclusive statistics on how people move and behave throughout the day. In this particular case we use an extension of the wrapped Gaussian model we introduced for location recognition, which is tuned to finding movement trajectories [4]. This model is a hidden Markov model (HMM), which is capable of tracking sequences of location based on phase estimates. This HMM can be used for unsupervised clustering, which then reveals the most common movement paths of people, even when the camera cannot track them.

Finally, we can also employ our localization capabilities to assist cameras with wide (or panoramic) lenses. To do so we highlight the visual sections that correspond to the areas where we detect sound coming from. This can help a human operator focus on an area of significance in an otherwise visually busy scene.

5.2.6   Conclusion

We have presented some of our work on assisting cameras with audio input. We have provided a basic description of the operations involved and presented some applications that we have used this work for that highlight the value of audio sensing for cameras. The technical descriptions given were simple so as to help easily construct a working model. Deployment in a real-life environment often necessitates more complex and robust methods. Some of these extensions are straightforward and some are complex, while others are still open research projects. Our intention was to expose the reader to the types of problems that audio-assisted cameras can deal with. We obviously have only started scratching the surface of what is possible, and we hope that we have managed to convey our excitement about this research direction and helped to expand the existing research base.

5.3   Acoustic Doppler Sensors for Gait Recognition

Gait is defined as a person's manner of walking. A person's gait often carries information beyond that present in the mere physiognomy of the individual—it is often possible to identify a person from a distance by his or her gait even when the person is too far away to distinguish facial or other features. In her seminal studies, Murray [5] proposed that the totality of a person's gait, including the entire cycle of motion, is a unique and identifying characteristic. In another early study, Johansson [6] attached lights to the joints of human subjects and demonstrated that observers could recognize the subjects from just the moving light display.

Traditionally, the identification of gait has been considered a visual phenomenon—it is a characteristic of a person that must be seen to be recognized. Correspondingly, most current algorithms for automatic identification of gait work from visual imagery, such as video or image data. The procedure typically begins by spotting, tracking, and obtaining a silhouette of the subject in the field of view. A set of features is then obtained from the silhouettes. *Model-based* feature extraction methods impose humanlike structures that are represented as stick models (e.g., [7], [8]), ellipse-based representations (e.g., [9]), and so forth. Physiologically motivated features such as joint angles and limb spacing are then derived from the models. *Model-free* feature extraction methods simply attempt to derive features from the silhouettes through a variety of dimensionality reduction methods, for example, through self-similarity plots [10], Euclidean distances from key frames [11], direct normalization of the silhouettes [12], and extraction of additional spatiotemporal representations [13]. Recognition of the subject is performed using classifiers such as K-nearest neighbors, HMMs, and so forth.

In the visual paradigm, gait is captured through a sequence of images, each of which represents an instantaneous capture of *static pose*, rather than movement. Any measurements of movement must be deduced from the sequence of poses. An alternative approach to gait recognition does not use visual information but captures motion through a continuous wave radar [14, 15]. A high-frequency EM tone is incident on the walking subject. The motion of the walker induces a Doppler shift in the frequency of the reflected tone, which is detected at a sensor.

Thus, rather than instantaneous pose, the CW radar captures instantaneous *velocities* of moving body parts.

Conventional wisdom is that high frequencies, such as those used in radars, are required for Doppler-based measurement of the low velocities that comprise gait-related motion. In this chapter, we present an *acoustic* Doppler-based gait recognition system that works from low-frequency ultrasonic tones. The device can be built at very low cost using off-the-shelf acoustic devices and acoustic-range sampling. We demonstrate that the proposed mechanism can provide highly accurate gait recognition, even using only simple signal processing schemes that are conventionally employed for analysing audio data and an equally simple Bayesian classifier.

In Section 5.3.1 we describe the Doppler principle underlying the proposed ultrasonic sensing mechanism for gait. Section 5.3.2 briefly describes the acoustic Doppler sensor we use, Section 5.3.3 describes the signal processing used to derive measurements from the signal captured by the sensor, and Section 5.3.4 presents experimental evidence of the effectiveness of the device.

Apart from differing fundamentally in the sensing mechanism, Doppler- and vision-based gait recognition also differ in their abilities. These differences, various restrictions and limitations of both mechanisms, and avenues of future work are presented in Section 5.3.5.

5.3.1 The Doppler Effect and Gait Measurement

The Doppler effect is the phenomenon by which the frequency perceived by a listener who is in motion, relative to a signal emitter, is different from that emitted by the source. Specifically, if the emitter emits a frequency f that is reflected by an object moving with velocity v with respect to the emitter, the reflected frequency $\hat{f}$ sensed at the emitter is shifted with respect to the original frequency f, and is given by

$$\hat{f} = \frac{v_s + v}{v_s - v} f \approx \left(1 + \frac{2v}{v_s}\right) f, \tag{5.1}$$

where v_s is the velocity of sound in the medium. The approximation to the right in (5.1) holds true if $v \ll v_s$.

If the signal is reflected by multiple objects moving with different velocities, the signal that is sensed will contain multiple frequencies, one from each object.

A human body is an articulated object, composed of a number of rigid bones connected by joints. Figure 5.2 illustrates this through an articulated model of various poses of a human walking. During the act of walking, the structure moves cyclically from peak stride, where most parts are moving at the overall velocity of the walker to midstride, where the arms and legs may be moving in opposition to each other. From the perspective of an observer directly ahead of the walker, the velocities of the various parts would appear to be cyclic, where the amplitude of the cycles depends on the distance of the part from its joint. For instance, the velocity of the foot would be observed to undergo larger swings than a point just below the knee.

When a continuous tone is incident on a walking person, the reflected signal contains a spectrum of frequencies arising from the Doppler shifts of the carrier tone by moving body parts. Figure 5.3 shows the spectrum of the reflected signal

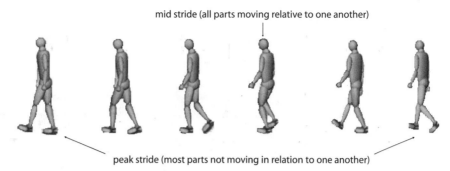

Figure 5.2   Articulated model of a walking person.

obtained at peak and mid strides, when a 40-kHz ultrasonic tone is incident on a walker. As expected, at peak stride the reflected signal shows a relatively narrow band of frequencies with a distinct peak representing the average velocity of the walker. A second peak representing the reflection of the carrier frequency from other static objects is also observed. At mid-stride all parts of the body are moving with respect to one another, and the reflected signal shows a wider range of frequencies without a clear peak beside the reflected carrier signal.

Figure 5.4 shows spectrographic representations of the reflections of an incident tone of 40 kHz captured from two walkers as they walk toward and away from the sensor. In the figures, the x-axis represents time, the y-axis represents frequency, and the color of the picture at any point represents the energy in that frequency at that instant of time. We observe a cyclic variation of frequency patterns in all spectrograms, representing the cyclic variation of the velocities of the limbs. As may be expected, the cyclic frequency patterns are mainly at frequencies higher than the carrier frequency (40 kHz) when the walker approaches the sensor, whereas they are below 40 kHz when the walker moves away from the sensor.

The patterns observed in the spectrogram are typical of the walker. For instance, it may be observed that the patterns in the left panels of Figure 5.4 are

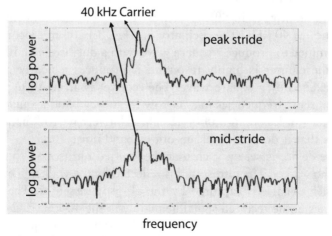

Figure 5.3   Doppler spectra of peak and mid-stride.

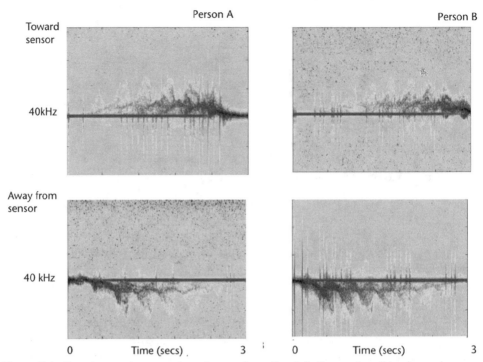

Figure 5.4 Doppler spectrograms from two walkers. Top left: Person A walking toward sensor. Bottom left: Person A walking away from sensor. Top right: Person B walking toward sensor. Bottom right: Person B walking away from sensor.

different from those in the right panels, although superficially both exhibit similar cyclic patterns. It is these differences that enable us to recognize the walker from the reflected signals.

5.3.2 The Acoustic Doppler Sensor for Gait Recognition

Figure 5.5 shows the acoustic Doppler device we use for gait recognition. It comprises an ultrasound emitter, an oscillator, and an ultrasonic sensor. In our setup, the emitter is an off-the-shelf MA40A3S ultrasound emitter with a resonant frequency of 40 kHz. The oscillator is a PIC10F206 microcontroller that has been programmed to produce a square wave with a duty cycle of 50% on one of its pins. Since the emitter is highly resonant, it produces a nearly perfect sinusoid even when excited by a square wave. The receiver is an MA40A3R ultrasound sensor that can sense frequencies in a narrow frequency band around 40 kHz, with a 3dB bandwidth of less than 3,000 Hz. The relatively narrow bandwidth of the sensor ensures that it does not pick up out-of-band noise.

The device is set up such that the emitted ultrasound tone is incident directly upon the walker. The ultrasound is reflected by the walker. The reflected signal (the "Doppler signal") is captured by the ultrasound sensor. The Doppler-shifted freoquencies in the received signal lie mainly in the 40-kHz $\pm$ 2-kHz range. Since the receiver is highly resonant, it automatically attenuates frequencies outside this range. This frequency range is sufficient to represent most velocities related to walking.

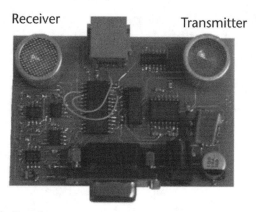

Figure 5.5   The acoustic Doppler sensor.

5.3.3   Signal Processing and Classification

The complete process of gait-based recognition of subjects involves two steps. In the first, a set of features is derived from the Doppler signal. In the second, these features are input to a Bayesian classifier. We describe both steps below.

5.3.3.1   Signal Processing

The spectrographic representations shown in Figures 5.3 and 5.4 can, in fact, be viewed as *velocity spectra*—each frequency represents the Doppler shift of the carrier frequency caused by a particular velocity. The energy in the frequency component is the sum energy of reflections from all components moving at that velocity.

The signal sensed at the Doppler receiver is a sum of the signals reflected by the various moving parts of the walker and is given by:

$$d(t) = \sum_i a_i \sin\left(2\pi f_c\left(t + \frac{2}{v_s}\int_0^t v_i(\tau)d\tau\right) + \phi_i\right) \tag{5.2}$$

Here f_c is the frequency of the emitted tone. Equation (5.2) models the walker's body as a discrete collection of moving parts (although, in reality, it is a continuum). The summation is over all moving parts. a_i is the amplitude of the signal reflected by the ith component and is related to its distance from the sensor. ϕ_i is a phase term representing the relative phase differences between the signals reflected by the various body parts in motion.

Equation (5.2) represents the sum of multiple frequency modulated (FM) signals operating on a carrier frequency f_c. To extract features, we first demodulate the signal $d(t)$ by differentiating it, followed by amplitude demodulation (i.e., multiplication with a sinusoid and low-pass filtering). This process yields $y(t)$:

$$
\begin{aligned}
y(t) &= \mathrm{LPF}\left(\sin(2\pi f_c t)\frac{d}{dt}d(t)\right) \\
&= -\sum_i 2\pi a_i f_c\left(1 + \frac{2v_i(t)}{v_s}\right)\sin\left(\frac{2\pi f_c}{v_s}\int_0^t v_i(\tau)d\tau + \phi_i\right)
\end{aligned} \tag{5.3}
$$

where LPF represents the low-pass-filtering operation.

The demodulated signal $y(t)$ is downsampled to 16,000 samples/sec and segmented into frames of 64 ms. The frames are relatively wide due to the slow varying nature of the signal. Adjacent frames overlap by 50%. Each frame is tapered by a Hamming window. A 1,024-point FFT is computed from the tapered frames, and a 513-point power spectrum is derived from it. The power spectrum is compressed by a logarithm in order to enhance the relative representation of low-energy events in the spectrum. The log-spectral vectors so obtained are finally decorrelated by applying a discrete cosine transform (DCT) to obtain a *cepstral* vector. Only the first 40 coefficients of the cepstral vectors are retained. The 40 cepstral coefficients are augmented by their first order difference terms to result in an 80-dimensional vector representing each frame of the signal.

5.3.3.2 Classifier

We model the sequence of cepstral vectors derived from the Doppler signal as IID. The distribution of the cepstral vectors for any walker w is assumed to be a Gaussian mixture of the form:

$$P(X|w) = \sum_i c_{w,i} N(X; \mu_{w,i}, \sigma_{w,i}) \qquad (5.4)$$

where X represents a cepstral vector, $P(X|w)$ represents the distribution of cepstral vectors for walker w, $N(X; \mu, \sigma)$ represents the value of a multivariate Gaussian with mean μ and variance σ at a point X, and $\mu_{w,i}$, $\sigma_{w,i}$ and $c_{w,i}$ represent the mean, variance, and mixture weight respectively of the ith Gaussian in the distribution for walker w. The parameters of the distribution are learned from a small set of training recordings for the walker.

The spectra, and consequently the cepstral vectors, for a walker are distinctly different for when they are approaching the sensor versus when they are walking away from the sensor. A single Gaussian mixture distribution may be learned for both cases. Alternately, the two cases may be treated independently and separate Gaussian mixtures may be learned for each.

Classification is performed using a simple Bayesian classifier. Let $\{X\}$ represent the set of cepstral vectors obtained from a recording of a subject. The subject is recognized as a walker $\hat{w}$ according to the rule:

$$\hat{w} = \text{argmax}_w P(w) \prod_{X \in X} P(X|w) \qquad (5.5)$$

where $P(w)$ represents the a priori probability of walker w. Typically, $P(w)$ is assumed to be uniform across all subjects, since it may not be reasonable to make any assumptions about the identity of the walker a priori.

5.3.4 Experiments

Experiments were conducted to evaluate the proposed acoustic Doppler-based gait recognition technique. The Doppler sensor described in Section 5.3.2 was mounted

Ultra sonic sensor

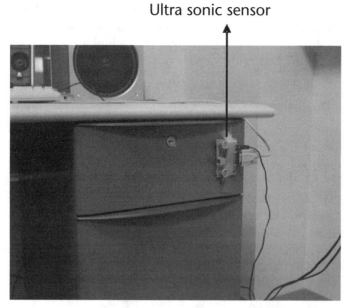

Figure 5.6   Mounting of acoustic Doppler sensor for experimental setup.

on a table as shown in Figure 5.6. A total of 30 subjects were asked to walk toward
and away from the sensor. The distance walked by the subjects was approximately
5m in each direction. Roughly half of the subjects were female and half were male.
Subjects varied in height from 1.5m to 1.93m in height.

A total of 20 recordings were obtained from each subject, ten each of the
subject approaching and walking away from the sensor. Each recoding was
approximately 3 seconds long. Five recordings each of the subject approaching
(*approach*) and departing from (*away*) the sensor were used as training data, while
the rest were used as test data. Separate Gaussian mixture distributions were
trained for the approach and away cases from the corresponding five training
recordings (representing about 15 seconds of training data for each case).

A number of different classification experiments were conducted. In the first,
we attempted to identify the subject. In subsequent experiments, we attempted to
distinguish between the approach and away cases and finally between female and
male subjects. The results of the classification are shown in Table 5.1.

The acoustic Doppler sensor is observed to be highly accurate at identifying
walkers from their gait, identifying the subject correctly well over 90% of the time.
It is also able to determine if the walker is approaching the sensor or moving away
from it over 96% of the time. This accuracy can be improved to 100% if we also

Table 5.1   Classificaton Results for Acoustic Doppler Sensor

Experiment	Number Evaluated	Correctly Classified	Percent Correct
Walker identification	300	275	91.66
Approach versus away	300	289	96.33
Male versus female	300	242	80.66

consider the energy contours of the Doppler signal–energy contours have a decreasing trend for departing walkers and an increasing one for approaching walkers. It is interesting to note that the technique is also significantly better than random at identifying the gender of the walker from the gait.

5.3.5 Discussion

Gait is by nature a very variable phenomenon, affected by such varied factors as the attire of the subject, the speed of walking, or even the subject's state of mind [16], making it generally difficult to recognize. Even in the best scenario, where the subject is walking with a measured regular gait, gait recognition algorithms, both vison-based and movement-based, are highly affected by the viewpoint and the orientation of the sensor. Here, there is a fundamental distinction between the performance of vision- and Doppler-based systems–vision-based systems work best when the subject is walking in a plane perpendicular to the vector from the subject to the camera, so that the entire range of movement of arms and legs can be captured. Doppler-based systems are most effective when the subject is moving directly toward the sensor—velocities perpendicular to the vector to the sensor are undetectable.

Vision-based algorithms are highly dependent on the ability to extract and track the silhouette of the subject accurately. Even when the only moving object in the video is the subject, various phenomena such as shadows, the layout of the background, and so forth, can all affect accurate tracking. On the other hand, Doppler-based devices are relatively insensitive to constant background effects, as these can usually be eliminated as clutter. The performance of both visual and Doppler-based sensors would be affected by the presence of secondary moving objects besides the subject. However, while it is theoretically still possible to track one subject among many in an image, it is currently not possible to do so in the Doppler-based mechanisms. On the other hand, the Doppler-based system can operate in the dark while vision-based systems are critically dependent on the presence of light.

Another important factor in gait recognition is distance. The acoustic Doppler sensor presented here can sense Doppler shifts from subjects at a distance of up to 10m from the sensor. Geisheimer et al. [14] state that they are able to recognize a gait at distances of over 100m using an EM radar. Vision-based mechanisms, on the other hand, are restricted primarily by the resolution of the images analyzed— as the distance to the subjects increases, the corresponding silhouettes become increasingly difficult to separate out in the captured images.

Possibly the biggest advantage to the proposed acoustic Doppler sensor for gait recognition may be cost. The complete cost of lab-manufacture of the device described in Section 5.3.2 is about $20 (at the time of this writing). If manufactured in bulk, it may be expected that the cost will be much less, possibly lower than $5 (at the time of this writing). As a result, it becomes sufficiently inexpensive for it to be used as a *complementary* device to other more accurate sensing mechanisms, with possibilities of ubiquitous deployment.

Several avenues remain for improving the performance of the acoustic Doppler-based gait recognition. Gait is a pseudocyclic phenomenon with a cadence that is displayed as temporal patterns in the spectrograms of the Doppler signal. However, we do not currently model these temporal patterns. We expect that significant improvement in classification accuracy may be obtained by appropriate modeling of these patterns. The proposed algorithm is unable to perform effectively when there are multiple subjects walking within the range of the sensor. However, it is possible to separate out known gait patterns of multiple subjects from a recording using methods such as convolutive nonnegative matrix factorization [17]. The proposed mechanism is also highly sensitive to the angle of approach of the subject—if the subject is walking at an angle that is significantly away from the direct vector to the sensor, it becomes difficult to identify the walker. We expect that it will be possible to normalize for the angle of approach through the use of multiple spatially separated sensors. These and other related issues are topics of future research.

5.4 Conclusions

In the preceding sections we provided a sampler on our work on audio-assisted cameras and Doppler sensing. We have made various arguments on why audio analysis can be a useful complement to audio cameras, and we have shown how Doppler sensing can be applied for gait recognition at a very low cost. Although the two subjects were treated separately, they are by no means meant to be kept as such. Future plans involve the convergence of these two projects, which will allow the enhancement of audio and camera systems by Doppler sensing, and vice versa, to help improve the Doppler estimates by multimodal inputs.

References

[1] Duda, R., P. Hart, and D. Stork, *Pattern Classification*, New York: Wiley-Interscience, 2000.

[2] Smaragdis, P., and P. Boufounos, "Position and Trajectory Learning for Microphone Arrays," *Proc. of the IEEE International Conference on Acoustics, Speech, and Signal Processing (ICASSP)*, January 2007.

[3] Birchfield, S., and D. Gillmor, "Fast Bayesian Acoustic Localization," *Proc. of the IEEE International Conference on Acoustics, Speech, and Signal Processing (ICASSP)*, May 2002.

[4] Smaragdis, P., and P. Boufounos, "Learning Source Trajectories Using Wrapped-Phase Hidden Markov Models," *IEEE Workshop on Applications of Signal Processing to Audio and Acoustics (WASPAA)*, October 2005.

[5] Murray, M., "Gait as a Total Pattern of Movement," *American Journal of Physical Medicine*, Vol. 46, No. 1, 1997, pp. 290–332.

[6] Johansson, G., "Visual Perception of Biological Motion and a Model for Its Analysis," *Perception and Psychophysics*, Vol. 14, 1973, pp. 201–211.

[7] Niyogi, S., and E. Adelson, "Analyzing and Recognizing Walking Figures in XYT," *Proc. of the IEEE Conf. Computer Vision and Pattern Recognition*, 1994, pp. 469–474.

[8] Bobick, A., and A. Johnson, "Gait Recognition Using Static, Activity-Specific Parameters," *IEEE Computer Society Conference on Computer Vision and Pattern Recognition (CVPR'01)*, 2001, pp. 423–435.

[9]   Lee, L., and W. Grimson, "Gait Analysis for Recognition and Classification," *Proc. of the IEEE Conference on Face and Gesture Recognition*, 2002, pp. 155–161.

[10]   BenAbdelkader, C., R. Cutler, and L. Davis, "Motion-Based Recognition of People in Eigengait Space," *Proc. of 5th International Conference on Automatic Face and Gesture Recognition*, 2002, pp. 254–259.

[11]   Kale, A., N. Cuntoor, and V. Krüger, "Gait-Based Recognition of Humans Using Continuous HMMs," *Proc. of the 5th IEEE International Conference on Automatic Face and Gesture Recognition*, 2002, pp. 321–326.

[12]   Collins, R., R. Gross, and J. Shi, "Silhouette-Based Human Identification from Body Shape and Gait," *Proc. of the 5th IEEE International Conference on Automatic Face and Gesture Recognition*, 2002, pp. 351–356.

[13]   Han, J., and B. Bhanu, "Individual Recognition Using Gait Energy Image," *IEEE Trans. on Pattern Analysis and Machine Intelligence*, Vol. 28, No. 2, 2006, pp. 316–322.

[14]   Geisheimer, J., W. Marshall, and E. Greneker, "A Continuous-Wave Radar for Gait Analysis," *Conference Record of the 35th Asilomar Conference on Signals, Systems and Computers*, Vol. 1, 2001, pp. 834–838.

[15]   Otero, W., "Application of a Continuous Wave Radar for Human Gait Recognition," *Signal Processing, Sensor Fusion, and Target Recgonition XIV, Proc. of the SPIE*, I. Kadar (ed.), No. 5809, 2005, pp. 538–548.

[16]   Lemke, M., et al., "Spatiotemporal Gait Patterns During Over Ground Locomotion in Major Depression Compared with Healthy Controls," *Journal of Psychiatric Research*, Vol. 34, 2000, pp. 277–283.

[17]   Smaragdis, P., "Discovering Auditory Objects Through Non-Negativity Constraints," *Proc. ISCA ITRW on Statistical and Perceptual Audio (SAPA 2004)*, 2004.

PART II

Multimodal Fusion Algorithms

CHAPTER 6

Audiovisual Speech Recognition

Stephen M. Chu and Thomas S. Huang

6.1 Introduction

There has been a growing interest in incorporating visual information into automatic speech recognition in recent years. A number of research systems have been implemented. Among these systems, some are designed to show that speech recognition is possible using only the visual information. Most of the recent research attempts, however, are concentrated on building bimodal systems that take advantage of both the audio and the visual channels, especially in the context of robust speech recognition.

The recognizers presented in the literature vary widely in the choice of visual features set, recognition engine, and fusion strategy. Significant efforts have been devoted to the extraction of visual features. Systems range from using raw images of the mouth region to parameters of a full-fledged three-dimensional face model of the speaker. Studies that compare the effectiveness of the various feature sets have been carried out. The recognition engine of the recognizer also takes many forms among the various research groups. Essentially, there is little fundamental difference between the bimodal systems and conventional audio-only ASR in term of recognition engine. Some earlier systems were based on dynamic time warping (DTW); since then, there have been a number of systems that use neural-network architectures and an increasing number of systems that rely on HMMs. The least understood aspect of audiovisual ASR is the fusion strategy. Although a number of fusion methods have been proposed in the literature, none has been shown to be consistently superior to others. Interestingly, here we see a clear parallel between the research in audiovisual ASR and the studies in bimodal human speech perception. To date, a majority of the various approaches to fusion fall into one of two categories: early integration or late integration. Only a very small number of researchers have explored the intermediate integration strategy. In the end, there is no clear consensus on what the optimal approach to fusion should be.

In the following sections, we will survey some of the representative works among the various audiovisual speech recognition systems found in the literature. The emphasis of the discussion will be on the choice of visual features and the fusion strategy, because firstly these issues are specific to audiovisual ASR, and secondly, they represent two of the most important decisions one has to make when considering a bimodal speech recognition system.

6.1.1 Visual Features

The main methods for extracting visual speech information from image sequences can be grouped into pixel-based and geometry-based approaches. In the basic pixel-based approach, the pixel-level data of the mouth region of the speaker is either taken directly or after some preprocessing to form the feature vector. The features obtained in this fashion are generally regarded as low-level features. Features of a slightly higher level can be extracted by a variation of the basic method that relies on the visual motion field of the region of interest. The geometry-based approaches usually apply more sophisticated computer vision techniques to get some forms of the geometry measurements of the talker's lips. The geometry features are high-level compared to the pixel-based features. In practice, these methods require accurate automatic lip-tracking in order to be successful. Some methods put markers on or paint the lips to facilitate tracking and feature extraction, while a growing number of methods take the model-based approach. In the model-based approach, a parameterized geometric model of the lip contours is the basis of the tracking algorithm. An advantage of the model-based methods is that the parameters of the lip model can be used directly as a compact representation of the visual features.

6.1.1.1 Examples of Pixel-Based Approach

An example of the basic pixel-based approach can be found in the bimodal system presented by Yuhas et al. [1]. In this system, the gray-level image of the mouth area was subsampled to lower resolution, and the pixel values of the resulting image were used to form the visual feature vector. An obvious disadvantage of using the image as a feature is the high dimension and high redundancy in the feature vector. A different pixel-based method based on vector quantization (VQ) was developed by Silsbee and Bovik [2]. In this method, the codebook consists of 17 vectors representing various mouth shapes that were selected by hand. Another approach to reduce the dimensionality of the feature space is to use the so-called Eigenlips [3, 4] based on principal component analysis (PCA). Feature extraction methods based on the discrete cosine transform (DCT) [5] and the discrete wavelet transform [6] have also been reported.

A common drawback of these pixel-based methods is that they are sensitive to variations in lighting conditions and noise. Furthermore, the extracted features account for both shape variability and intensity variability, and therefore might be less discriminative for a large population of speakers with different appearances and mouth shapes.

Mase and Pentland described another low-level visual feature extraction method using optical flow [7]. The algorithm computes the average optical flow vectors in four windows around the mouth region in hopes of capturing the muscle activities involved in speech production. Visual motion information is likely to be tolerant to speaker variability. However, the optical flow features, like the basic pixel-based features, still suffer from some common variations present in the input video signal, such as scaling, 3-D rotation, and illumination change.

6.1.1.2 Examples of Geometry-Based Approach

Petajan developed one of the first audiovisual speech recognition systems [8]. In this system, binary images are analyzed to derive a set of geometric features, which comprises the mouth opening area, the perimeter, the height, and the width. Goldschen later used this visual analysis system to build a continuous speech recognition system [9]. He found that the best recognition performance was achieved when the derivative information was dominant in the feature set. Similar conclusions were also reached by Mase and Pentland in their experiments.

Finn and Montgomery [10] confirmed that physical dimensions of the mouth could give good recognition performance. In their system, visual features were obtained by measuring 14 distances with the help of reflective markers put on the speaker's mouth. A system developed by Stork et al. [11] took a similar approach to extract visual features using markers. Instead of using markers, some researchers highlighted the lips with color to facilitate lip tracking and the extraction of geometric visual features. An example is the audiovisual system developed by Adjoudani and Benoît [12].

Robust tracking of the lips is indeed a difficult problem. The use of markers and lip colors may simplify the task. However, they are not feasible solutions for real applications. A number of model-based methods have been proposed to tackle the problem. Both the deformable templates [13] and active contour models (*snakes*) [14] fit into this category. The basic idea is to form an energy function that relates a parameterized model to an image. This energy function can be minimized using any standard optimization technique to obtain a parameter set. An application of deformable template to lip-tracking was described in [15]. The lip contours are modeled by a set of polynomials hand-tuned to match the outline of the lips. Several research groups have applied the *snakes* to lip-tracking [16]. Rao and Mersereau [17] proposed a tracking algorithm that combines the deformable templates with statistical color modeling. This method was used to extract visual features by Huang and Chen [18]. Both the deformable templates method and the *snakes* perform tracking in 2-D. Tao and Huang [19] described a method based on the Bézier volume deformation model to track facial motion in 3-D. This algorithm has been applied by Zhang to extract visual features for an audiovisual speech recognizer.

The model-based methods are able to give compact feature sets for visual speech. However, the substantial reduction in data may also result in the loss of speech-relevant information. For example, the intensity information of the mouth region may reflect the 3-D shape of the lips that is not represented in the 2-D geometric features. Efforts that aim to integrate the low-level and the high-level features have been made. Luettin et al. [20] used the active shape models for tracking. The visual features were comprised of both the shape information represented by the parameters of the shape model and the intensity information in the neighborhood of the lip contours. Potamianos et al. [6] combined geometric features and the wavelet coefficients of the mouth image to form a joint visual feature vector. Neti et al. [5] compared Luettin's feature sets with a low-level approach based on DCT. It was found that the shape-intensity joint feature set performs slightly better in a visual-only speech-reading experiment.

6.1.2 Fusion Strategy

Perhaps the most challenging issue in bimodal speech processing and recognition is how to integrate the audio and visual information in an optimal way. As stated in the discussion of the studies of human speech perception in Chapter 1, there are two groups of hypotheses that attempt to model the sensory fusion in humans: the EI models and the LI models. The various approaches to fusion found in the audiovisual ASR literature can largely be categorized in a similar fashion. One class of methods performs fusion at the feature level. These are referred to as *early integration* or *feature fusion*. In a system based on *early integration*, the fusion takes place before the classification stage. Therefore, only one classifier is necessary. The other class of methods carries out fusion in the decision level. These are referred as *late integration* or *decision fusion* methods. These methods combine the outputs of single-modality classifiers to recognize audiovisual speech. A small number of methods fit into neither the *early integration* nor the *late integration* models. In these methods, the fusion of the audio and visual information takes place within the classification process. We shall call these the *intermediate integration* methods. The schematics of the three classes of approaches are shown in Figure 6.1.

One could argue that early integration is the most general method, and both late and intermediate integration are special cases of it. Given suitable system architecture and proper learning algorithm, a system that uses early integration should perform at least as well as one that integrates at a later stage. Here we limit our definition of *early integration* to a more restrictive sense: the classifier in *early*

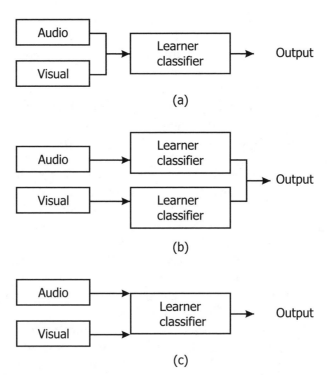

Figure 6.1 Different approaches to audiovisual fusion: (a) early integration, (b) late integration, and (c) intermediate integration.

integration is unaware of the bimodal nature of the feature vector it sees and just takes it as unity.

6.1.2.1   Examples of Early Integration

The simplest way to integrate the audio and visual information is to concatenate the audio and visual features to form a composite feature vector [3, 12, 21–24]. Since the integration of features takes place before classification, the method is independent of the choice of recognition engine. Yuhas described an interesting way to realize feature integration in [1]. The visual information was first mapped to the frequency envelope of the acoustic domain and then combined with the acoustic spectrum. The motivation of this approach was to integrate the audio and visual data in a common feature space. Neti et al. proposed an early integration method that seeks to reduce the size of the concatenated audiovisual feature vector. This is achieved by projecting the composite vector onto a lower dimensional space by linear discriminative analysis (LDA), followed by maximum likelihood linear transform (MLLT) [5].

An assumption that the *early integration* methods make implicitly is that the acoustic channel and the visible channel of speech production are always tightly synchronized. In reality, the visual cues usually precede the audible speech with a variable lead-time. The variability of this desynchronization becomes more prominent at the word boundaries. Studies of human audiovisual speech perception have shown that human bimodal speech recognition does not require tight synchronization between the two modalities. The *early integration* methods are not able to capture the bimodal cues that are embedded in the asynchrony, such as the VOT.

6.1.2.2   Examples of Late Integration

A *late integration* approach was used in the first audiovisual speech recognition system developed by Petajan [8]. In this system, the acoustic speech recognizer and the visual classifier are combined in a serial fashion. First, the acoustic speech recognizer gives a list of candidate words. Then the visual classifier takes the word list as input and makes a decision based on the visual information. Adjoudani and Benoît [25] studied two *late integration* methods. In both methods, the acoustic speech recognizer and the visual speech recognizer are running in parallel. In the first method, the two recognizers compete with each other, and the winner is chosen based on a confidence measure. In the second method, the likelihood scores outputted by the audio and visual recognizers are weighted and combined to give the recognition result. In addition, they compared the performance of the two *late integration* methods with *early integration* using concatenation. It was demonstrated that their *late integration* methods give higher recognition accuracy, especially at low SNR. Stork et al. [11] implemented both early and late integration strategies using neural networks. They found that the *late integration* strategy produced slightly better results.

The *late integration* methods treat the audio and visual modalities as conditionally independent. This is in contradiction with the fact that the acoustic speech and visual speech represent two temporally coupled information channels of the same underlying speech production process. Studies have concluded that the fusion of the two modalities in human speech perception occurs before phonetic characterization, and the information of the two modalities is continuous rather than discrete. The *late integration* methods will not be able to capture this low-level influence between the two modalities. Furthermore, temporal information across the audio and visual channels is also lost.

6.1.2.3   Examples of Intermediate Integration

The *intermediate integration* methodology presents a fresh approach to the fusion problem. Conceptually, an *intermediate integration* method employs a single classifier. This classifier is responsible for both the fusion and classification of the audiovisual speech. Furthermore, the architecture of the classifier must include elements that explicitly model the low-level intermodal influence between the audio and the visual channels.

Very few integration methods in existence fit this description. Hennecke et al. [26] pointed out that *intermediate integration* can be achieved using neural networks. Potamianos and Graf proposed an HMM-based method [27]. In this system, a two-stream HMM architecture is used to carry out fusion and classification of bimodal speech. Stream exponents that weight the output probabilities of the audio and visual modalities were introduced and learned through discriminative training. The stream exponents can be introduced at the state level, or they can be introduced at the model level or the global level. Good results were reported for using global exponents.

When the exponents are state-specific, the two-stream HMM is in fact performing *intermediate integration,* because varying the stream exponents will not only affect the total output probabilities of the states, but also will be likely to change the state-occupancy path, which reflects the internal dynamics of the HMM. One limitation posed by the two-stream HMM method is that synchronized progression of the audio and visual modalities is assumed. This is due to the fact that both information channels share a single state machine. Therefore, like in early *integration*, this architecture is not able to model asynchronies between the two modalities.

In the next chapter, we present a novel *intermediate integration* approach to audiovisual speech fusion based on the CHMM framework. The CHMM architecture allows controlled asynchronies between information channels. Furthermore, the proposed fusion scheme is more general than the two-stream HMM method. We will show later that the latter can be viewed as a special case of the former.

6.2   Sensory Fusion Using Coupled Hidden Markov Models

The fusion of audio and visual speech is an instance of the general sensory fusion problem. The sensory fusion problem arises in the situation when multiple channels carry complementary information about different components of a system. In the case of audiovisual speech, the two modalities manifest two aspects of the same underlying speech production process. From an observer's perspective, the audio channel and the visual channel represent two interacting stochastic processes. We seek a framework that can model the two individual processes as well as their dynamic interactions.

6.2.1 Introduction to Coupled Hidden Markov Models

It is a fundamental problem to model stochastic processes that have structure in time. A number of frameworks have been proposed to formulate problems of this kind. Among them is the HMM, which has found great success in the field of ASR. The theory of the HMM is well established, and there exist efficient algorithms to accomplish inference and learning of the model. Figure 6.2 shows an HMM with four discrete states in the conventional state machine representation. In this form of representation, each node denotes one of the discrete hidden states. The arrows indicate the transition probabilities among the states.

In recent years, a more general framework, dynamic Bayesian networks (DBNs), has emerged as a powerful and flexible tool to model complex stochastic processes [28]. The DBNs and the HMMs both observe the Markov assumption that the future is conditionally independent of the past, given the current state. The DBNs generalize the hidden Markov models by representing the hidden states as state variables and allow the states to have complex interdependencies. Under the DBN framework, the conventional HMM is just a special case with only one state variable in a time slice. DBNs are commonly depicted graphically in the form of probabilistic inference graphs. An HMM can be represented in this form by rolling out the state machine in time, as shown in Figure 6.3. Under this representation, each vertical slice represents a time step. The circular node in each slice is the multinomial state variable, and the square node in each slice represents the observation variable. The directed links signify conditional dependence between nodes.

It is convenient to use probabilistic inference graphs to represent the independence structure of a distribution over random variables. An important property of the graph is the *directed Markov* property, which states that a variable is conditionally independent of its nondescendants, given its parents:

$$X \perp nd(X) | \, pa(X) \tag{6.1}$$

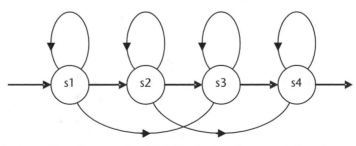

Figure 6.2 State machine diagram of an HMM (output nodes are not shown).

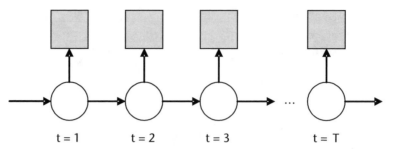

Figure 6.3 DBN representation of an HMM. The circular node denotes state variable, and the square node represents the observation variable.

In the case of HMM, each node has only one parent. It can be easily read from the graph that the future state is independent of the past, given the present; moreover, the observation variable is only dependent on the current state.

It is possible to just use HMM to carry out the modeling and fusion of multiple information sources. This can be accomplished by attaching multiple observation variables to the state variable, and each observation variable corresponds to one of the information sources. Figure 6.4 illustrates the fusion of audio and visual information using this scheme. In fact, this is precisely the underlying model that the two-stream HMM approach discussed in the previous chapter is based on. Because both channels share the single state variable, this approach in effect assumes that the two information sources always evolve in lockstep. Therefore, it is not possible to model asynchronies between the two channels using this approach.

An interesting instance of the DBNs is the so-called coupled hidden Markov models (CHMMs). The name CHMMs comes from the fact that these networks can be viewed as parallel rolled-out HMM chains coupled through cross-time and cross-chain conditional probabilities. It is this particular type of coupling that makes CHMMs a promising framework for learning and modeling multiple interacting processes [29]. In the perspective of DBNs, an *n*-chain CHMM has

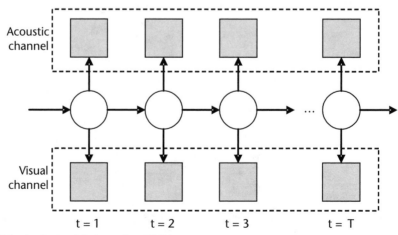

Figure 6.4 Audiovisual fusion using HMM.

n hidden nodes in a time slice, each connected to itself and its nearest neighbors in the next time slice. For the purpose of audiovisual speech modeling, we considered the case of $n = 2$, or the two-chain CHMMs. Figure 6.5 shows the inference graph of a two-chain CHMM.

There are two state variables in the graph. The state of the system at a certain time slice is jointly determined by the states of these two multinomial variables. More importantly, the state of each state variable is dependent on both of its two parents in the previous time slice: one parent from the same chain, the other from the opposite chain. This configuration essentially permits unsynchronized progression of the two chains, while encouraging the two subprocesses to assert temporal influence on each other's states. Note that the Markov property is not jettisoned by introducing the additional state variable and the directed links. Given the current state of the system, the future is conditionally independent of the past. Furthermore, given its two parents, a state variable is also conditionally independent of the other state variable. This property is important for our later development of the inference algorithms.

In addition to the two state variables, there are two observation variables in each time slice. Each observation variable is a private child of one of the state variables. The observation variables can be either discrete or continuous. It is possible with this framework that one of the state variables is continuous and the other one is discrete.

In the context of audiovisual speech fusion, the audio and visual channels are associated with the two state variables, respectively, through the observable nodes. Interchannel asynchrony is allowed. The overall dynamics of the audiovisual speech are determined by both modalities. The temporal dependencies of the two modalities are learned from data through training.

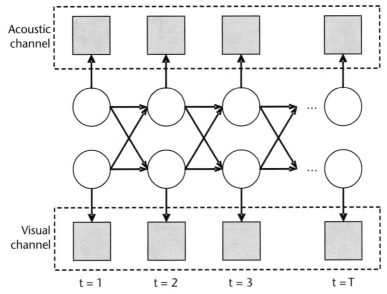

Figure 6.5   Audiovisual fusion using CHMM.

In general, the time complexity of exact inference in DBNs is exponential in the number of state variables per time slice. For systems with a large number of state variables, exact inference quickly becomes computationally intractable. Consequently, much attention in the literature has been paid to approximation methods that aim to solve the general problem. Existing approaches include the *variational methods* [30] and the *sampling methods* [31]. However, these methods usually exhibit nice computational properties in an asymptotical sense. When the number of states is very small, the computational overhead embedded in the approximation method is often large enough to offset the theoretical reduction in time complexity. In this situation, the approximation becomes superfluous and exact inference becomes more desirable. The HMM is an example of a DBN in which exact inference is tractable. In fact, there exist efficient algorithms to carry out inference and learning in HMM. In the following section, we develop an exact inference algorithm for CHMM based on the HMM approach. Following the development, the correctness of the algorithm is verified through a series of experiments using simulated data. The modeling capabilities of the CHMM will also be empirically assessed, especially in the context of maximum likelihood classification of temporal patterns.

6.2.2 An Inference Algorithm for CHMM

6.2.2.1 Task Definition

Given the form of CHMM shown in Figure 6.5, we are concerned with the following two basic problems:

1. The inference problem: given the parameters of the model, evaluate the probability of some given events. Specially, given the observation sequence $\mathbf{O} = (\mathbf{o}_1\mathbf{o}\ldots\mathbf{o}_T)$, and a model λ, we want to compute $P(\mathbf{O}|\lambda)$, the probability of the observation sequence given the model.
2. The learning problem: find the model parameters λ to maximize $P(\mathbf{O}|\lambda)$.

For conventional HMMs, $P(\mathbf{O}|\lambda)$ can be efficiently computed using the forward procedure, which reduces the problem to inductively solving the forward variable α_t defined as

$$\alpha_t = P(\mathbf{o}_1\mathbf{o}_2\ldots\mathbf{o}_1, q_t|\lambda) \tag{6.2}$$

where q_t is the value of the state variable at time t. Similarly, the backward variable is defined as

$$\beta_t = P(\mathbf{o}_{t+1}\mathbf{o}_{t+2}\ldots\mathbf{o}_T| q_t, \lambda) \tag{6.3}$$

Using the forward and backward variables as basic statistics, the learning task can be achieved by applying the expectation-maximization (EM) algorithm. In the next section, we show how to extend the forward procedure to the two-chain CHMMs.

6.2.2.2 Algorithm Definition

Let us denote the two hidden state variables q_t^1 and q_t^2. Then a two-chain CHMM λ is described by the conditional probability tables $P(q_t^1 | q_{t-1}^1, q_{t-1}^2)$ and $P(q_t^2 | q_{t-1}^1, q_{t-1}^2)$, specifying the state transition probabilities, together with the output distribution table $P(o_t | q_t^1, q_t^2)$. Suppose each state variable can take Q possible discrete values, the state transition tables will then have $Q \times Q \times Q$ entries, and the output distribution table will have $Q \times Q$ entries.

Define

$$\alpha_t = P(o_1 o_2 \ldots o_t, q_t^1, q_t^2 | \lambda) \tag{6.4}$$

which is the probability of the joint event of the partial observation up to t and the two hidden states being in $\{q_t^1, q_t^2\}$ at time t. At the next time slice $t+1$, the probability of the joint event of observing the partial observation up to t and the two hidden states reaching $\{q_{t+1}^1, q_{t+1}^2\}$ from $\{q_t^1, q_t^2\}$ is

$$P(o_1 o_2 \ldots o_t, q_t^1, q_t^2, q_{t+1}^1, q_{t+1}^2 | \lambda) = \alpha_t \cdot P(q_{t+1}^1, q_{t+1}^2 | q_t^1, q_t^2) \tag{6.5}$$

Using the conditional independence implied from the graph in Figure 6.1, we have

$$P(o_1 o_2 \ldots o_t, q_t^1, q_t^2, q_{t+1}^1, q_{t+1}^2 | \lambda) = \alpha_t \cdot P(q_{t+1}^1 | q_t^1, q_t^2) \cdot P(q_{t+1}^2 | q_t^1, q_t^2) \tag{6.6}$$

Now, marginalize out q_t^1 and q_t^2

$$P(o_1 o_2 \ldots o_t, q_{t+1}^1, q_{t+1}^2 | \lambda) = \sum_{\{q_t^1, q_t^2\}} \alpha_t \cdot P(q_{t+1}^1 | q_t^1, q_t^2) \cdot P(q_{t+1}^2 | q_t^1, q_t^2) \tag{6.7}$$

Multiply both sides of (6.6) by the output density $p(o_{t+1} | q_{t+1}^1, q_{t+1}^2)$

$$P(o_1 o_2 \ldots o_t, q_{t+1}^1, q_{t+1}^2 | \lambda) \cdot p(o_{t+1} | q_{t+1}^1, q_{t+1}^2) = P(o_1 o_2 \ldots o_{t+1}, q_{t+1}^1, q_{t+1}^2 | \lambda) \tag{6.8}$$

$$\alpha_{t+1} = \left[\sum_{\{q_t^1, q_t^2\}} \alpha_t P(q_{t+1}^1 | q_t^1, q_t^2) P(q_{t+1}^2 | q_t^1, q_t^2) \right] \cdot p(o_{t+1} | q_{t+1}^1, q_{t+1}^2) \tag{6.9}$$

We have derived the update formula for the forward variable α_t. Finally,

$$P(O | \lambda) = \sum_{\{q_T^1, q_T^2\}} \alpha_T \tag{6.10}$$

Clearly, this procedure performs exact inference in the two-chain CHMM, and its time complexity is $O(TQ^4)$. The backward variable and the backward procedure can be similarly derived (omitted here for brevity).

Once we have the ability to compute the forward and backward variables, many inference problems in the two-chain CHMMs can be solved efficiently. Furthermore, the learning can be carried out following the well-known Baum-Welch reestimation formulas (EM).

One objective of this study is to evaluate the capabilities of the CHMMs in the context of maximum likelihood classification of temporal patterns. Given a set of K models $\{\lambda_k\}_{1 \leq k \leq K}$ and observation sequence $\mathbf{O}$, we want to know which model is the most likely to have generated the observation

$$\hat{k} = \arg\max_k [P(\mathbf{O}|\lambda_k)] \qquad (6.11)$$

6.2.2.3   Performance Expectations

The inference algorithm presented in the previous subsection is an exact algorithm. Therefore, we expect the inference outcome to be the true probability of the event being inferred. On the other hand, the EM algorithm used in training is known to give locally maximized parameters, thus it is not guaranteed to produce the optimal model for a given training set. Nevertheless, when the structure of the hypothesized model used in EM matches the structure of the true underlying distribution that generates the examples, we can expect the EM algorithm to give satisfactory results. In fact, we ensure this matched-structure condition in the subsequent experiments. Furthermore, the number of training examples has a direct effect on the training result. The expectation is that increasing the number of training examples will give a resulting hypothesis that is closer to the true distribution that the examples were drawn from.

In the classification experiment, it is expected that given a set of models that is close enough to the true models generating the observations, the error rate should be low. In the next section, we aim to empirically evaluate these expectations.

6.2.3   Experimental Evaluation

6.2.3.1   Methodology

The experiments are carried out in two phases. The objectives of the first phase are: (1) to verify the correctness of the algorithm, and (2) to study its behavior under different conditions. The algorithm is employed to learn the parameters of a two-chain CHMM given a set of temporal observations. The training examples are generated by a predefined CHMM through random walk. We vary the number of allowable states in each chain and the number/length of training examples, and compare the parameters of the resulting CHMM with the true model that generated the data. The error is measured using the l_1 distance between the corresponding parameters. For example, for the conditional probability table (CPT)

$$\theta_1 = [\{P(q_{t+1}^1 | q_t^1, q_t^2)\}] \in \Re^{Q^3} \qquad (6.12)$$

the error is defined as

$$err_{\theta_1} = |\hat{\theta}_1 - \theta_1| \qquad (6.13)$$

Notice that this is a rather naïve distance measure, because statistical equivalence of two models can occur without them necessarily having similar parameters. Nonetheless, the above error does provide a useful measure of how close two models are when their parameters are similar.

In the second phase of the experiments, the algorithm is evaluated on a simulated temporal pattern classification task. Starting with the two-class case, we first create two CHMMs by assigning the parameters randomly according to the uniform distribution. These models are used to generate both the training and the testing examples. Supervised training is then carried out to produce two estimated CHMMs. During testing, we seek the model that gives the higher likelihood given an observation. The classification is considered correct when the class of the selected model matches the label of the example.

To assess whether any classification error is the result of inadequate learning or is due to other limitations, we perform the test on the same test set again, this time using the two true CHMMs that generated the data. This can be viewed as the situation when the learning is perfect, thus giving us an upper bound on the performance given the kind of stochastic processes we are trying to model and the available observation.

In addition to the two-class case, a ten-class classification task is also carried out following the same procedure.

6.2.3.2   Results and Discussion

A two-chain CHMM is created with the following configuration. Each of the two state variables can take three discrete values; the output nodes are binary. We use this model to randomly generate 5 training sets, containing 10, 20, 50, 100, and 200 examples, respectively. All the examples are ten time-slices in duration. Training typically converges in two to four EM iterations, except for the 10-element set, where it does not converge after ten iterations. The algorithm, which is implemented in MATLAB, takes 121 seconds to process 50 examples on a SUN Ultra 10 workstation. The error of the state transition CPT and the output CPT between the estimated model and the true models is recorded in Table 6.1.

It is evident that the learned model becomes closer to the true model as the number of training examples increases. This trend can be even more clearly observed in Figure 6.6, in which the error incurred for the CPT $P(q_{t+1}^1 | q_t^1, q_t^2)$ is plotted against the size of the training set.

Let us take a closer look at the error for the CPT $P(q_{t+1}^1 | q_t^1, q_t^2)$ of the model trained using 100 examples. The CPT contains $3^3 = 27$ entries, and the error defined in (6.13) is the sum of the errors of the individual entries. Therefore, the average error for each entry in the table is $0.2164/27 \approx 28$. Considering that each entry in the CPT is a probability, which is a real number between 0 and 1, the error is indeed quite small. This indicates that the learning is successful.

The two-class classification experiment is first carried out using two CHMMs that have the same topological configuration as in the learning experiment, each

Table 6.1   Distance Between the Learned Parameters and the True Parameters

Training Set Size	10	20	50	100	200
Err $P(q_{t+1}^1 | q_t^1, q_t^2)$	5.1644	1.7636	0.8078	0.2164	0.1181
Err $P(q_{t+1}^2 | q_t^1, q_t^2)$	5.2886	1.1683	0.6776	0.1267	0.1724
Err $P(o_t^1 | q_t^1)$	1.1221	0.4242	0.0879	0.0619	0.0409
Err $P(o_t^2 | q_t^2)$	0.5053	0.1018	0.0695	0.0201	0.0531

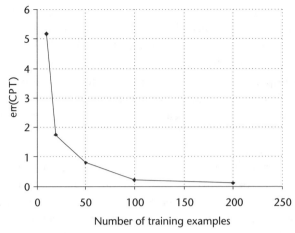

Figure 6.6   Error for CPT $P(q_{t+1}^1|q_t^1, q_t^2)$ decreases as a larger training set is used.

trained by 100 examples. The length of the samples T, measured in number of time slices, is 10. During testing, we randomly pick one of the CHMMs and use it to produce a test sequence. A total of 100 test cases are generated. The number of correct classifications out of 100 is shown in Table 6.2.

The tests are performed on three kinds of observations. The first is the regular test where the two state variables are hidden, and only the output nodes are observed. In the second case, in addition to the output nodes, one of the state variables is also visible. In the last case, all nodes in a time slice are visible.

We see that the performance of the learned CHMMs is virtually the same as the true models at all three conditions. This implies that the learning is successful. However, it is rather unexpected to see that even the true CHMMs perform poorly in the regular test. This seems to suggest that even when perfect learning is achieved, it may still be difficult to infer the correct class from the given observations. Results from the second and the third case show that additional information about the true state of the network can greatly improve the chance of making the correct classification. Notice that in the third case, the problem becomes virtually deterministic.

The tests were repeated several times, and similar results are observed. Table 6.3 shows the results for CHMMs with $Q = 5$.

Table 6.2   Two-Class Classification Results for 100 Test Cases ($Q = 3$, $T = 10$)

CHMMs Used in Testing	Learned	True
All state variables hidden	73	76
One state variable visible	87	87
All state variables visible	100	98

Table 6.3   Two-Class Classification Results for 100 Test Cases ($Q = 5$, $T = 10$)

CHMMs Used in Testing	Learned	True
All state variables hidden	68	67
One state variable visible	95	93
All state variables visible	100	100

Table 6.4   Two-Class Classification Results Using Samples with Different Duration Ts

T	50	100	200	500	1,000
N Correct	76	85	88	93	96

Increasing the duration of the examples T from 10 time-slices to 100 leads the number of correct classification in regular test on learned CHMMs to improve from 68 to 83. This trend continues when even larger Ts are used. The classification results using test cases of $T = 50$, 100, 200, 500, and 1,000 are summarized in Table 6.4.

The number of correct classifications versus the duration of the test samples is plotted in Figure 6.7. The results clearly indicate that longer test sequences are more likely to be classified correctly by the classifier. And the low classification accuracy observed in the earlier experiments with $T = 10$ is because the short samples do not carry sufficient statistical evidences in order to be reliably distinguished from each other.

This observation may provide important understanding of the capabilities of CHMMs in real world applications. A major promise of the framework is its ability to model and fuse two temporally interacting processes, for example, audiovisual speech. Now, let us consider a hypothetical isolated digit recognition experiment using CHMMs. A 10-ms frame rate is a typical choice in speech processing. If the CHMMs with $Q = 5$ were used to model the bimodal speech, according to the results shown in Figure 6.7, reliable classification would have required at least 500 frames, or 5 seconds of sample duration, which is considerably longer than a normal utterance of digits.

This complication arises because of the exponential state space resulted from coupling. Compared with a Q-state conventional HMM, a CHMM whose two state variables can respectively take Q_1 and Q_2 discrete values has $Q_1 \times Q_2$ possible states. In the hypothetical example, with the addition of the second state variable, the amount of observation is doubled, while the number of possible states

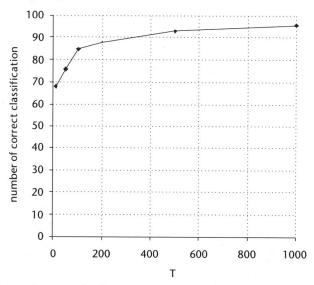

Figure 6.7   Number of correct classifications versus the duration of the test samples.

increases fivefold when comparing with a five-state conventional HMM. As a result, the large and more complex state space of the CHMMs requires considerably more evidence in order for the classifier to be statistically reliable.

Therefore, when applying the CHMM framework in real-world applications, it is necessary to carefully consider the Q values of the state variables to avoid excessively large state space. In the above example, one way to resolve the problem is to make the Q value of one of the state variables smaller.

Finally, a 10-class classification task was carried out. The correct rate is 58 out of 100 for $T = 10$. Closer examination of the log-likelihood lists reveals that for the misclassified instances, the model representing the correct class almost always ranks in the top three. Increasing T to 500 improves the number of correct classifications to 81.

6.3 Audiovisual Speech Recognition System Using CHMM

The challenges of developing a mature audiovisual speech recognition system are manifold. Two of the core issues that are specific to an audiovisual system are: (1) visual feature extraction and (2) bimodal sensory fusion. The latter is the main focus of this work. However, from the perspective of building a complete speech recognition system, many more important components are involved. For example, modern audio-only speech recognizers rely extensively on language modeling to improve recognition performance. In addition, the training of acoustic models usually engages several incremental steps to refine the models. Various tying and clustering schemes are necessary for robust estimation of model parameters. Moreover, the decoding engines in the recognizers often employ sophisticated searching algorithms to achieve higher efficiency. Lastly, speaker adaptation is usually crucial for good performance in speaker independent systems.

Each of these components presents its own challenge in the development of a bimodal system. As a result, most of the existing research systems on audiovisual speech recognition tend to be very simple. Many are designed around small-vocabulary isolated-word experiments. Although these simple systems serve their proof-of-concept purpose well, it often remains unknown whether the concept can be successfully transferred to a larger setup and how the system will perform when scaled up. Furthermore, the bare-bones nature of these systems makes it difficult to compare the performance of an audiovisual system with the performance of the state-of-the-art audio-only speech recognizers.

In this chapter, we present a novel approach that facilitates the realization of a CHMM-based audiovisual speech recognition system using the facilities found in most of the advanced acoustic-only ASR systems. The central part of this approach is a strategy to transform a CHMM to an equivalent HMM through appropriate tying of parameters and manipulation of the state transition matrix. The strategy is fairly general and can be applied to any HMM-based ASR packages that support parameter tying and multiple data streams. Using this approach, existing speech recognition systems can be modified easily to take advantage of the CHMM framework, thus making sophisticated audiovisual systems capable of continuous speech recognition tasks.

6.3.1   Implementation Strategy of CHMM

6.3.1.1   Required Facilities

The proposed scheme of implementing CHMM makes use of the training and decoding facilities of an existing HMM-based system. In particular, the scheme requires the HMM package to have the following two capabilities:

1. Support for multiple observation streams. In most continuous-density HMM systems, the output distributions are represented by *Gaussian mixture densities*. In a multistream HMM, each observation vector at time t is split into S data streams $o_{s,t}$. The output distribution $\{b_j(o_t)\}$ usually takes the following form [32]:

$$b_j(o_t) = \prod_{s=1}^{S} \left[\sum_{m=1}^{M_s} c_{j,s,m} N(o_{s,t}, \mu_{j,s,m}, \Sigma_{j,s,m}) \right]^{\gamma_s} \qquad (6.14)$$

where M_s is the number of mixture components in stream s, $c_{j,s,m}$ is the weight of the mth component and $N(.,\mu,\Sigma)$ is a multivariate Gaussian with mean vector μ and covariance matrix Σ. The exponent γ_s is a stream weight that can be used to give a particular stream more emphasis.

   Multiple data streams are used to enable separate modeling of multiple information sources. In acoustic speech modeling, multiple streams are usually used to separately model static coefficients, the delta coefficients, and the acceleration coefficients.

2. Support for parameter tying. In particular, the HMM package should allow intramodel parameter tying at the state-stream level. When two or more parameter sets are tied, the same set of parameter values are shared by all the *owners* of the tied set. During training, when the values of a tied parameter set are reestimated, all of the data that would have been used to estimate each individual untied parameter is effectively pooled. In acoustic ASR systems, parameter tying is often used to reduce the number of parameters in the models so that the estimation can be more robust, given limited training data.

In the proposed implementation scheme for CHMM, the multistream and parameter tying capabilities of an HMM systems are exploited from somewhat unconventional perspectives. Combined, they give rise to an interesting way to transform CHMMs to HMMs that exhibit analogous inferential characteristics. Thus, the training and decoding of the CHMMs can be undertaken entirely within the HMM domain.

6.3.1.2   CHMM Transformation

Recall the discussion in Chapter 3 that the state of a two-chain CHMM is jointly determined by the two state variables in the parallel chains. If the two state variables can take Q_1 and Q_2 discrete values, respectively, then the CHMM in effect has $Q_1 \times Q_2$ possible states. The same state space can also be represented by a conventional HMM that has $Q_1 \times Q_2$ hidden states. Moreover, in CHMM, the output distribution of a joint state can be obtained by taking the product of the two output densities of the two individual state variables; similarly, in a two-stream HMM, the output distribution of a state is the product of the two stream-dependent densities. Hence, it is also possible to represent the output configurations of a two-chain CHMM with a two-stream HMM that has an equivalent state space. However, the observable nodes of a $Q_1 \times Q_2$ CHMM are fully specified by a table containing $Q_1 + Q_2$ entries. On the other hand, an unconstrained two-stream HMM with $Q_1 \times Q_2$ hidden states has $2 \times Q_1 \times Q_2$ distinct output densities. This difference arises because in the CHMM an output node is only dependent on its single parent, while in the state-equivalent HMM the output is effectively conditioned on both state variables in the original CHMM. Fortunately, this discrepancy can be readily resolved through tying the appropriate output densities in the two-stream HMM according to the mapping from CHMM states to HMM states. This state mapping and parameter tying procedure is easy to visualize graphically.

Figure 6.8 illustrates the state-machine diagram of two-stream HMM obtained by transforming a two-chain CHMM with $Q_1 = 3$ and $Q_2 = 2$. The state space of the original CHMM is represented by the six hidden states in the HMM. This mapping is explicitly depicted in the diagram. For example, the state 3 in the HMM is equivalent to the state $\{q_1 = 2, q_2 = 1\}$ in the CHMM. The output densities of the HMM are tied according to the mapping. In Figure 6.8 the observation nodes with the same color shade are tied. For example, the output densities modeling the lower stream in states 2, 4, and 6 are tied because they all correspond to the entry $p(o_t | q_2 = 2)$ in the CPT of the CHMM.

The allowed state transition in the HMM is also derived from the state space mapping. In this example, it is assumed that the conditional probabilities concerning the two state variables in the CHMM satisfy the following condition:

$$P(q_{t+1}^i | q_t^1, q_t^2) = 0 \text{ if } q_{t+1}^i \neq q_t^i \text{ and } q_{t+1}^i \neq q_t^i + 1 \tag{6.15}$$

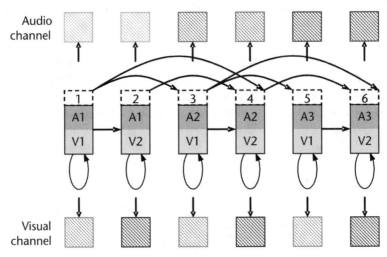

Figure 6.8 Transforming CHMM to HMM through state mapping and parameter tying.

This condition essentially enforces the left-to-right no-skip policy in the sense of conventional HMM for the two state variables in the CHMM, which is commonly used in audio-only speech recognizers. For example, a possible state path in the CHMM could be $\{q_1 = 1, q_2 = 1\} \rightarrow \{q_1 = 2, q_2 = 1\} \rightarrow \{q_1 = 3, q_2 = 2\}$; this is equivalent to the allowed state path $1 \rightarrow 3 \rightarrow 6$ in the HMM.

Other meaningful model configurations can be obtained through manipulating the allowed state transitions. For instance, it might be reasonable to model the dynamics of the lip motion using an ergodic state variable, that is, no restriction on the possible state transitions for this variable. The resulting HMM is shown in Figure 6.9.

It is worthy to note that the two-stream HMM approach to audiovisual fusion found in [27] can be considered as a special case of the CHMM-based fusion architecture. In that case, the number of the audio states must be equal to the number visual states, and the two-state variables always progress in lock step, in other words, $Q_1 = Q_2$, and q_t^1, q_t^2 for all t. The CHMM-based fusion architecture permits a much richer space for modeling interactions between the two modalities.

The model transformation strategy described above facilitates the realization of the CHMM framework using advanced HMM-based ASR packages. The approach is a general one and thus can be applied in many existing systems. In the following section, an implementation using HTK [32] is described.

6.3.2   Audiovisual Speech Recognition Experiments

An audiovisual speech recognition system using the proposed CHMM-based fusion architecture was developed and evaluated. The realization of CHMMs was carried out through the model transformation strategy discussed in the previous section using HTK. HTK is a powerful toolkit for building HMMs. It supports multistream HMMs and allows flexible parameter tying. More importantly, HTK is primarily designed for building HMM-based large vocabulary continuous speech recognition systems. Therefore, extensive infrastructure support for various

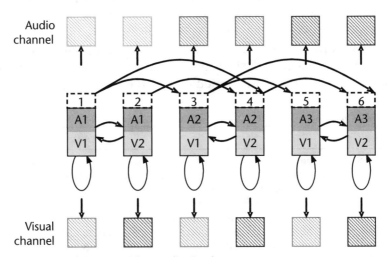

Figure 6.9   Transformed CHMM with ergodic visual states.

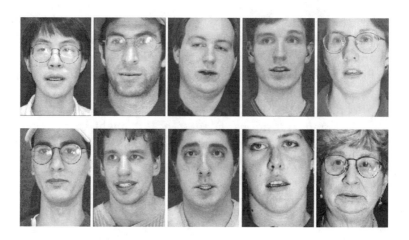

Figure 6.10 The 10 subjects in the dataset.

components of a modern ASR system is provided within the toolkit. Experimental evaluation of the bimodal speech recognition system was performed on an audiovisual speech dataset collected by Chen [22] at the Carnegie Mellon University.

6.3.2.1 CMU Audiovisual Speech Dataset

The CMU audiovisual speech dataset contains 10 subjects. Seven of the subjects are male and three are female, as shown in Figure 6.10. The vocabulary is designed around scheduling applications, with a total of 78 words. The words are listed in Table 6.5.

During taping, the subjects were asked to put a short pause between words. Every word in the vocabulary was repeated 10 times by each subject.

A major obstacle in the audiovisual ASR research is the lack of a standard corpus like the TIMIT dataset in acoustic-only ASR. Without a standard database, it is often difficult to make meaningful comparisons among systems from different research groups. The CMU dataset is one of the more recent collections of audiovisual speech data, and it is publicly available. Therefore, using this dataset allows us to directly compare the performance of our bimodal system with the system developed at CMU, and with results from other researchers using the same data. Furthermore, in addition to the raw data, the dataset also contains

Table 6.5 Vocabulary of the CMU Audiovisual Dataset

Category	Words
Numbers	one, two, three, four, five, six, seven, eight, nine, ten, eleven, twelve, thirteen, fourteen, fifteen, sixteen, seventeen, eighteen, nineteen, twenty, thirty, forty, fifty, sixty, seventy, eighty, ninety, hundred, thousand, million, billion
Months	January, February, March, April, May, June, July, August, September, October, November, December
Days	Monday, Tuesday, Wednesday, Thursday, Friday, Saturday, Sunday
Miscellaneous	morning, noon, afternoon, night, midnight, evening, a.m., p.m., now, next, last, yesterday, today, tomorrow, ago, after, before, from, for, through, until, till, that, this

the lip-tracking data that can be used to derive visual features. This is a thoughtful feature, because it enables a focused comparison between the fusion architectures of different systems when they use the same audiovisual feature set.

6.3.2.2   Audiovisual Speech Features

The acoustic speech was processed using a 25-ms Hamming window, with the frame period set at 10 ms. For each frame, 12 MFCC coefficients were calculated from the result of filterbank analysis using 26 channels. Delta coefficients were also computed and then appended to the static features, resulting in a 24-dimensional acoustic feature vector.

The visual features were derived from the lip-tracking data provided with the bimodal speech dataset. The tracking algorithm [22] that generated this data was based on deformable templates and statistical color modeling using Gaussian mixture models. As shown in Figure 6.11, the deformable template was constructed by two parabolas to fit the outer contours of the upper and lower lips. Note that this deformable template assumes that the mouth is symmetric. The tracking results were given by six parameters: (x_1, y_1) and (x_2, y_2) specify the 2-D coordinates of the left and the right mouth corners; h_1 and h_2 measure the vertical openings of the upper and lower lips.

The frame rate of the tracking data is 29.97 fps. To align the visual features with the acoustic features, the data were first up-sampled to 100 fps using piecewise cubic interpolation. The primary visual features considered in the experiments are composed of h_1, h_2, and the distance between the two mouth-corners, w. Delta features were also included, thus the actual visual feature vector is six-dimensional.

6.3.2.3   Experimental Results

There are two main objectives behind this round of experiments. The first is to evaluate the improvement in noise robustness brought by the bimodal approach to ASR. The second is to compare the performance of the proposed fusion architecture with other fusion techniques.

To fulfill the first objective, we built an acoustic speech recognizer as the baseline system. The recognizer was trained using clean speech. The noisy condition of a particular SNR level was simulated by adding a controlled amount of white Gaussian noise to the clean speech samples. The performance of the acoustic-only

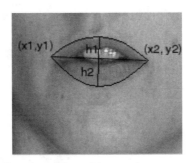

Figure 6.11   Lip-tracking results using deformable templates.

system was assessed on both the clean and noisy speech. The same acoustic feature sets were also used in the audio channel of the bimodal system. As in the acoustic-only case, training was performed on noise-free data, and evaluation involves both the clean and the noisy testing sets. However, it is assumed that the visual channel is not affected by any additional noise during testing. A visual-only recognizer was built and used as a benchmark.

To achieve the second objective, we implemented the most common form of the early integration approach, that is, fusion by concatenating the audio and visual feature vectors. Comparisons between this fusion method and the CHMM-based method were made through recognition experiments at different SNR levels.

Because the vocabulary size is relatively small, we constructed the acoustic and the audiovisual speech models at the word level. Recognition was performed in the connected-word mode without the help of any grammatical constraints. A cross-validation scheme was used in the evaluations due to the limited amount of data. Specifically, the recognizers were trained on a subset containing 90% of the available data and tested on the remaining 10%; this process was repeated until all data had been covered in testing. The results are summarized in Table 6.6.

The configurations of the systems that gave the above results are as follows. The audio-only system is based on HMMs with nine states, left-to-right topology, and no skips. The HMMs used in the visual-only system have a similar topology, but with only five states. HMM configuration identical to the audio-only system is used in the early integration bimodal system. The CHMM-based bimodal system uses five states to model the audio channel and three states for the visual channel. The allowed state transitions follow the policy specified in (6.15). Figure 6.12 shows the recognition results in graphical form.

As shown in Figure 6.12, it is evident that both of the bimodal systems demonstrate improved noise robustness in comparison to the audio-only system. However, at 10 dB, the gain in robustness achieved by the early integration system is very limited. On the other hand, the CHMM approach managed to give a clear improvement in performance at the same SNR level. At 30 dB, which is the SNR of the clean speech data, the recognition accuracy of the CHMM-based system is slightly worse than both the audio-only recognizer and the early integration bimodal system.

The CHMM configuration with ergodic visual state variable mentioned earlier in this chapter was also implemented. In addition, we experimented with globally tying the output distribution of the visual states in hope of getting more robust estimation of the dynamics in the visual channel. However, in both cases, the performance was inferior to that of the system with forward CHMMs.

Table 6.6 Summary of Recognition Results (Measured in Percentage Word Accuracy)

SNR	10 dB	20 dB	30 dB
A	4.03	43.61	99.10
V	42.95	42.95	42.95
A+V	10.58	72.79	99.74
CHMM	35.32	86.58	93.32

A indicates the audio-only system; V indicates the visual-only system; A+V indicates the bimodal system using early integration; and CHMM indicates the CHMM-based system.

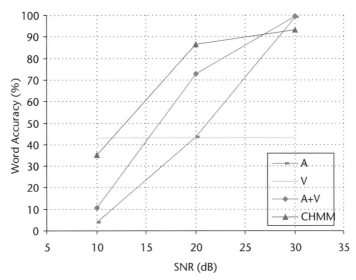

Figure 6.12   Comparing the recognition results from the audio-only, visual-only, early integration, and CHMM fusion systems.

An important cue the visual modality provides in bimodal speech perception is the information about boundary locations of the speech units within an utterance. It would be interesting to see if this effect can be observed in our audiovisual ASR system. We computed forced alignment of a speech segment in the 20-dB test set using both the acoustic-only recognizer and the CHMM-based bimodal recognizer. The results are illustrated in Figure 6.13.

Figure 6.13 covers a 10-second segment of the alignment result. The two subplots on the bottom show the word boundaries superimposed with the speech waveform. The upper one is the alignment obtained using audiovisual CHMMs; the lower one shows the alignment obtained using acoustic-only HMMs. The three subplots on the top display the static visual features used in the bimodal system. All five plots are time-aligned so that the correspondence among them can be visualized.

From the plot, we see that the audio-only recognizer almost always gives the incorrect end-of-word boundary at this noise level. In contrast, the bimodal system was able to precisely determine the end boundaries in six out of seven cases. On the front edge of a word, the acoustic speech recognizer had no difficulty detecting the boundary in the audio signal. It is interesting to observe that the bimodal recognizer consistently introduced a lead-time before the audible starting point of a word. This observation is in agreement with the finding from human speech perception that the visual speech usually leads the acoustic speech by a varying time window. The duration of the visual lead-in shown in Figure 6.13 ranges from about 40 ms to 150 ms.

6.3.3   Large Vocabulary Continuous Speech Experiments

In the previous section, we have discussed the audiovisual speech recognition experiments on a small vocabulary connected-word dataset. The experimental results clearly validate the theorized improvement in recognition performance

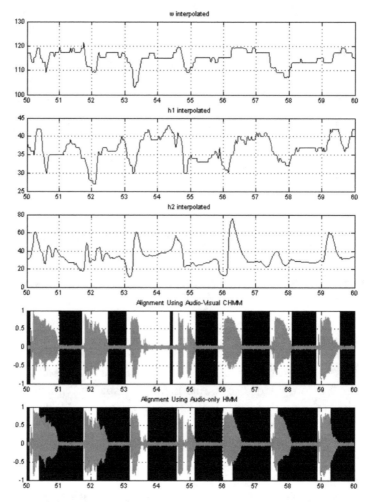

Figure 6.13   Forced alignment using audio-only HMM and audiovisual CHMM.

brought by the bimodal approach. In particular, we have shown that the proposed CHMM-based fusion architecture implemented through the model transformation strategy is able to attain even greater performance gain in noisy conditions compared with fusion via feature concatenation. There is strong evidence indicating that the CHMMs indeed successfully capture the asynchrony between the audio and visual channels, which gives the framework the added modeling strength as observed in the experiments.

In this section, we extend the experiments to a large vocabulary continuous speech recognition (LVCSR) task. Similar to the earlier experiments, the recognition systems are implemented using HTK, and comparison is made between the straightforward feature concatenation fusion and the proposed fusion scheme using CHMMs. In contrast to the small vocabulary experiments, the LVCSR experiments described in this section involve a considerably more elaborate training process. The recognition systems are trained and evaluated using the audiovisual speech corpus provided by the IBM T. J. Watson Research Labs.

6.3.3.1 IBM Audiovisual Speech Dataset

The audiovisual speech data used in the LVCSR experiments is derived from the IBM ViaVoice audiovisual database. The original database is a large corpus designed for speaker-independent audiovisual LVCSR. It consists of 290 subjects and is approximately 49 hr in duration. The utterances in the corpus are drawn from the ViaVoice training scripts, read continuously with verbalized punctuations, in a typical dictation style. The size of the vocabulary is 10.5K. The visual channel contains frontal-view full-face video of the subjects, similar to the CMU dataset. Figure 6.14 shows five examples of the video frames taken from the database. Notice that the variations in terms of lighting conditions, visual background, and head pose are minimal across the corpus. The acoustic speech is sampled at 16 kHz, and the SNR is measured as 19.5 dB.

As a part of the original design of the database, the corpus is partitioned into three subsets as listed in Table 6.7. The training set covers 17,111 utterances and 239 subjects. The test set contains 1,038 utterances spoken by 26 subjects. In addition to the training set and the test set, another held-out set with 2,277 utterances and 25 subjects is also provided for the tuning of various parameters in the recognition systems, for example, word-insertion and language model penalties. Note that the three partitions have mutually exclusive subject sets.

The IBM ViaVoice audiovisual database is a proprietary corpus and is not available in the public domain. The dataset granted to us is a limited version of the original corpus and is specifically intended for carrying out experiments focused in the fusion problem. The original audio and visual data files are not provided. Instead, the dataset consists of audio channel and visual channel features extracted from the bimodal speech data. Furthermore, only a single noisy acoustic condition is considered in the dataset, for which the audio signal in the original database is artificially corrupted by adding speech babble noise to achieve an average SNR of 8.5 dB.

The audio channel front end processing includes the subsequent steps. First, 24-dimensional MFCCs are extracted from the noisy audio signal, followed by utterance-level cepstral mean subtraction and energy normalization. Then, every nine consecutive feature frames in the MFCC stream are concatenated to form a 216-dimesional vector. The resulting vectors are projected to a 60-dimensional space using linear discriminant analysis (LDA). Finally, maximum likelihood linear transform (MLLT) is applied to rotate the feature vectors so that they can be better modeled by multivariate Gaussian densities with diagonal covariance matrices.

The visual features provided in the dataset are pixel-based. Discrete cosine transform (DCT) is applied directly to the gray-level pixel values of a rectangular

Figure 6.14 Pictured are 5 of the 290 subjects from the IBM ViaVoice audiovisual speech corpus. The visual data consists of frontal-view full-face video recorded under constant lighting conditions and uniform background.

Table 6.7 The IBM Audiovisual Speech Corpus Is Partitioned into Three Subsets, with Different Sets of Subjects

Set	Number of Utterances	Duration (hh:mm)	Number of Subjects
Train	17,111	34:55	239
Test	1.038	02:32	26
Held-out	2,277	04:47	25

region that encloses the speaker's mouth. Similar to the audio front end, the DCT coefficients are further processed using LDA and MLLT to obtain visual feature vectors in a 41-dimensional space. In both the audio and the visual feature extraction, the frame rate is set at 100 Hz, so that the feature streams in the two channels can be synchronized.

In addition to the feature files, the dataset also includes the transcriptions of the utterances at the word level, a dictionary file, and trigram lattices to facilitate LVCSR decoding.

6.3.3.2 System Implementation

The LVCSR systems based on context-dependent triphone models are implemented using HTK Version 2.2 on an 850-MHz Pentium III computer running the Windows XP operating system. Four different recognition systems are considered in the experiments: audio-only, visual-only, direct feature fusion, and CHMM fusion. The audio and visual feature vectors provided in the IBM dataset are first preprocessed to yield a composite audiovisual feature, which takes the form of the two-stream feature configuration following the HTK convention. This allows the implementations of the first three recognition systems to share a common baseline platform for training and testing. Furthermore, the two-stream composite features are also perfectly compatible with the model transformation strategy proposed for CHMM implementation.

The training of the context-dependent triphone models involves a sequence of steps. The basic idea is to build the systems incrementally. Starting with an initial set of context-independent monophone HMMs, a system can be refined systematically by cycles of re-estimation while gradually increasing the level of sophistication of the model set configuration. The development of the baseline system and the CHMM-based system both follow a similar path in terms of model refinement, which is illustrated in Figure 6.15.

The training starts with monophone models with Gaussian output densities. The densities are first identically initialized to the global mean and variance of the entire training set. With the phonetic transcriptions of the training utterances generated from the word-level transcriptions and the dictionary file, the models are re-estimated using the embedded EM algorithm. Additional treatment is then given to the silence model and the short pause model, to ensure that they have the special topologies necessary for absorbing the various voids in speech. After two more passes of embedded reestimation, the monophone model set is used to generate a new phonetic transcription by realigning the training data. Using the new transcription, the models are reestimated twice more. At this point, a well-trained set of single-Gaussian monophone models has been created.

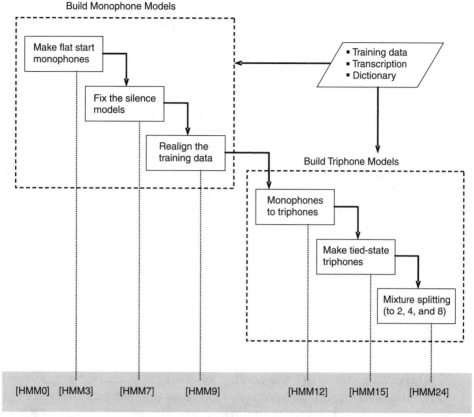

Figure 6.15 The triphone models of the LVCSR systems are built through a sequence of stages. Each refinement step involves the modification of the model definitions followed by several re-estimation passes. The entire training process begins at monophone-set HMM0 and results in the final tied-state context-dependent triphone-set HMM27.

The next stage of the training process is to build context-dependent triphone models. Starting from the monophone model set obtained in the previous stage, context-dependent triphones can be made by cloning the monophones and then reestimating using triphone transcriptions. To ensure robust parameter estimation with limited data, states within the triphone set are clustered using the decision tree–based clustering method in HTK, and similar states are tied to reduce the number of free parameters. Subsequently, the tied-state triphone models are reestimated twice. The final step of model building is to refine the output densities. The number of Gaussian mixtures is iteratively increased to two, four, and eight through mixture splitting. Each splitting is followed by two passes of embedded reestimation.

As metioned earlier, the audio-only, visual-only, and direct feature fusion systems are developed on a common baseline platform, which is built on two-stream HMMs with three emitting states and left-to-right topology. A particular system can be derived from the baseline by appropriately setting the global exponent stream weights γ^a and γ^v, which weight the audio observation stream and the visual observation stream in the composite audiovisual features, respectively. For instance, the visual-only system can be configured by setting γ^a and γ^v a priori.

Training is then carried out following the procedure described above. The CHMM-based audiovisual system is built upon CHMMs with three audio states and two visual states. The models are first initialized using the model transformation strategy depicted in Figure 6.8, then trained and refined in the same sequence of stages as the other systems. Note that the parameter tying established in the model transformation is only preserved until the clustering of states when building the tied-state triphone models.

6.3.3.3   Experimental Results and Discussions

Audiovisual LVCSR systems with direct feature fusion and CHMM-based fusion, as well as systems configured as audio-only and visual-only, are trained and evaluated on the IBM dataset. In addition to the phonetic models, a bigram language model is also constructed from the transcriptions of the training utterances. The selection of training set and test set follows the recommended partition shown in Table 6.7. The held-out set is used for optimizing the language model scaling factor and the word insertion penalty in decoding. The recognition performance is measured on the word error rate (WER) metric,

$$WER = \frac{D + S + I}{N} \times 100\% \tag{6.16}$$

where D, S, and I are the number of deletion errors, substitution errors, and insertion errors, respectively; and N is the total number of labels in the reference transcriptions. The experimental results are summarized in Table 6.8. Figure 6.16 shows the WERs with the corresponding 95% confidence error-bars.

It is evident in the recognition results that with an overall WER of 40.91%, audio-only speech recognition is fairly unreliable at the given noise level. One the other hand, the visual-only recognizer is only able to correctly recognize approximately 20% of the words. However, although recognition based on the visual channel alone performs poorly, when combining with the audio channel, the limited visual information is able to give a clear improvement in the performance. Integrating the two modalities using direct feature fusion reduces the WER to 36.70%, which translates to a 10.3% reduction of word errors over the audio-only recognition result. The WER decreases further when the audiovisual speech is modeled and fused using the CHMMs. The CHMM-based bimodal recognizer achieves the lowest WER among the four systems. At 34.33% WER, the relative improvement obtained by CHMM fusion is 6.46% compared with direct feature fusion, and 16.1% compared with the audio-only recognizer.

Table 6.8   Experimental Results of the LVCSR Systems: The Evaluations Are Carried Out on the IBM Audiovisual Speech Dataset, Which Considers a Single Noisy Acoustic Condition at 8.5-dB SNR

LVCSR System	WER
Audio only	40.91%
Visual only	79.69%
Audiovisual: feature fusion	36.70%
Audiovisual: CHMM fusion	34.33%

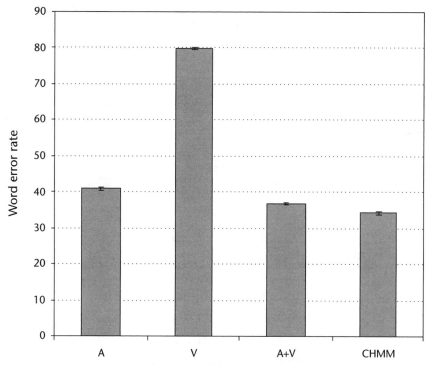

Figure 6.16   LVCSR results in term of WER. The corresponding 95% confidence error bars are shown.

Notice that the differences in recognition performance among the audio-only, direct feature fusion, and CHMM fusion in the LVCSR experiments are of a lower magnitude than those observed in the small vocabulary experiments on the CMU database. Apart from the obvious differences in terms of the scope and setup of the two recognition tasks, one important factor that vindicates the smaller performance gaps is that in the LVCSR experiments, the acoustic condition of the training set matches the condition of the testing set, whereas in the small vocabulary experiments, clean speech is used in training. The matched condition essentially gives an upper limit of recognition performance achievable by the audio channel alone. Thus, even though the relative improvements are small in the LVCSR experiments, they are nonetheless significant because it clearly shows that the fusion algorithms have indeed captured additional discriminative information embedded in the visual modality that is beyond the capability of audio-only speech recognizers.

One limitation of the IBM dataset is that it does not include the original audio files of the utterances, thus making it impracticable to perform the alignment analysis to visually confirm the asynchronies captured by the CHMMs as in the small vocabulary experiments with the CMU dataset. Also, the dataset only provides audio-channel features at a single noisy condition (8.5-dB SNR). Without the audio files, it is difficult to simulate additional acoustic conditions. Although noise can be injected directly to the feature files, or, more generally, introduced through manipulations in the feature domain, the implication of such operations in the original signal domain is usually not transparent, and their relevance to real-world ambient conditions is likely to be weak.

Nevertheless, experiments on the IBM dataset indeed give statistically significant measurements of the recognition performance of the baseline systems and the CHMM-based bimodal system, thus allowing us to evaluate the proposed fusion algorithms in the LVCSR setting. In spite of the limitations associated with the experimental data as discussed above, it is prudent to conclude that the relative performance ranking of the different systems observed in the small-vocabulary studies is upheld in the LVCSR experiments. In particular, it has been empirically demonstrated once again that the CHMM-based intermediate integration scheme utilizes the information in the visual channel more effectively than early integration through direct feature fusion. The experiments confirm that model transformation strategy for CHMM implementation is viable in both the small-vocabulary connected-word recognizers with word-level models as well as the LVCSR systems based on subword models. Furthermore, it is shown that the audiovisual system can give higher recognition accuracy than audio-only speech recognizers even when the acoustic condition of the training set perfectly matches the testing environment.

6.4 Conclusions

This study has considered the fundamental problem of multimodal fusion in the context of pattern recognition tasks in HCI. We proposed a novel sensory fusion method based on the CHMMs for audiovisual speech modeling. The CHMM framework allows the fusion of two temporally coupled information sources to take place as an integral part of the statistical modeling process. An important advantage of the CHMM-based fusion method lies in its ability to model asynchronies between the audio and visual channels. We described two approaches to carry out inference and learning in CHMMs. The first is an exact algorithm derived by extending the forward-backward procedure used in HMM inference. The second method relies on the model transformation strategy that maps the state space of a CHMM onto the state space of a classic HMM, and therefore facilitates the development of sophisticated audiovisual speech recognition systems using existing infrastructures.

Audiovisual speech recognition experiments on the CMU dataset and the IBM dataset validate the proposed fusion approach. In both evaluations, it is clearly demonstrated that the CHMM-based intermediate integration scheme can utilize the information in the visual channel more effectively than the early integration methods in noisy conditions. Moreover, the LVCSR results show that the audiovisual systems can achieve higher recognition accuracy than audio-only systems even when matched acoustic conditions are ensured.

Although visual feature extraction is not the focus of this work, the visual front end constitutes an important challenge in audiovisual speech recognition and deserves no less attention than the fusion problem. In fact, reliable face/feature detection and tracking that is robust to pose, lighting, and speaker variations remains a difficult task. Significant progress in the relevant research areas needs to be made before the visual channel can be consistently incorporated into mainstream speech recognition applications.

Looking at the multimodal fusion problem in HCI from a broader perspective, many more research areas can benefit from the joint processing of audiovisual information, for example, biometrics, speech synthesis, speech enhancement, event detection, video indexing, and information retrieval. Combined, these topics form a solid and promising vector for future research.

References

[1] Yuhas, B., et al., "Neural Network Models of Sensory Integration for Improved Vowel Recognition," *Proc. IEEE*, Vol. 78, No. 10, 1990, pp. 1658–1668.

[2] Silsbee, P., and A. Bovik, "Computer Lipreading for Improved Accuracy in Automatic Speech Recognition," *IEEE Trans. on Speech and Audio Processing*, Vol. 4, No. 5, 1996, pp. 337–351.

[3] Bregler, C., and Y. Konig, "Eigenlips for Robust Speech Recognition," *Proc. IEEE ICASSP*, Adelaide, Australia, 1994, pp. 669–672.

[4] Brooke, N., and S. Scott, "PCA Image Coding Schemes and Visual Speech Intelligibility," *Proc. Institute of Acoustics*, Vol. 16, No. 5, 1994, pp. 123–129.

[5] Neti, C., et al., "Final Workshop 2000 Report," in *Audio-Visual Speech Recognition*, Center for Language and Speech Processing, The Johns Hopkins University, Baltimore, MD, 2000.

[6] Potamianos, G., H. Graf, and E. Cosatto, "An Image Transformation Approach for HMM Based Automatic Lipreading," *Proc. IEEE Intl. Conf. Image Processing*, Chicago, IL, 1998, pp. 173–177.

[7] Mase, K., and A. Pentland, "Automatic Lipreading by Optical Flow Analysis," *Systems and Computers in Japan*, Vol. 22, No. 6, 1991, pp. 67–75.

[8] Petajan, E., "Automatic Lipreading to Enhance Speech Recognition," Ph.D. dissertation, University of Illinois at Urbana-Champaign, Urbana, IL, 1984.

[9] Goldschen, A., "Continuous Automatic Speech Recognition by Lipreading," Ph.D. dissertation, George Washington University, Washington, D.C., 1993.

[10] Finn, K., and A. Montgomery, "Automatic Optically Based Recognition of Speech," *Pattern Recognition Letters*, Vol. 8, No. 3, 1988, pp. 159–164.

[11] Stork, D., G. Wolff, and E. Levine, "Neural Network Lipreading System for Improved Speech Recognition," *Proc. Intl. Joint Conf. Neural Networks*, Vol. 2, 1992, pp. 289–295.

[12] Adjoudani, A., and C. Benoît, "Audio-Visual Speech Recognition Compared Across Two Architectures," *Proc. European Conf. Speech Communication and Technology*, 1995, pp. 1563–1566.

[13] Yuille, A., P. Hallinan, and D. Cohen, "Feature Extraction from Faces Using Deformable Templates," *International Journal of Computer Vision*, Vol. 8, No. 2, 1992, pp. 99–111.

[14] Kass, M., A. Witkin, and D. Terzopoulos, "Snakes: Active Contour Models," *International Journal of Computer Vision*, Vol. 1, No. 4, 1987, pp. 321–331.

[15] Hennecke, M., K. Prasad, and D. Stork, "Using Deformable Templates to Infer Visual Speech Dynamics," *28th Annual Asilomar Conference on Signals, Systems, and Computers*, Pacific Grove, CA, November 1994.

[16] Bregler, C., and S. Omohundro, "Nonlinear Manifold Learning for Visual Speech Recognition," *Proc. IEEE International Conference on Computer Vision*, Cambridge, MA, 1995, pp. 494–499.

[17] Rao, R., and R. Mersereau, "On Merging Hidden Markov Models with Deformable Templates," *Proc. of the Intl. Conf. Image Processing*, Washington, D.C., 1995, pp. 556–559.

[18]   Huang, F., and T. Chen, "Real-Time Lip-Synch Face Animation Driven by Human Voice," *IEEE Workshop on Multimedia Signal Processing*, Los Angeles, CA, 1998.

[19]   Tao, H., and T. Huang, "Bézier Volume Deformation Model for Facial Animation and Video Tracking," in *Modeling and Motion Capture Techniques for Visual Environments*, New York: Springer-Verlag, 1998, pp. 242–253.

[20]   Luettin, J., N. Thacker, and S. Beet, "Speechreading Using Shape and Intensity Information," *Proc. Intl. Conf. Spoken Language Processing*, Philadelphia, PA, 1996, pp. 58–61.

[21]   Luettin, J., "Visual Speech and Speaker Recognition," Ph.D. dissertation, University of Sheffield, Western Bank, Sheffield, U.K., 1997.

[22]   Chen, T., "Audiovisual Speech Processing," *IEEE Signal Processing Magazine*, Vol. 18, No. 1, 2001, pp. 9–21.

[23]   Zhang, Y., "Information Fusion for Robust Audio-Visual Speech Recognition," Ph.D. dissertation, University of Illinois at Urbana-Champaign, Urbana, IL, 2000.

[24]   Tomlinson, M., M. Russell, and N. Brooke, "Integrating Audio and Visual Information to Provide Highly Robust Speech Recognition," *Proc. IEEE ICASSP*, Atlanta, GA, 1996, pp. 821–824.

[25]   Adjoudani, A., and C. Benoît, "On the Integration of Auditory and Visual Parameters in an HMM-Based ASR," in *Speechreading by Humans and Machines*, Vol. 150 of *NATO ASI Series F: Computer and Systems Sciences,* D. Stork and M. Hennecke (eds.), Berlin: Springer-Verlag, 1996, pp. 461–471.

[26]   Hennecke, M., D. Stork, and K. Prasad, "Visionary Speech: Looking Ahead to Practical Speech Reading Systems," in *Speechreading by Humans and Machines*, Vol. 150 of *NATO ASI Series F: Computer and Systems Sciences,* D. Stork and M. Hennecke (eds.), Berlin: Springer-Verlag, 1996, pp. 331–349.

[27]   Potamianos, G., and H. Graf, "Discriminative Training of HMM Stream Exponents for Audio-Visual Speech Recognition," *Proc. IEEE ICASSP*, Seattle, WA, 1998, pp. 3733–3736.

[28]   Grahramani, Z., "Learning Dynamic Bayesian Networks," in *Adaptive Processing of Sequences and Data Structures*, C. Giles and M. Gori (eds.), Lecture notes in Computer Science, Vol. 1387, London: Springer-Verlag, 1997, pp. 168–197.

[29]   Brand, M., "Coupled Hidden Markov Models for Modeling Interacting Processes," MIT Media Lab, Learning and Common Sense Technical Report 405, 1997.

[30]   Jordan, M., et al., "An Introduction to Variational Methods for Graphical Models," *Machine Learning*, Vol. 37, No. 2, 1999, pp. 183–233.

[31]   Mackay, D., "Introduction to Monte Carlo Methods," in *Learning in Graphical Models*, M. Jordan, (ed.), Boston, MA: Kluwer Academic Publishers, 1998, pp. 175–204.

[32]   Young, S., et al., *The HTK Book* (for HTK Version 2.2), Cambridge, U.K.: Cambridge University Engineering Department Speech Group and Entropic Research Laboratory Inc., 1999.

CHAPTER 7

Multimodal Tracking for Smart Videoconferencing and Video Surveillance

Dmitry N. Zotkin, Vikas C. Raykar, Ramani Duraiswami, and Larry S. Davis

Many applications such as interactive multimedia, videoconferencing, and surveillance require the ability to track the 3-D motion of the subjects. Particle filters represent an attractive solution for the tracking problem because they do not require solution of the inverse problem of obtaining the state from the measurements and because the tracking can naturally integrate multiple modalities. We build a framework for multimodal tracking using multiple cameras and multiple microphone arrays. In order to calibrate the resulting distributed multisensor system, we propose a method to automatically determine the 3-D positions of all microphones in the system using at least five loudspeakers. Our method does not require knowledge of the loudspeaker positions but assumes that for each loudspeaker there exists a microphone very close to it. We derive the maximum likelihood estimator, which reduces to the solution of the nonlinear least squares problem. A closed-form approximate solution that can be used as an initial guess is derived. We also derive an approximate expression for the estimator covariance using the implicit function theorem and Taylor series expansion. Using the estimator covariance matrix, we analyze the performance of the estimator with respect to the positions of the loudspeakers; in particular, we show that the loudspeakers should be as far away from each other, and the microphones should lie within the convex hull formed by the loudspeakers. We verify the correctness and robustness of the multimodal tracker and of the self-calibration algorithm both with Monte Carlo simulations and on real data from three experimental setups. We also present practical details of system implementation.

7.1 Introduction

Our perception of the environment is a strong function of our relative location to objects being perceived. It changes with our orientation, body posture, whether we are seated or standing, and so forth, and it becomes apparent that multiple modalities are merged in the brain to produce space-time perception [1]. Many applications thus require an ability to track the motion of a person; for example, in surveillance it is necessary to focus the system's attention on the action taking place in the system's field of view; virtual and augmented reality (VR/AR) applications that seek to create convincing experiences need to adapt the presentation according to the person's position and orientation; and applications that seek to provide telepresence (such as videoconferencing) need to determine the spatial distribution of people in an environment. There are multiple modalities that are used to acquire such positional information, including multiperspective video, active audio transmitters placed on moving objects, speech-based audio, magnetic tracking of installed tags, and so forth. Integrating information obtained from multiple sensors can lead to improvements in both the accuracy and the sampling rate and to a practical design for the development of a surveillance, augmented reality, or smart videoconferencing system. Indeed, the development of multimodal sensor fusion algorithms has seen many applications recently, including multisensor vehicle navigation system where computer vision, sonar, and laser and microwave radar sensors are used together [2]; audiovisual person identification using support vector machine classifier [3]; multimodal speaker detection using Bayesian networks [4]; multimodal tracking using inverse modeling techniques from computer vision, speech recognition, and acoustics [5]; and discourse segmentation using gesture, speech, and gaze cues [6].

One of the problems faced by the developers of the multisensor systems is the calibration of sensor's positions. Automatic camera calibration algorithms are relatively well-developed ([7–11], to cite a few); therefore, in this chapter we will focus on presenting a novel algorithm for autocalibration of the audio subsystem of the tracker. Most multimicrophone array processing algorithms need to know the positions of the microphones very precisely (although some algorithms, e.g., blind beamforming, do not need to know the microphone positions); for example, in the case of source localization, even relatively small uncertainties in sensor location could make substantial, often dominant, contributions to overall localization error [12]. Manual calibration employing a tape or laser range device is error-prone and has to be repeated every time the geometry of the system changes either because of adhoc redeployment or accidentally. Thus, automatic position calibration of multiple microphones is essential and has many applications such as sound source localization, audio tracking, hands-free voice communication, beamforming, speech enhancement, and speech recognition, where distributed arrays of multiple microphones are widely used.

We propose a method to automatically determine the three-dimensional positions of multiple microphones. The approach we follow is to use a few loudspeakers, to play known calibration signals from each of them, and to measure the time taken by the sound to reach the microphones and thus determine the distances between each loudspeaker and all the microphones. From these distances, the maximum likelihood (ML) estimate of the microphone positions can be derived. This approach was taken before by several authors, who assumed known loudspeaker positions (e.g., [13] describes an experimental setup for automatic calibration of a large-aperture microphone array using acoustic signals from transducers whose locations are known). However, we wish to consider the more complex case when the loudspeaker positions are unknown as well and must also be estimated from measurements (related theoretical work can be found in [12, 14, 15]). The ML estimate in this case turns out to be a nonlinear optimization problem, which needs a very good initial guess for solution. To obtain it, we derive a novel closed-form solution for the microphones and loudspeakers positions under the assumption that each loudspeaker has a microphone that is very close (i.e., attached) to it. In practice we achieve this by placing a microphone right next to the loudspeaker. (This scenario is also realized practically in the case of mobile phones, handheld computers, and laptops.) The closed-form solution is further refined by the nonlinear minimization procedure. In order to study the ML estimator, we derive the approximate mean and covariance of the implicitly defined estimator using the implicit function theorem and Taylor series expansion and analyze the estimator accuracy with respect to the loudspeaker positions. In particular, we show that the loudspeakers should be placed as far away from each other and all the microphones should be in the convex hull formed by the loudspeakers as opposed to the setup in [13] where all the loudspeakers are close to each other.

We then propose a multimodal information fusion algorithm for audio and video measurements obtained from multiple calibrated cameras and microphone arrays using sequential Monte Carlo methods (also known as particle filters [16]). One advantage of our proposed tracker is its ability to seamlessly handle temporary absence of some measurements (e.g., camera occlusion or silence). Another advantage is the ability to track self-configuration parameters of a changing system (e.g., when a sensor is located on a mobile platform and is moving itself) together and in the same framework as object tracking. This is done by including those parameters into the state vector of the system. We describe a particular setup in which experimental results are obtained, which is an extension of the videoconferencing setup in [17]; analyze tracking performance using synthetic data; and show results of several successful tracking experiments in different environments, including a person moving with an ultrasonic sound source in an anechoic room, an echolocating bat in flight in an anechoic room, and a speaker moving in a typical office room. We also show that our algorithm is capable of sensor self-motion recovery together with object tracking and of successful handling of the temporary absence of some measurements (the target being occluded from one or both cameras, or absence of audio data).

The rest of this chapter is organized as follows. Section 7.2 is devoted to theoretical derivation of the framework for automatic microphone array calibration. We formulate the problem and the ML estimator for the microphone and loudspeaker positions and then derive a closed-form approximate solution that can be used as an initial guess for the nonlinear minimization routine. We also derive the theoretical mean and covariance of the estimated parameters, validate those using Monte Carlo simulations, and analyze the dependency of the estimator covariance on the location of speaker-microphone pairs. In Section 7.3, a discussion of the issues involved in designing a practical system is given, and the possible error sources are outlined. Section 7.4 provides a brief introduction to particle filters. Section 7.5 describes the particulars of the multimodal setup used for the experiments as well as mapping of audio and video measurements into the particle filter framework. In Section 7.6, we show the experimental evaluation of the multimodal tracker using synthetic and real input data. Finally, Section 7.7 concludes the chapter with a summary of the work presented.

7.2   Automatic Calibration of Multimicrophone Setup

In this section, we present a systematic analysis of the problem of self-calibrating a microphone array of unknown geometry using at least five loudspeakers. We have used the technique to calibrate the microphone array, which was used later in particle filter tracking experiments. However, the method is obviously not restricted to such scenarios and can be used in any application, such as smart conference rooms [18], hands-free voice communication [19], speech enhancement [20], acoustic surveillance [21], and many others. In related work [22], we present an alternative, somewhat more flexible method, which can recover more complex microphone and loudspeaker configurations and can handle the case of no common timebase among all stations (a station is defined as a unit that includes at least one microphone and at least one loudspeaker, such as a mobile phone). However, it is more expensive and requires more measurements than the method described here.

7.2.1 ML Estimator

Given a set of M microphones and S loudspeakers in unknown locations, our goal is to estimate their 3-D coordinates. Each loudspeaker is excited using a known calibration signal (such as maximum length sequence or chirp), and the signal is captured by each of the microphones. The time of flight (TOF) is estimated from the captured audio signal. The TOF for a given microphone and speaker pair is defined as the time taken by the acoustic signal to travel from the speaker to the microphone. We assume that the signals emitted from each of the speakers do not interfere with each other (i.e., each signal can be associated with a particular speaker). This can be achieved by confining the signal at each speaker to disjoint frequency bands or time intervals. Alternately, we can use coded sequences so that the signal due to each speaker can be extracted at the microphones and correctly attributed to the corresponding speaker. The $M \times S$ TOF measurements constitute our observations, based on which we have to estimate the microphone and speaker positions.

7.2.1.1 Formulation

Let $\mathbf{m}_i = [mx_i : my_i : mz_i]^T$ and $\mathbf{s}_j = [sx_j : sy_j : sz_j]^T$ be the three-dimensional vectors representing the x, y, and z coordinates of the ith microphone and jth loudspeaker, respectively. We excite each of the S speakers one at a time and measure the TOF at each of the M microphones. The TOF_{ij} for the ith microphone and the jth speaker is defined as the time taken for the acoustic signal to travel from the jth speaker to the ith microphone. Let $TOF_{ij}^{estimated}$ and TOF_{ij}^{actual} be the estimated and the actual TOF respectively for the ith microphone and jth speaker. The actual TOF can be written as

$$TOF_{ij}^{actual} = \frac{\|\mathbf{m}_i - \mathbf{s}_j\|}{c} \tag{7.1}$$

where $\|\ \|$ is the Euclidean norm and c is the speed of the sound.

Assuming a Gaussian noise model for our observations, we can derive the ML estimator as follows. Let Θ be a vector of length $P \times 1$ representing all the unknown nonrandom parameters to be estimated (microphone and speaker coordinates), Γ a vector of length $N \times 1$ representing noisy estimated TOF measurements, and

$T(\Theta)$ a vector of length $N \times 1$ representing the actual value of the TOF observations. Then our model for the observations is $\Gamma = T(\Theta) + \eta$, $\eta = N(0, \Sigma)$ (i.e., η is the $N \times 1$ Gaussian noise vector with zero mean and covariance matrix Σ). The likelihood function of Γ in vector form can be written as

$$p(\Gamma|\Theta) = (2\pi)^{-\frac{N}{2}}|\Sigma|^{-\frac{1}{2}}\exp\left(-\frac{1}{2}[\Gamma - T(\Theta)]^T\Sigma^{-1}[\Gamma - T(\Theta)]\right) \qquad (7.2)$$

The ML estimate $\hat{\Theta}_{ML}$ of Θ is defined as

$$\hat{\Theta}_{ML}(\Gamma) = \arg\{\max_{\Theta} p(\Gamma|\Theta)\} \qquad (7.3)$$

Maximizing $p(\Gamma|\Theta)$ is equivalent to maximizing $\log p(\Gamma|\Theta)$. The log-likelihood function is given by

$$\log p(\Gamma|\Theta) = -\frac{N}{2}\log(2\pi) - \frac{1}{2}\log|\Sigma| - \frac{1}{2}[\Gamma - T(\Theta)]^T\Sigma^{-1}[\Gamma - T(\Theta)] \qquad (7.4)$$

The ML estimator can be written as

$$\hat{\Theta}_{ML}(\Gamma) = \arg\{\max_{\Theta} F(\Theta, \Gamma)\}$$
$$F(\Theta, \Gamma) = -\frac{1}{2}[\Gamma - T(\Theta)]^T\Sigma^{-1}[\Gamma - T(\Theta)] \qquad (7.5)$$

Assuming that each of the TOFs are independently corrupted by zero-mean additive white Gaussian noise of variance σ_{ij}^2, the ML estimate can also be formulated as a nonlinear least squares problem (in which case Σ becomes a diagonal matrix):

$$\hat{\Theta}_{ML} = \arg_{\Theta}\min\sum_{i=1}^{M}\sum_{j=1}^{S}\frac{(TOF_{ij}^{estimated} - TOF_{ij}^{actual})^2}{\sigma_{ij}^2} \qquad (7.6)$$

We estimate the TOF using generalized cross correlation (GCC) [23]. The estimated TOF is corrupted due to ambient noise and room reverberation. For high SNR, the delays estimated by the GCC can be shown to be normally distributed with zero mean [23].

7.2.1.2   Reference Coordinate System

As the solution depends only on pairwise TOFs, any translation, rotation, and reflection of the global minimum found will also be a global minimum. In order to make the solution invariant to rotation and translation, we select three arbitrary nodes to lie in a plane such that the first is at $(0, 0, 0)$, the second at $(x_1, 0, 0)$, and the third at $(x_2, y_2, 0)$. To eliminate the ambiguity due to reflection along the z-axis, we specify one more node to lie in the positive-Z half-space. Also the reflections along the x and y axes can be eliminated by assuming that $x_1 0$ and $y_2 0$. In practice, microphones belonging to the speaker-microphone pairs are used as the reference nodes. We prefer microphones as reference points because they are usually smaller than loudspeakers, hence the reference coordinate system can be known more precisely. This would be beneficial if we wanted to subsequently transform this reference coordinate system to the room coordinate system.

7.2.1.3   Nonlinear Least Squares

The ML estimate for the node coordinates of the microphones and loudspeakers is implicitly defined as the minimum of the nonlinear function given in (7.5). This function has to be minimized using numerical optimization methods. The Levenberg-Marquardt method [24] is a popular method for solving nonlinear least squares problems. It is a compromise between steepest descent and Newton's methods. The steepest descent method potentially has a very slow convergence but can converge from any starting point. Newton's method converges fast but requires a good initial guess and computation of the Hessian matrix inverse. (For more details on nonlinear minimization, please refer to [24]). Appendix 7A gives the nonzero partial derivatives needed for the minimization routines (the actual routines are available in many software products such as MATLAB and Numerical Recipes). A significant problem is that the Levenberg-Marquardt minimization routine will not converge to the global minimum unless we have a very good initial guess. In the next section, we derive an approximate closed-form solution, which can be used as an initial guess for the minimization routine.

Also, the total number of observations should be greater than or equal to the total number of parameters to be estimated. In our case $MS \geq 3(M + S) - 6$. If $M = S = K$ then $K \geq 5$. Thus we need a minimum of five speaker-microphone pairs. We will use this result later to get a closed-form solution for the microphone coordinates.

7.2.2   Closed-Form Solution

Given the pairwise Euclidean distances between N nodes, their relative positions can be determined by means of metric or classical multidimensional scaling (MDS) [25]. MDS is a popular technique in psychology and denotes a set of data-analysis techniques for the analysis of proximity data on a set of stimuli for revealing the hidden structure underlying the data [26]. The proximity data refers to some measure of pairwise dissimilarity. Given a set of N stimuli along with their pairwise dissimilarities p_{ij}, MDS arranges them as points in a multidimensional space in a way that the distances between any two points are a monotonic function of the corresponding dissimilarity. MDS is widely used to visually study the structure of proximity data. If proximity data are based on the Euclidean distances, then classical metric MDS [25] can exactly recreate the configuration.

7.2.2.1   Classical MDS

Given a set of N points in 3-D space, let X be an $N \times 3$ matrix where each row represents the 3-D coordinates of each point. Then the $N \times N$ matrix $B = XX^T$ is called the dot product matrix. By definition, B is a symmetric positive definite matrix, so the rank of B (i.e., the number of positive eigenvalues) is equal to the dimensionality of space in which the data points lie (i.e., 3 in this case). Starting with a matrix B (possibly corrupted by noise), it is possible to factor it to get the matrix of coordinates X. One method to factor B is to use singular value decomposition (SVD) [27], that is, decompose $B = U\Sigma U^T$ where Σ is a $N \times N$ diagonal matrix of singular values. The diagonal elements are arranged as $s_1 \geq s_2 \geq s_r s_{r+1} = \ldots\ldots = s_N = 0$, where r is the rank of the matrix B. The columns of U are the corresponding singular vectors. We can write $X' = U\Sigma^{1/2}$ and take the first three columns of X' to get X. If the elements of B are not corrupted by noise, then all the other columns are zero. It can be shown that SVD factorization minimizes the matrix norm $\| B - XX^T \|$.

In practice, we can estimate the distance matrix D, where the ijth element is the Euclidean distance between the ith and the jth points. This distance matrix D must be converted into a dot product matrix B before MDS can be applied. We need to choose some point as the origin of our coordinate system in order to form the dot product matrix. Any point can be selected as the origin, but Togerson [25] recommends the choice of the centroid of all the points. If the distances have random errors, then such choice minimizes the errors, as they would tend to cancel each other. We can convert the distance matrix into a dot product matrix using simple geometry; please refer to Appendix 7B for a derivation.

In our case of M microphones and S speakers, we cannot use MDS directly as we cannot measure some of the pairwise distances (e.g., the distance between two microphones). Figure 7.1 shows an example consisting of seven microphones and four speakers, with each speaker attached to one of the microphones forming in

	s1	s2	s3	s4	m1	m2	m3	m4	m5	m6	m7
s1	?	?	?	?	X	X	X	X	X	X	X
s2	?	?	?	?	X	X	X	X	X	X	X
s3	?	?	?	?	X	X	X	X	X	X	X
s4	?	?	?	?	X	X	X	X	X	X	X
m1	X	X	X	X	?	?	?	?	?	?	?
m2	X	X	X	X	?	?	?	?	?	?	?
m3	X	X	X	X	?	?	?	?	?	?	?
m4	X	X	X	X	?	?	?	?	?	?	?
m5	X	X	X	X	?	?	?	?	?	?	?
m6	X	X	X	X	?	?	?	?	?	?	?
m7	X	X	X	X	?	?	?	?	?	?	?

Figure 7.1   Pairwise distance matrix for four loudspeakers and seven microphones. Four microphones are attached to the loudspeaker forming four speaker-microphone pairs. The measured quantities are shown as X and the unknown distances are shown as ?.

effect four speaker-microphone pairs and three single microphones. The cells marked X and $?$ show available and unavailable measurements, respectively.

7.2.2.2   Forming Speaker-Microphone Pairs

In our practical setup, for every loudspeaker there is a microphone attached to it. Such speaker-microphone pairs are considered as one entity (i.e., we assume that the distance between them is zero). Based on this approximation, the distance d_{ij} between the ith and jth speaker-microphone pair is given by

$$d_{ij} \approx 0 : if : i = j,$$
$$d_{ij} \approx \frac{c(TOF_{ij} + TOF_{ji})}{2} : if : i \mathrel{/}= j \qquad (7.7)$$

where c is the speed of the sound. Once all the pairwise distances are obtained, classical MDS is performed to get the approximate positions of the speaker-microphone pairs. The position estimate from MDS is with respect to an arbitrary centroid and orientation, and hence it is converted into the reference coordinate system as described in Section 7.2.1.2.

The approximate locations of the speaker-microphone pairs are slightly perturbed to obtain the initial guess for the microphone and speaker locations. We use this as an initial guess for the nonlinear minimization routine and obtain the exact locations of the microphones and loudspeakers in each speaker-microphone pair. As discussed before, for the ML estimation procedure we need a minimum of five speaker-microphone pairs.

7.2.2.3   Closed-Form Solution for Microphone Positions

From the previous step we have obtained the locations of five speaker-microphone pairs. If the location of four speakers are known, then by triangulation the positions of remaining (single) microphones can be determined analytically. (Given its distance from one loudspeaker, the microphone can lie anywhere on a sphere centered at that loudspeaker. With two loudspeakers, the unknown microphone can lie on a circle, as two spheres intersect at a circle. With three, the set of solutions is reduced to two points, and with four loudspeakers a unique location is determined). As the estimated distances are corrupted by noise and further five (instead of four) distances are available, the intersection in general need not to be a unique point. Hence, we solve the problem in a least square sense.

As before, assume that we have S loudspeakers. Let $\mathbf{s}_j = [sx_j : sy_j : sz_j]^T$ be the x, y, and z coordinates of the jth speaker. The locations of the loudspeakers are determined as discussed in the previous sections. Let $\mathbf{m}_i = [mx_i : my_i : mz_i]^T$ be the unknown microphone coordinates, which we have to determine. For the ith microphone, we have S TOF measurements

$$c^2 TOF_{ij}^2 = \| \mathbf{m}_i - \mathbf{s}_j \|^2 :: j = 1 \ldots S \qquad (7.8)$$

In order to write a closed-form solution for m_i, we take the difference of every pair of equations:

$$\| \mathbf{m}_i - \mathbf{s}_j \|^2 - \| \mathbf{m}_i - \mathbf{s}_k \|^2 = c^2 TOF_{ij}^2 - c^2 TOF_{ik}^2 \tag{7.9}$$

Expanding, we can write,

$$mx_i(sx_k - sx_j) + my_i(sy_k - sy_j) + mz_i(sz_k - sz_j)$$
$$= \frac{c^2 TOF_{ij}^2 - c^2 TOF_{ik}^2 - \| \mathbf{s}_j \|^2 + \| \mathbf{s}_k \|^2}{2} \tag{7.10}$$

or simply

$$(\mathbf{s}_k - \mathbf{s}_j)^T \mathbf{m}_i = b_{jk}^i \tag{7.11}$$

where

$$b_{jk}^i = \frac{c^2 TOF_{ij}^2 - c^2 TOF_{ik}^2 - \| \mathbf{s}_j \|^2 + \| \mathbf{s}_k \|^2}{2} \tag{7.12}$$

Each pair of speakers generates one equation in three unknowns; for S speakers there are $S(S-1)/2$ equations, and we need a minimum of four speakers to determine the position of one microphone. For $S4$ speakers we define the following matrix A and the vector $\mathbf{b}^i$

$$A = \begin{bmatrix} (\mathbf{s}_1 - \mathbf{s}_2)^T \\ (\mathbf{s}_1 - \mathbf{s}_3)^T \\ \vdots \\ (\mathbf{s}_1 - \mathbf{s}_S)^T \\ (\mathbf{s}_2 - \mathbf{s}_3)^T \\ \vdots \\ (\mathbf{s}_k - \mathbf{s}_j)^T \\ \vdots \\ (\mathbf{s}_{S-1} - \mathbf{s}_S)^T \end{bmatrix} \quad \mathbf{b} = \begin{bmatrix} b_{21}^i \\ b_{31}^i \\ \vdots \\ b_{S1}^i \\ b_{32}^i \\ \vdots \\ b_{jk}^i \\ \vdots \\ b_{S(S-1)}^i \end{bmatrix} \quad A\mathbf{m}_i = \mathbf{b}^i \tag{7.13}$$

The least squares solution can be written as

$$\mathbf{m}_i = (A^T A)^{-1} A^T \mathbf{b}^i \tag{7.14}$$

The closed-form solution for the microphone coordinates is further refined via a final ML estimation of all the unknown parameters (i.e., the positions of speaker-microphone pairs and of single microphones).

The following summarizes the complete algorithm.

Algorithm 7A

Assume that we have $M+S$ microphones and $S \geq 5$ loudspeakers. Attach one microphone to each of the loudspeakers so that we have S speaker-microphone pairs and M microphones. Place the speakers such that the microphones are in the convex hull formed by the speakers.

- **STEP 0**: *Select three loudspeakers to form a reference coordinate system: the first as the origin, the second to define the positive X-axis, and the third to form the positive XY-plane. Also select a fourth one to define the positive-Z half-space. The three reference speakers should be chosen such that they are as far away as possible from each other.*
- **STEP 1**: *Measure the $(M+S) \times S$ TOF matrix by exciting each of the loudspeakers using an appropriate signal.*
- **STEP 2**:
  (a) *Form the approximate distance matrix D between the S speaker-microphone pairs using equation (7.7).*
  (b) *Convert the distance matrix D to the dot product matrix B (Appendix 7B).*
  (c) *Get the approximate positions of the speaker-microphone pairs using metric MDS.*
  (d) *Slightly perturb the coordinates to get approximate separate initial guesses for the microphones and speakers positions.*
  (e) *Minimize the TOF-based error function using the Levenberg-Marquardt method to get the final positions of the S loudspeakers forming the reference coordinate system.*
  (f) *Translate, rotate, and mirror the coordinates to the coordinate system specified in Step 0.*
- **STEP 3**: *Get the closed-form solution for the M microphones using the reference coordinate system formed using (7.14).*
- **STEP 4**: *Refine all the values by performing the final ML estimation using the Levenberg-Marquardt method.*

Figure 7.2 shows an example in 2-D with 10 microphones (shown as x) and 3 speaker-microphone pairs. (In 2-D we select two nodes to lie on a line, the first at $(0, 0)$ and the second at $(x_1, 0)$, and specify one more node to lie in the positive-Y area to eliminate the reflection ambiguity). Using MDS, we obtain the approximate locations of the three speaker-microphone pairs, shown as filled squares in the figure. This approximate position is refined using ML estimation procedure to obtain the actual (separate) locations of the microphone and the loudspeaker in each speaker-microphone pair (no longer assuming that the distance between those is zero). Using the obtained loudspeaker locations, we compute a closed-form solution for the microphone locations, shown as squares in the figure. In the final ML estimation, we refine the closed-form solution to obtain the exact location of the microphones (shown as circles).

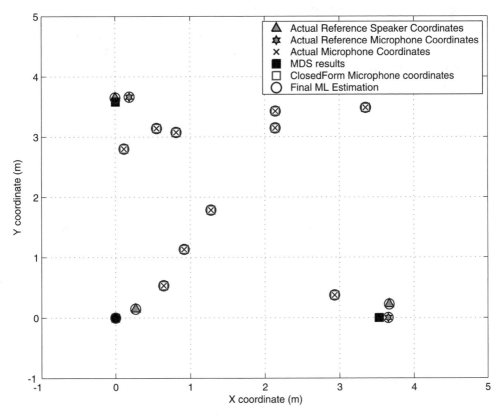

Figure 7.2 Result of the proposed algorithm in 2-D consisting of 10 microphones and 3 speaker-microphone pairs.

7.2.3   Estimator Bias and Variance

The properties of the ML estimator can be studied in terms of its bias and error variance. The error variance depends on the noise standard deviation, on the microphone array geometry, and on the positions of the reference speaker-microphone pairs. One way to study it is to perform extensive Monte Carlo simulations for various geometries and positions of the reference speaker-microphone pairs. However, if we could construct an analytical expression for the estimator bias and variance, then such simulation studies can be carried out much more quickly.

The ML estimate for the microphone and the speaker positions is defined implicitly as the minimum of a certain error function (7.5). Hence it is not possible to obtain exact analytical expressions for its mean and variance. However, by using the implicit function theorem and Taylor series expansion, it is possible to derive approximate expressions for those, similarly to derivations in [28–30]. Also, it is possible to derive the Cramér-Rao lower bound on the error covariance matrix of any unbiased estimator [31]; however, we cannot determine whether our estimator is unbiased.

7.2.3.1 Estimator Covariance Matrix

The ML estimate $\hat{\Theta}$ of Θ is the one that maximizes the likelihood ratio. In our case, the ML estimator is implicitly defined by (7.5). The maximum can be found by setting the first derivative to zero:

$$\nabla_{\Theta} F(\Theta, \Gamma)|_{\Theta = \hat{\Theta}} = 0 \qquad (7.15)$$

where ∇_{Θ} is a $P \times 1$ column gradient operator defined as

$$\nabla_{\Theta} F(\Theta, \Gamma) = \left[\frac{\partial F(\Theta, \Gamma)}{\partial \theta_1}, \frac{\partial F(\Theta, \Gamma)}{\partial \theta_2}, \ldots, \frac{\partial F(\Theta, \Gamma)}{\partial \theta_P} \right]^T \qquad (7.16)$$

As discussed earlier, the function can have multiple global minima. However, by defining a reference coordinate system and removing the assumed parameters, we can make sure that the function has only one global minimum. The implicit function theorem guarantees that (7.15) implicitly defines a vector-valued function

$$\hat{\Theta} = h(\Gamma) = [h_1(\Gamma), h_1(\Gamma), \ldots, h_P(\Gamma)]^T \qquad (7.17)$$

that maps the observation vector Γ to the parameter vector $\hat{\Theta}$. As such, (7.15) can be written as

$$\nabla_{\Theta} F(h(\Gamma), \Gamma) = 0 \qquad (7.18)$$

It is not possible to find an analytical expression for $h(\Gamma)$; however, we can approximate the covariance using the first-order Taylor series expansion for $h(\Gamma)$. Let $\bar{\Gamma}$ be the mean of Γ. Expanding $h(\Gamma)$ around $\bar{\Gamma}$ gives

$$h(\Gamma) \approx h(\bar{\Gamma}) + [\nabla_{\Gamma} h(\Gamma)^T|_{\Gamma = \bar{\Gamma}}]^T (\Gamma - \bar{\Gamma}) \qquad (7.19)$$

where

$$\nabla_{\Gamma} = \left[\frac{\partial}{\partial \gamma_1}, \frac{\partial}{\partial \gamma_2}, \ldots, \frac{\partial}{\partial \gamma_N} \right]^T \qquad (7.20)$$

is an $N \times 1$ column gradient operator. Taking the covariance on both sides of (7.19), we obtain

$$Cov[h(\Gamma)] \approx [\nabla_{\Gamma} h(\Gamma)^T|_{\Gamma = \bar{\Gamma}}]^T Cov(\Gamma) [\nabla_{\Gamma} h(\Gamma)^T|_{\Gamma = \bar{\Gamma}}] \qquad (7.21)$$

Note we do not know $h(\Gamma)$, but the dependence is only through the first-order partial derivatives of $h(\Gamma)$. Differentiating (7.18) with respect to Γ and evaluating it at $\bar{\Gamma}$ yields

$$\nabla_{\Theta} \nabla_{\Theta} F(h(\bar{\Gamma}), \bar{\Gamma}) [\nabla_{\Gamma} h(\bar{\Gamma})^T]^T + \nabla_{\Theta} \nabla_{\Gamma} F(h(\bar{\Gamma}), \bar{\Gamma}) = 0 \qquad (7.22)$$

Assuming that $\nabla_\Theta \nabla_\Theta F(h(\bar{\Gamma}), \bar{\Gamma})$ is invertible, we can write

$$[\nabla_\Gamma h(\bar{\Gamma})^T]^T = -[\nabla_\Theta \nabla_\Theta F(h(\bar{\Gamma}), \bar{\Gamma})]^{-1} \nabla_\Theta \nabla_\Gamma F(h(\bar{\Gamma}), \bar{\Gamma}) \qquad (7.23)$$

At $\Gamma = \bar{\Gamma}$, the vector derivatives involved can be shown to be

$$\begin{aligned}
\nabla_\Theta \nabla_\Theta F(\Theta, \Gamma) &= -J^T \Sigma^{-1} J \\
\nabla_\Theta \nabla_\Gamma F(\Theta, \Gamma) &= J^T \Sigma^{-1} \\
\nabla_\Gamma \nabla_\Theta F(\Theta, \Gamma) &= \Sigma^{-1} J \\
\nabla_\Gamma \nabla_\Gamma F(\Theta, \Gamma) &= -\Sigma^{-1}
\end{aligned} \qquad (7.24)$$

where J is an $N \times P$ Jacobian matrix of partial derivatives of $T(\Theta)$,

$$[J]_{ij} = \frac{\partial T_i(\Theta)}{\partial \Theta_j} \qquad (7.25)$$

Please refer to Appendix 7A for full derivations. Substituting the vector derivatives, we obtain

$$[\nabla_\Gamma h(\bar{\Gamma})^T]^T = -[-J^T \Sigma^{-1} J]^{-1} J^T \Sigma^{-1} \qquad (7.26)$$

Substituting into the covariance (7.21), we finally arrive at the following:

$$Cov\hat{\Theta} \approx Cov[h(\Gamma)] \approx [J^T \Sigma^{-1} J]^{-1} \qquad (7.27)$$

If we assume that all the observations have the same variance σ^2 (i.e., $\Sigma = \sigma^2 I$), we get

$$Cov\hat{\Theta} = \sigma^2 [J^T J]^{-1} = \sigma^2 F^{-1} \qquad (7.28)$$

where $F = J^T J$. If we assume that all the microphone and source locations are unknown, F is rank deficient and hence not invertible. This happens because the solution to the ML estimation problem as formulated is not invariant to rotation and translation. In order to make F invertible, we remove the rows and columns corresponding to the known parameters.

7.2.3.2 Estimator Mean

Taking the expectation of the first order Taylor series expansion in (7.16), we get

$$E(h(\Gamma)) \approx h(\bar{\Gamma}) = h(T(\Theta)) \qquad (7.29)$$

We see that the mean is the value given by the estimation procedure when applied to the actual noise-free measurements $T(\Theta)$. A more accurate expression for the mean can be derived using the second order Taylor series expansion. However, it involves third-order vector derivatives and generally cannot be stated in a simple form comparable to (7.27).

7.2.3.3   Monte Carlo Simulations

In order to validate the derived expression for the estimator variance, we performed a series of Monte Carlo simulations with 20 microphones randomly placed in a room of dimensions $4.0\text{m} \times 4.0\text{m} \times 4.0\text{m}$. We placed five speaker-microphone pairs so that all 20 microphones were within the convex hull formed by those pairs. Based on the geometry of the setup, the actual TOF values were calculated and then corrupted with additive white Gaussian noise (mean zero, variance σ^2) in order to model the room ambient noise and reverberation. The Levenberg-Marquardt method was used as the minimization routine. The results were averaged over 200 trials for each noise variance value. Figures 7.3(a) and 7.3(b) show the total estimator variance (sum of estimated variances of each parameter) and the total estimator bias (sum of estimate biases for each parameter) of all unknown microphone coordinates plotted against the noise standard deviation σ. The theoretical estimator variance is also shown. From Figure 7.3(a) we can see that experimentally computed estimator variance closely tracks the theoretical one. Figure 7.3(b) shows that the estimator is unbiased for low-noise variance; however, as the noise variance increases, the estimator starts showing an increasing bias.

7.2.3.4   Implications for Placement of Loudspeakers

In our evaluation of the covariance matrix, we have assumed that we know the positions of some nodes (i.e., we have fixed three loudspeakers to reside in the $z = 0$ plane). The covariance matrix has significant dependence on how those known nodes are arranged. Figure 7.4 shows the 95% uncertainty ellipses for a regular 2-D array containing 25 microphones and 4 loudspeakers for different positions of the loudspeakers. The microphones are represented as dots (.) and the loudspeakers as crosses ($\times$). The position of one loudspeaker and the x coordinate of another one are assumed to be known (shown in bold). For the second fixed loudspeaker, only the variance in the y direction is shown (as its x coordinate is fixed). For TOF estimation, the noise variance was assumed to be 10^{-9} in order to properly visualize the uncertainty ellipses.

In Figure 7.4(a), all four loudspeakers are placed at one corner of the grid. It can be seen that the farther the estimation is performed from the known nodes, the wider is the uncertainty ellipse. The uncertainty in the direction tangential to the line joining the microphone and the loudspeaker cluster is much larger than along the line. The same can be seen in Figure 7.4(b), where all loudspeakers are placed at the grid center. A simple geometric explanation can be provided; assume that we know the locations of two speakers [as shown in Figure 7.4(d)]. Each circular band represents the uncertainty in the distance estimation. The intersection of two bands corresponding to two speakers gives the uncertainty region for the microphone position, which widens if bands are intersecting far away from the speakers because of the decrease in curvature. From this reasoning, we can deduce that to minimize the uncertainty ellipse area, one should place loudspeakers as far away from each other as possible so that they enclose the area containing microphones, as in Figure 7.4(c), where substantially smaller uncertainty ellipses are seen. As such, in order to minimize the error due to Gaussian noise, we should choose the three reference nodes as far apart as possible.

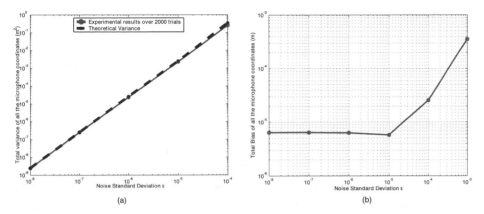

Figure 7.3 (a) The total variance and (b) the total bias of all the microphone coordinates for increasing noise standard deviation σ. The network consists of 20 microphones and 5 speaker-microphone pairs. The theoretical variance is also shown.

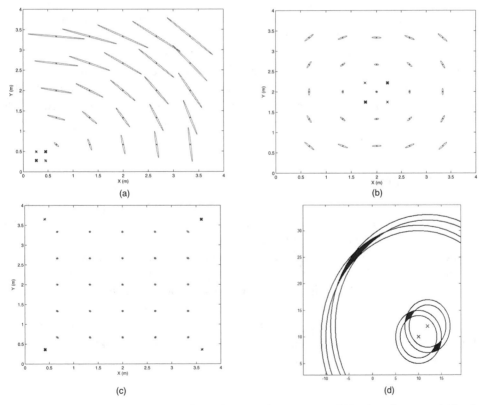

Figure 7.4 95% uncertainty ellipses for a regular 2-D array of 25 microphones and 4 loudspeakers. Noise variance for all cases is $\sigma^2 = 10^{-9}$. The microphones are represented as dots (.) and the loudspeakers as crosses (×). The position of one loudspeaker and the x coordinate of another is assumed to be known. In (a) and (b) the loudspeakers are close to each other, and in (c) they are spread out one at each corner of the grid. Drawing (d) explains the shape of the uncertainty ellipses.

7.3 System Autocalibration Performance

In this section we discuss some of the practical issues of our autocalibration implementation, such as the type of calibration signal, the TOF estimation procedure, and other design choices.

7.3.1 Calibration Signals

In order to measure the TOF accurately, the calibration signal has to be selected and tuned properly for the particular setup. ML sequences and frequency sweeps (chirps) are the two most popular choices for the calibration signal. A linear chirp signal is a short pulse in which the signal frequency varies linearly between two preset frequencies. The cosine linear chirp signal of duration T with the instantaneous frequency varying linearly between f_0 and f_1 is given by

$$s(t) = A\cos\left(2\pi\left(f_0 + \frac{f_1 - f_0}{T}t\right)t\right), 0 \le t \le T \tag{7.30}$$

In our system, we used a chirp signal of 512 samples at 39.0625 kHz as our calibration signal. The instantaneous frequency varied linearly from 5 kHz to 10 kHz. The initial and final frequencies were chosen to lie in the common passband of the microphone and the speaker frequency responses. Convolution of the chirp signal sent by the loudspeaker with the room impulse response results in signal spreadout and distortion. Figure 7.5(a) shows the chirp signal as sent out by a loudspeaker. Figure 7.5(b) shows the signal recorded by the microphone attached directly to the same loudspeaker. Figure 7.5(c) is the same signal recorded by another microphone. Changes in the signal shape are due to the speaker, microphone, and room response.

7.3.2 Time Delay Estimation

TOF estimation is the most crucial part of the algorithm and a potential source of error as well. Hence a lot of care is required to obtain the TOF accurately in noisy and/or reverberant environments. The time delay between two signals can be found by locating the peak in the cross-correlation of those; however, this method is not robust to noise and reverberations. Knapp and Carter [23] developed an ML estimator for determining the time delay between signals received at two spatially separated sensors in the presence of uncorrelated noise. They introduce the generalized cross correlation (GCC) function, which is the cross-correlation of the filtered versions of the received signals; the delay is still estimated by locating a peak in the GCC. The GCC function $R_{x_1 x_2}(\tau)$ is computed as

$$R_{x_1 x_2}(\tau) = \int_{-\infty}^{\infty} W(\omega) X_1(\omega) X_2^*(\omega) e^{j\omega\tau} d\omega \tag{7.31}$$

where $X_1(\omega)$ and $X_2(\omega)$ are the Fourier transforms of the microphone signals $x_1(t)$ and $x_2(t)$, respectively, and $W(\omega)$ is the weighting function. Two of the most

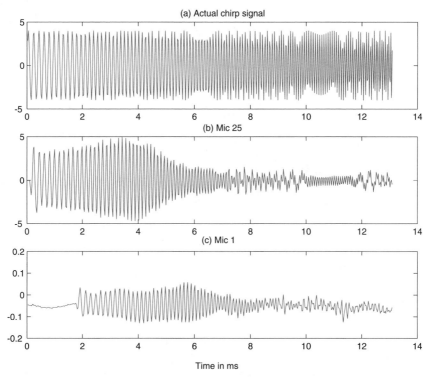

Figure 7.5   (a) The actual chirp signal used in our setup. (b) The chirp signal received by the microphone directly attached to the speaker. (c) The delayed chirp signal received by another microphone.

commonly using weighting functions are the ML and the Phase Transform (PHAT) weighting. The ML weighting function accentuates the signal passed to the correlator at frequencies for which the signal-to-noise ratio is the highest and simultaneously suppresses the noise power [23]. It performs well for low room reverberation levels. As the room reverberation increases, this method shows severe performance degradation [32]. Due to the spectral characteristics of the received signal being modified by the multipath propagation in a room, the GCC function can be made more robust by deemphasizing the frequency dependence. The PHAT weighting takes this idea to the extreme and flattens the magnitude spectrum. It is given by

$$W_{PHAT}(\omega) = \frac{1}{|X_1(\omega)X_2^*(\omega)|} \tag{7.32}$$

As a result, the GCC peak corresponds to the dominant delay. The disadvantage of the PHAT weighting is that it equally emphasizes both the low and high SNR regions and hence works well only when the noise level is low. However, it is generally observed that in typical in-room microphone array scenarios, reverberation is more detrimental to TOF determination than noise [32–35]; therefore, we use PHAT weighting for our experiments in real setup. Figure 7.6 shows the calibration signal, the received signal, and the corresponding GCC-PHAT function. The TOF is the position of the peak in the correlation function.

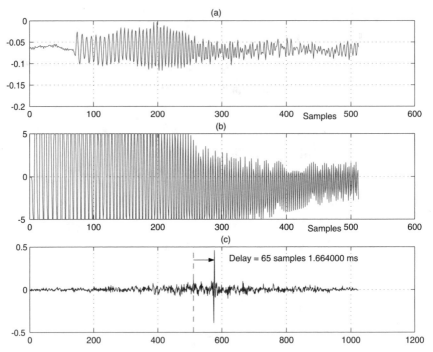

Figure 7.6 (a) The delayed chirp signal received by a microphone. (b) The chirp signal received by the microphone attached directly to the speaker. (c) The GCC-PHAT function.

7.3.3 Speed of Sound

Inaccuracy in knowledge of the speed of sound in the air can lead to errors in microphone position estimations. The speed of sound depends on the air temperature and is given by $c = (331 + 0.6T)$ m/s, where T is the temperature in degrees Celsius. In practice, we assume that c is known and constant. However, we can also estimate the speed of the sound along with the positions of the microphones and loudspeakers [13].

7.3.4 Synchronization Error

All the derivations in this chapter assume that all the microphones and loudspeakers are sharing a common clock source (e.g., are interfaced via a single or multiple but synchronized data acquisition boards). If that is not the case, TOFs will obviously contain errors due to lack of common time base. The method described in this chapter can be extended to handle such a case and recover time shifts between multiple sound production and acquisition units, but becomes significantly more complicated and requires more measurements; for full treatment, please refer to our related work in [22].

7.3.5 Testbed Setup and Results

We have set up a microphone array with 32 elements (Knowles Electronics model FG-3629). The array, shown in Figure 7.7(a), was built and is currently used for

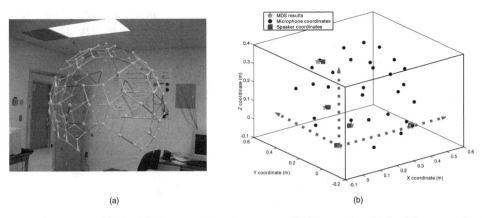

(a)   (b)

Figure 7.7 (a) The 32-element microphone array. (b) The results obtained from our algorithm.

fast measurement of the head-related transfer function (HRTF) of human subjects [36]. A microphone is placed at each node of the structure. In order to calibrate this array, we placed five microloudspeakers (Knowles Electronics model ED-9689) next to five microphones of the array to form the speaker-microphone pairs. The first four speakers define the coordinate system, and all positions are computed with respect to that coordinate system. Figure 7.7(b) shows the results obtained by the proposed algorithm for our microphone array; asterisks indicate the speaker positions, and circles indicate the microphone positions. In order to validate our results, we have measured the actual microphone coordinates with a Polhemus tracker and found that results are in good agreement.

7.4 The Tracking Algorithm

The particle filtering tracker, known also as a CONDENSATION tracker, was first introduced in the computer vision area in the work of Isard and Blake [16]. Various improvements of the technical nature were provided by Isard and Blake [37], Carpenter et al. [38], MacCormick and Blake [39], Li and Chellappa [40], Philomin et al. [41], and Qian and Chellappa [42]. The algorithm has seen application to multiple aspects of both computer vision and signal processing and was extended to track multiple objects, and a book on this topic [43] describes many different applications in signal detection and estimation. The mathematical framework of the tracker assumes that there exists a state vector X_s that describes the state of the tracked object and includes parameters of interest (coordinates, velocities, Euler angles, color histogram, and so forth). There also exists a measurement vector (sometimes called the observation vector) X_m that consists of the measurement values obtained from the sensors, which are related to and carry some information about the underlying state of the object. The (unknown) true state vector for any given time corresponds to point in a *state space*. The *probability distribution function* is defined on the state space and represents the uncertainty in knowledge of the state vector. For example, the extended Kalman filter assumes that the distribution is Gaussian and thus implicitly creates a probability distribution function (PDF) on the state space by keeping its mean and variance, which is sufficient to define a Gaussian. In contrast, the CONDENSATION tracker maintains an explicit approximate PDF by computing the PDF value at a set of randomly selected sample points (called *particles*) in the state space, which allows it to work well when Kalman filtering fails due to the underlying PDF being non-Gaussian. A further practical advantage of the technique is that it allows one to mix modalities and measurements during the tracking relatively easily, as it is not necessary to construct inverse solutions explicitly. Such multiple-modality tracking is the focus of the current chapter.

7.4.1 Algorithm Overview

The particle set update algorithm used in this paper is very similar to the original algorithm [16]. In the simple algorithm used in this paper, each particle $\{x_i\}$, $i = 1...N$, in the state space X has a weight π_i associated with it. This set of particles along with their weights is called *properly weighted* if it approximates the true PDF $P(x)$ correctly; this means that for every integrable function $H(x)$,

$$E_x(H(x)P(x)) = \lim_{N \to \infty} \frac{\sum_N H(x_i)\pi_i}{\sum_N \pi_i} \tag{7.33}$$

Given a properly weighted set of particles at time t with $\pi_i = \frac{1}{N}$ for every particle, it is possible to update the set reflecting the new measurements obtained at time $t + \delta t$ and end up, again, with a properly weighted set. The updated algorithm is as follows:

1. Propagate each particle x_i in time using the object *motion model*, which captures physical or any other knowledge about the relationship between the object's current state and that at subsequent time steps, obtaining particle set $\{x_i^*\}$.

2. Obtain a new measurement vector X_m and perform an evaluation of the *posterior probability density* π_i^* on set $\{x_i^*\}$ using the posterior probability estimation function $\pi_i^* = p(x_i^*|X_m)$, which essentially answers the question, "How likely is it that the object state is x_i^*, given that the current measurement vector is X_m?" In this way, the hidden internal state of the object is recovered from the observations. This probability cannot be computed in closed form, as it would require one to invert the measurement equations, often resulting in a nonlinear iterative process. Therefore, it is instead expanded using Bayes's rule:

$$p(x_i^*|X_m) = \frac{p(X_m|x_i^*)p(x_i^*)}{p(X_m)} \tag{7.34}$$

in which $p(X_m)$ is the prior probability of measurement, which is assumed to be known (e.g., assumed constant), and $p(x_i^*) = \frac{1}{N}$ is the weight of each particle at time t. Thus, $p(x_i^*|X_m) = Kp(X_m|x_i^*)$ for some constant K. Thus the expression is obtained for π_i^* using only $p(X_m|x_i^*)$, which *can* be computed without inversion of the measurement equations.

3. Resample from set $\{x_i^*\}$ with probabilities π_i^*, generating a new *properly weighted* set $\{x_i'\}$ with equal weights $\frac{1}{N}$ associated with every particle. The algorithm can be repeated now for the next time step, and so forth.

The algorithm does not keep an explicit representation of the currently most probable object state; however, it can be computed at any time by performing the numerical integration of a desired function of state variables over the state PDF approximately represented by the particle set.

7.4.2 Instantiation of the Particle Filter

The particle filter framework as described previously is very general; in this section, we describe the state vector, the audio and the video measurement vectors, the motion model, and the posterior probability estimation functions used in this work. The state vector X_s for the system consists of the coordinates and the velocities of the tracked object. Thus, $X_s = [x\,y\,z\,\dot{x}\,\dot{y}\,\dot{z}]$. The motion model that is used to propagate object state in time is given by

$$
\begin{aligned}
x(t + \delta t) &= x(t) + \dot{x}(t)\delta t \\
\dot{x}(t + \delta t) &= \dot{x}(t) + F\delta t
\end{aligned}
\tag{7.35}
$$

and similar expressions for $y, z, \dot{y}, \dot{z}$. F is the random excitation force applied to the particle. It generally depends on the expected range of object velocity and was set to an empirical value of $10\ \mathrm{ms}^{-2}$ for all test runs.

The observation vector X_m is built from audio and video measurements. The video part consists of the pairs $(\hat{u}_i, \hat{v}_i)$ of *image coordinates* of feature points on the tracked object for every camera in the system. Thus, N video cameras produce $2N$ components of the observation vector for each feature point. The transformation $(u_i, v_i) = \Psi_i(X_s) = \Psi_i(x, y, z)$ that converts the world coordinates into the image coordinates is precomputed during a camera calibration procedure (described later). This transformation is used to project the state vector to the image coordinate (observation) space and compute the posterior probability of a state vector given the observation vector by measuring the distance between the projection of a state vector and the observed image coordinates. The audio part of the observation vector consists of the values of *time differences of arrivals* (TDOA) $\hat{\tau}_{ij}$ of the acoustic source signal between different microphone pairs in the microphone array; for M microphones, the number of such observations is equal to C_2^M. The corresponding transformation from the state space to the observation space $\{\tau_{ij}\} = \Phi(X_s) = \Phi(x, y, z), i, j = 1 \ldots M$ is easy to compute and is also described later.

The final component of the tracker is the specific form of the posterior probability function, which is used to measure how likely it is that the particular observation vector X_m at time t is caused by some candidate object state X_s. We define the video and audio error distances and the posterior probability as

$$
\varepsilon_v^2(X_s, X_m) = \frac{1}{N} \sum_{i=1}^{N} [(u_i - \hat{u}_i)^2 + (v_i - \hat{v}_i)^2]
$$

$$(u_i, v_i) = \Psi_i(X_s), (\hat{u}_i, \hat{v}_i) \in X_m$$

$$\varepsilon_a^2(X_s, X_m) = \frac{1}{C_2^M} \sum_{i,j} (\tau_{ij} - \hat{\tau}_{ij})^2, \tau_{ij} = \Phi(X_s), \hat{\tau}_{ij} \in X_m$$

(7.36)

$$p(X_m|X_s) = \frac{\exp\left(-\frac{1}{2}\frac{\varepsilon_v^2(X_s, X_m)}{\sigma_v^2}\right)}{\sqrt{2\pi}\sigma_v} \frac{\exp\left(-\frac{1}{2}\frac{\varepsilon_a^2(X_s, X_m)}{\sigma_a^2}\right)}{\sqrt{2\pi}\sigma_a}$$

The parameters σ_v and σ_a in the PDF correspond to the standard deviation of the corresponding Gaussian. Loosely speaking, they play the same role as the variance of the measurement noise in the EKF and define how much trust is put on every individual measurement. Variation of these parameters affects the filter behavior. If the measurements are known to be inaccurate, larger values of σ should be used. On the other hand, if the value used is too large, the filter would be slow to learn the correct object motion.

Note that the $p(X_m|X_s)$ is a product of Gaussians formed from individual measurements. If some audio or video measurement is unavailable or unreliable at some time instant, then the part of the $p(X_m|X_s)$ corresponding to this measurement is simply set to a constant value, and the particle set update is performed using the marginalized values; and when the measurement becomes available again, it is put back into the framework.

7.4.3 Self-Calibration Within the Particle Filter Framework

The particle filter is usually employed for tracking the motion of an object. However, it can be used equally well to estimate the *intrinsic system parameters* or the sensor ego-motion. For example, in a videoconferencing framework there often exists an uncertainty in the position of the sensors. The position of a microphone array with respect to the camera can be measured with a ruler or determined from a calibrated video sequence; however, both methods are subject to measurement errors. These errors can lead to disagreement in audio and video estimations of the object position and ultimately to tracking loss. In another scenario, a multimodal tracking system with some sensors moving independently (either by requirements or by design) requires estimation of sensor motion, which can be done simultaneously with tracking in the proposed framework. Such a system can include, for example, several moving platforms, each with a camera and a microphone array, or a rotating microphone array. To perform simultaneous tracking with parameter estimation, we simply include the sensor motion parameters into the state space. (Note that it is not correct to talk about object state space now, since the state space includes also the system parameters. In some sense, the parameters of a whole system including the object and the tracking system itself are estimated.) One should be careful, though, to avoid introducing too many free parameters, as this will boost the dimensionality of the state space (curse of dimensionality) and lead to poor tracking performance.

7.5   Setup and Measurements

We have implemented the tracking algorithms in two different naturally multimodal setups. Setup 1 was constructed in a large anechoic room (flight room) and was used for studying the hunting behavior of the *Eptesicus Fuscus* echolocating bat. The flight room is pictured in Figure 7.8. Setup 2 is a typical acoustically untreated office environment with two cameras and two microphone arrays, shown in Figure 7.9. In the following, we describe details of the audio and video hardware used in these setups.

7.5.1   Video Modality

For the flight room setup, the video hardware consisted of two Kodak MotionCorder digital infrared cameras used at the resolution of 640×480 pixels and a frame rate of 240 Hz, placed at two corners of the experimental room. There were no sources of visible light in the room during the recording to ensure that the bat navigates using echolocation calls only. The video stream was recorded at a digital video recorder with embedded timestamps, and then the parts of the recording corresponding to the audio activity were extracted. For the office setup, two color Sony EVI-D30 active cameras were used at a resolution of 320×240 pixels and the frame rate determined by the performance limit of the single computer doing both audio and video acquisition and data processing in real time. The recorded video shows a frame rate of approximately 7 fps. The frame grabbers used were Matrox Meteor II.

To convert the world coordinates to the image coordinates, we use *direct linear transformation* (DLT) [44]. The DLT is defined by a 3×4 camera calibration matrix P, which has 11 free parameters (the transformation is invariant to matrix

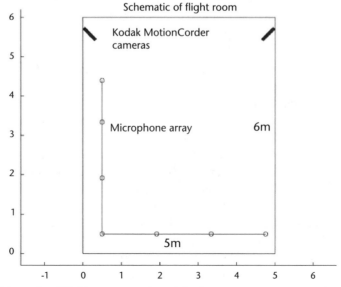

Figure 7.8   Schematic of flight room experimental setup.

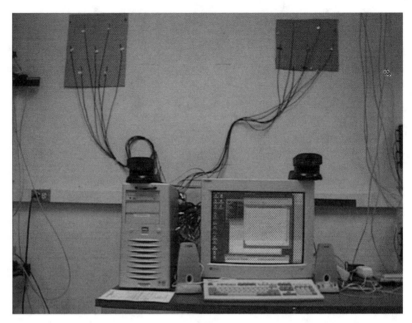

Figure 7.9 A two-camera, two-array setup used in office-room experiments.

scaling) and relates the world point coordinates (x, y, z) to the image coordinates (u, v) by the transformation

$$u = \frac{p_{11}x + p_{12}y + p_{13}z + p_{14}}{p_{31}x + p_{32}y + p_{33}z + 1}$$

$$v = \frac{p_{21}x + p_{22}y + p_{23}z + p_{24}}{p_{31}x + p_{32}y + p_{33}z + 1} \tag{7.37}$$

This is the transformation $(u_i, v_i) = \Psi_i(X_s) = \Psi_i(x, y, z)$ involved in the computation of the posterior probability). The matrix P_i for the ith camera has 11 parameters $\{p_{11}, \ldots, p_{14}, p_{21}, \ldots, p_{33}\}$, which in this model are assumed to be independent with $p_{34} = 1$. These parameters were estimated by using a calibration object of known geometry (Figure 7.10) placed in the field of view of both cameras with both camera pan and tilt set to zero. The calibration object consists of 25 white balls on black sticks arranged in a regular spatial pattern; the three-dimensional coordinates of the balls are known within 1 mm. The image coordinates of every ball is determined manually from an image of the calibration object, thus giving 25 relationships between (x_j, y_j, z_j) and (u_{ij}, v_{ij}), $j = 1 \ldots 25$ for the ith camera of the form above with the unknown parameters P. This overdetermined linear system of equations is then solved for P using least squares for $i = 1$ and $i = 2$, providing the DLT parameters for both cameras of the system.

To obtain the feature-point locations coordinates from the video streams captured by the cameras, we use different techniques for different setups. In the flight room, infrared light is used for imaging, and the room walls are covered by black audioabsorbing material; because of that, only a few bright spots can be seen in an infrared image, and a simple background subtraction technique works well to

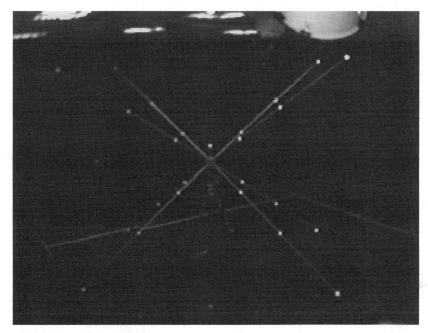

Figure 7.10   Video calibration object.

detect the bat as the only moving spot. The tracking is more complicated in case of a person moving in an office environment. The size of the image of a head of a person standing 3m or 4m from the camera is at most 15×15 pixels, so the level of detail is insufficient to find facial features. We used a simple automatic tracker based also on background subtraction to roughly locate the head of the person and then refined the results manually by hand-clicking on the person's nose (it is easier to identify in a low-resolution picture than the lip area) to provide video coordinates for the tracker. We were able to achieve the agreement of separately computed video and audio trajectories within 50-mm error using this hand-tracking mode.

7.5.2   Audio Modality

The audio tracking setup used is also different for the flight room setup and the office room setup because the nature of the sound signals is different. The bat produces ultrasonic echolocation chirps with the duration ranging from 2 ms to 20 ms and with the signal frequency decreasing from 50 kHz to 20 kHz during each chirp. To capture the signal, seven Knowles Electronics FG-3329 microphones were used, with the microphones arranged on a horizontal plane in an L-shaped frame and the arms of the frame along two adjacent room walls. The microphones were connected to a custom-made preamplifier and digitized at 140 kHz per channel using an IoTech Wavebook board. The positions of the individual microphones for audio processing were obtained from the video images (the microphones are visible in the image as small bright dots).

The office room audio setup consists of two microphone arrays mounted on the wall. Two groups of seven Panasonic WM-60A button microphones are used;

each group includes one microphone at the center and six at the circumference of a circle of 0.30m diameter. The microphones are connected to a custom-made low-noise, low-distortion preamplifier based on an AD797 chip, and the signal is digitized at 22.05 kHz per channel using a PowerDAQ board. To ensure a good match between the coordinate systems used in audio and video processing, the calibration frame is set up with its axes parallel to the room walls, and the location of the origin of the audio coordinate system with respect to the central ball of calibration frame is obtained using a measuring tape.

The audio algorithms for estimating the TDOAs are based on generalized cross-correlation method [23] and have been described in our prior work [17, 45]. The TDOAs obtained by the cross-correlation constitute the audio measurements. The projection function from the state space to the measurement space is trivially obtained as

$$\tau_{ij} = \Phi(X_s) = \Phi(x, y, z) = (\chi_j - \chi_i)/c \qquad (7.38)$$

where χ_i is the distance between ith microphone and the sound source and c is the sound speed. To recover the audio trajectory from a set of TDOAs for tracker verification, the set of algorithms described in [17, 46] were used.

7.6 Tracking Performance

We performed evaluation of the developed multimodal tracker on several sets of synthetic and real data obtained in different conditions. The synthetic data were used to verify the algorithm performance when the ground truth data is available. The real data includes the tracking of an artificial moving sound source, a bat in the flight, and the speaking person in the room. Tracking in the flight room setup was done offline, and the real-time tracking system for the office environment was implemented on a Dell workstation with dual PIII-933 MHz PC using Windows NT 4.0 and MSVC++ 6.0, working at approximately 7 fps. We are able to show that the performance of the multimodal tracker is better that the performance of both the audio and the video trackers taken separately.

7.6.1 Synthetic Data

We created a set of synthetic data by simulating an object moving in the spiral motion over the trajectory given by $x = \sin(2\pi t)$, $y = 2.0 - t$, $z = \cos(2\pi t)$, $t \in [0,1]$. The frame rate was set to 240 fps, the discretization frequency to 140 kHz, and all geometric parameters of the system for the run were kept the same as the parameters for the real setup in the flight room. In every frame, we obtained true object coordinates in image frames and TDOA values. Then, the image coordinates and TDOA values were perturbed by Gaussian noise with zero mean and variances of 3 pixels and 10 samples, respectively. The tracker was initialized by the correct source position at $t = 0$ and zero velocity; the initial distribution of particles in the state space was chosen to be Gaussian around the correct source position with $\sigma_{init} = 0.2$m. The σ_v and σ_a for the tracker were set to 3 pixels and 10 samples, corresponding to the true value of measurement noise. Any change to those values

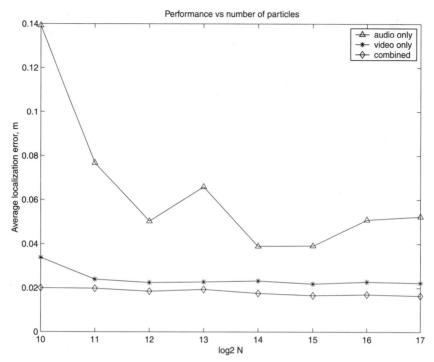

Figure 7.11   Performance versus log number of particles.

resulted in increase of the track estimation error, as can be expected. We performed several runs of the tracker with a different number of particles measuring the average distance between the estimated and the true object position at every time step; the results are shown in Figure 7.11.

It can be seen that the performance improves with the number of particles. The audio tracker performance alone is not very good; this can be attributed to the fact that all the microphones lie in the same horizontal plane, which decreases the accuracy of object height determination. The performance improves with the number of particles. The performance of the video tracker alone is better, and the performance for the combined tracker is improved even more (approximately by 15%). For a large number of particles, the average tracking error is approximately 16.5 mm, which is 2.5 times less than the error obtained by pure object detection in every frame (about 38.3 mm). This shows the effect of learning the object motion parameters by the filter.

The sensor motion recovery capability of the algorithm was also tested. We performed several experiments with synthetic data using one and two simulated planar microphone arrays rotating independently and one and two rotating cameras. We used two L-shaped microphone arrays placed on the ground, rotating with different speeds of 0.5 and 0.25 radian per second in opposite directions. The object is moving along the same spiral trajectory as before. The rotation was modeled by adding two rotation angles and two rotational velocities into the state of the system. The measurement vector was computed using true microphone coordinates and the object position. Then, random Gaussian noise with the same parameters as before was added to the measurement vector. The case we show here

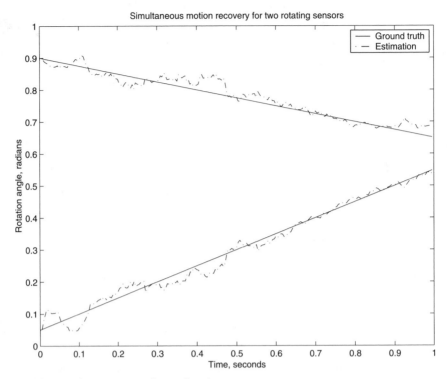

Figure 7.12   Rotating sensor motion estimation.

corresponds to the simultaneous tracking and sensor motion recovery using only one fixed camera. The algorithm succeeds in tracking, despite the fact that using any sensor alone is not sufficient to recover full object motion and the sensor's relative geometry is constantly changing. We show the plot of recovered sensor motion in Figure 7.12; the solid lines correspond to the true sensor rotation angles, and the dashed lines are the estimates computed by the tracking algorithm. The object tracking error for this set of experiments is only slightly increased (approximately 21.4 mm) compared to the case of two static arrays and two static cameras (16.5 mm). The same results were obtained for the case of two rotating cameras and one fixed microphone array, and in all cases where at least one sensor position is fixed, tracking with simultaneous parameter estimation succeeded in recovering both the object motion and the sensor motion. When all sensors are free to rotate, it is impossible to distinguish between sensor and object motion. Multipoint self-calibration should be used in this case.

7.6.2 Ultrasonic Sounds in Anechoic Room

To verify algorithm performance in controlled environment on real data, we acquired the multichannel audio and video recordings of an echolocating bat hunting for a mealworm prey [46] in an anechoic environment (the flight room). In addition, we performed several experiments with the person-carried sound source in the flight room. TDOA values for all microphone pairs and the object position in the frame for both cameras were used as an input to our multimodal CONDENSATION tracker.

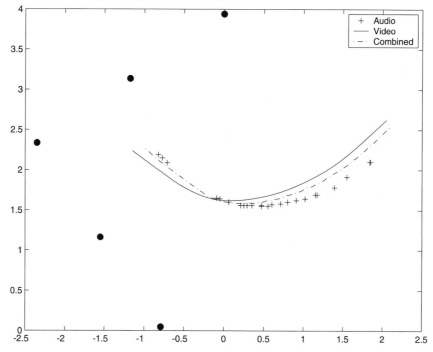

Figure 7.13   Bat flight (0.95 second) and the microphone setup (large black dots).

As TDOA values are not available in every frame (due to the bat being acoustically active only intermittently), linear interpolation is used to obtain missing values. (If audio information is used only when it is available, the trajectory essentially jumps back and forth between the video trajectory and audio points). In addition, the video trajectory and audio data points were determined from video and audio data independently. In Figures 7.13 and 7.14, the video-determined trajectory, audio data points corresponding to the individual echolocating bat calls, and the output of the tracker are plotted for two experiments.

The pictures show that the independently obtained video and audio trajectories are in fairly good agreement. The misalignment between them is likely to be due to the bias in determination of the microphone coordinates, which is done by video and can be inaccurate because the microphones lie far from the well-calibrated area where the calibration object was placed. The output of the multimodal tracker integrates the audio and video information and lies between the audio and video tracks, as expected. No ground truth data is available for these runs, so the comparison with ground truth cannot be done.

7.6.3   Occlusion Handling

We tested occlusion handling ability in the office environment setup. It is much more noisy and reverberant than the flight room. Using the setup described above, we performed real-time tracking of a single speaker moving in the field of view of a tracking system. The data was then processed off-line as well to recover the speaker trajectory. Low discretization frequency, relatively small intermicrophone distance

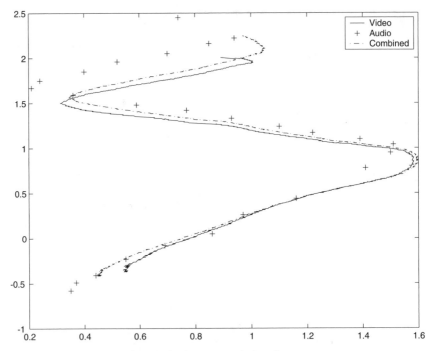

Figure 7.14 A 6.5-second walk with the bat trainer in hand.

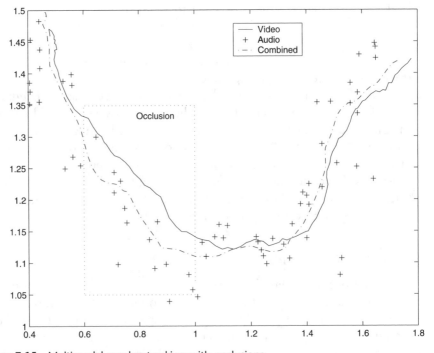

Figure 7.15 Multimodal speaker tracking with occlusions.

within an audio subarray, and large distance from the microphone array to the speaker all contribute to relatively low accuracy of audio data, so the video data is the primary source of information for the run. The audio error is, though, about 10 cm on average, which is sufficient to roughly localize the speaker (e.g, for camera pointing) using audio data alone.

We also tested the algorithm's robustness to occlusion. Normally, the speaker is visible to both of the tracking cameras. When only one camera can see the speaker due to occlusion or being out of the field of view, video information alone cannot be used to recover the speaker coordinates. The multimodal tracker, however, can continue to track the speaker because of the audio constraints. In Figure 7.15, we show the tracking results with a simulated occlusion. For the marked part of track, the video data from one of the cameras was omitted. Still, the track stays near the video trajectory, although it is influenced more by audio data now. When the video data is available again, the tracking error decreases back to the original value.

7.7 Conclusions

The described multimodal tracking algorithm provides a natural framework to integrate multimodal information and is robust to partial unavailability of input measurements (e.g., video occlusion or audio noise). The audio subsystem of the tracker is calibrated using the novel automatic microphone position calibration algorithm. The algorithm does not require precise placement of the calibration loudspeakers, and the only constraint we impose is that each loudspeaker is placed at or very close to one of the array microphones. We derive a closed-form approximate solution and further refine it by nonlinear minimization. We also derive and verify the expression for the variance of our estimator. The tracking and calibration algorithms are extensively validated on simulated and real data. We anticipate continued development of the system and achievement of real-time multimodal tracking ability on common hardware.

Acknowledgments

This chapter is an extended version of conference papers [47, 48]. Partial support of NSF awards 0086075 and 0205271 and of ONR grant N00014951021 is gratefully acknowledged. We would also like to thank Dr. Cynthia F. Moss and Kaushik Ghose (Auditory Neuroethology Laboratory, Neuroscience and Cognitive Science Program, Department of Psychology, University of Maryland) for providing us with the flight room experimental data used in this chapter.

References

[1]   Stein, B., et al., "Neural Mechanisms for Integrating Information from Multiple Senses," *Proc. IEEE ICME 2000*, New York, July 2000, pp. 567–570.

[2]   Cheok, K., G. Smid, and D. McCune, "A Multisensor-Based Collision Avoidance System with Application to a Military HMMWV," *Proc. IEEE Conf. Intelligent Transportation Systems*, Dearborn, MI, October 2000, pp. 288–292.

[3]   Ben-Yacoub, S., et al., "Audio-Visual Person Verification," *Proc. CVPR 1999*, Vol. 1, Fort Collins, CO, June 1999, pp. 1580–1585.

[4]   Pavlovic, V., et al., "Multimodal Speaker Detection Using Error Feedback Dynamic Bayesian Networks," *Proc. CVPR 2000*, Vol. 2, Hilton Head, SC, June 2000, pp. 34–41.

[5]   Pingali, G., G. Tunali, and I. Carlbom, "Audio-Visual Tracking for Natural Interactivity," *Proc. ACM Multimedia 1999*, Vol. 1, Orlando, FL, October 1999, pp. 373–382.

[6]   Quek, F., et al., "Gesture, Speech and Gaze Cues for Discourse Segmentation," *Proc. IEEE CVPR 2000*, Vol. 2, Hilton Head, SC, June 2000, pp. 247–254.

[7]   Faugeras, O., *Three-Dimensional Computer Vision: A Geometric Viewpoint*, Cambridge, MA: MIT Press, 1993.

[8]   Wei, G., and S. Ma, "Implicit and Explicit Camera Calibration: Theory and Experiments," *IEEE Trans. on Pattern Analysis and Machine Intelligence*, Vol. 16, No. 5, 1994, pp. 469–480.

[9]   Liebowitz, D., and A. Zisserman, "Metric Rectification for Perspective Images of Planes," *Proc. IEEE CVPR 1998*, Santa Barbara, CA, June 1998, pp. 482–488.

[10]   Triggs, B. "Autocalibration from Planar Scenes," *Proc. ECCV 1998*, Freiburg, Germany, June 1998, pp. 89–105.

[11]   Zhang, Z., "A Flexible New Technique for Camera Calibration," *IEEE Trans. on Pattern Analysis and Machine Intelligence*, Vol. 22, No. 11, 2000, pp. 1330–1334.

[12]   Rockah, Y., and P. Schultheiss, "Array Shape Calibration Using Sources in Unknown Locations. Part II: Near-Field Sources and Estimator Implementation," *IEEE Trans. on Acoustics, Speech, and Signal Processing*, Vol. ASSP-35, No. 6, 1987, pp. 724–735.

[13]   Sachar, J., H. Silverman, and W. Patterson III, "Position Calibration of Large-Aperture Microphone Arrays," *Proc. IEEE ICASSP 2002*, Vol. 2, Orlando, FL, May 2002, pp. 1797–1800.

[14]   Weiss, A., and B. Friedlander, "Array Shape Calibration Using Sources in Unknown Locations—A Maximum-Likelihood Approach," *IEEE Trans. on Acoustics, Speech, and Signal Processing*, Vol. ASSP-37, No. 12, 1989, pp. 1958–1966.

[15]   Ng, B., and C. See, "Sensor-Array Calibration Using a Maximum-Likelihood Approach," *IEEE Trans. on Acoustics, Speech, and Signal Processing*, Vol. ASSP-44, No. 6, 1996, pp. 827–835.

[16]   Isard, M., and A. Blake, "CONDENSATION Conditional Density Propagation for Visual Tracking," *Intl. J. Computer Vision*, Vol. 29, No. 1, 1996, pp. 5–28.

[17]   Zotkin, D., et al., "Smart Videoconferencing," *Proc. IEEE ICME 2000*, New York, August 2000, pp. 1597–1600.

[18]   Wang, C., et al., "Real-Time Automated Video and Audio Capture with Multiple Cameras and Microphones," *J. VLSI Signal Processing Systems*, Vol. 29, No. 1/2, 2001, pp. 81–99.

[19]   Oh, S., and V. Viswanathan, "Hands-Free Voice Communication in an Automobile with a Microphone Array," *Proc. IEEE ICASSP 1992*, San Francisco, CA, April 1992, pp. 281–284.

[20]   Omologo, M., M. Matassoni, and P. Svaizer, "Speech Recognition with Microphone Arrays," in *Microphone Arrays: Signal Processing Techniques and Applications*, M. S. Brandstein and D. Ward (eds.), Berlin, Germany: Springer-Verlag, 2001, pp. 331–353.

[21]   de Jesus, J., J. Calvo, and A. Fuente, "Surveillance System Based on Data Fusion from Image and Acoustic Array Sensors," *IEEE Aerospace and Electronic Systems Magazine*, Vol. 15, No. 2, 2000, pp. 9–16.

[22]   Raykar, V., I. Kozintsev, and R. Lienhart, "Position Calibration of Microphones and Loudspeakers in Distributed Computing Platforms," *IEEE Trans. on Speech and Audio Processing*, Vol. 13, No. 1, pp. 70–83.

[23]   Knapp, C., and G. Carter, "The Generalized Correlation Method for Estimation of Time Delay," *IEEE Trans. on Acoustics, Speech, and Signal Processing*, Vol. ASSP-24, No. 4, 1976, pp. 320–327.

[24]   Gill, P., W. Murray, and M. Wright, *Practical Optimization*, New York: Academic Press, 1982.

[25]   Torgerson, W., "Multidimensional Scaling. Part I: Theory and Method," *Psychometrika*, Vol. 17, 1952, pp. 401–419.

[26]   Steyvers, M., "Multidimensional Scaling," in *Encyclopedia of Cognitive Science*, London, U.K.: Nature Publishing Group, 2002.

[27]   Press, H., et al., *Numerical Recipes in C: The Art of Scientific Computing*, Cambridge, U.K.: Cambridge University Press, 1995.

[28]   Broida, T., and R. Chellappa, "Estimating the Kinematics and Structure of a Rigid Object from a Sequence of Monocular Images," *IEEE Trans. on Pattern Analysis and Machine Intelligence*, Vol. 13, No. 6, 1991, pp. 497–513.

[29]   Fessler, J., "Mean and Variance of Implicitly Defined Biased Estimators (Such as Penalized Maximum Likelihood): Applications to Tomography," *IEEE Trans. on Image Processing*, Vol. 5, No. 3, 1996, pp. 493–506.

[30]   Roy Chowdhury, A., and R. Chellappa, "Stochastic Approximation and Rate Distortion Analysis for Robust Structure and Motion Estimation," *Intl. J. Computer Vision*, Vol. 55, No. 1, 2003, pp. 27–53.

[31]   van Trees, H., *Detection, Estimation, and Modulation Theory*, Vol. 1, New York: John Wiley and Sons, 2001.

[32]   DiBiase, J., H. Silverman, and M. Brandstein, "Robust Localization in Reverberant Rooms," in *Microphone Arrays: Signal Processing Techniques and Applications*, M. Brandstein and D. Ward, (eds.), Berlin, Germany: Springer-Verlag, 2001, pp. 157–180.

[33]   Brandstein, M., and H. Silverman, "A Robust Method for Speech Signal Time-Delay Estimation in Reverberant Rooms," *Proc. IEEE ICASSP 1997*, Munich, Germany, 1997, pp. 375–378.

[34]   Omologo, M., and P. Svaizer, "Use of the Crosspower-Spectrum Phase in Acoustic Event Location," *IEEE Trans. on Speech and Audio Processing*, Vol. 5, No. 3, 1997, pp. 288–292.

[35]   Wang, H., and P. Chu, "Voice Source Localization for Automatic Camera Pointing System in Videoconferencing," *Proc. IEEE ICASSP 1997*, Munich, Germany, 1997, pp. 187–190.

[36]   Zotkin, D., et al., "Fast Head-Related Transfer Function Measurement Via Reciprocity," *J. Acoustical Society of America*, Vol. 120, No. 4, 2006, pp. 2202–2215.

[37]   Isard, M., and A. Blake, "ICONDENSATION: Unifying Low-Level and High-Level Tracking in a Stochastic Framework," *Proc. ECCV 1998*, Freiburg, Germany, June 1998, pp. 893–908.

[38]   Carpenter, J., P. Clifford, and P. Fearnhead, "An Improved Particle Filter for Non-Linear Problems," *IEEE Proc. Radar, Sonar, and Navigation*, Vol. 146, 1999, pp. 2–7.

[39]   MacCormick, J., and A. Blake, "Probabilistic Exclusion and Partitioned Sampling for Multiple Object Tracking," *Intl. J. Computer Vision*, Vol. 39, No. 1, pp. 57–71.

[40]   Li, B., and R. Chellappa, "Simultaneous Tracking and Verification Via Sequential Posterior Estimation," *Proc. CVPR 2000*, Vol. 2, Hilton Head, SC, June 2000, pp. 110–117.

[41]   Philomin, V., R. Duraiswami, and L. Davis, "Quasi-Random Sampling for CONDENSATION," *Proc. ECCV 2000*, Dublin, Ireland, June 2000, pp. 134–149.

[42]   Qian, G., and R. Chellappa, "Structure from Motion Using Sequential Monte-Carlo Methods," *Proc. ICCV 2001*, Vancouver, Canada, July 2001, pp. 614–621.

[43]   Doucet, A., N. de Freitas, and N. Gordons (eds.), *Sequential Monte-Carlo Methods in Practice*, New York: Springer, 2001.

[44]   Hartley, R., and A. Zisserman, *Multiple View Geometry in Computer Vision*, Cambridge, U.K.: Cambridge University Press, 2000.

[45]   Zotkin, D., et al., "An Audio-Video Front End for Multimedia Applications," *Proc. IEEE SMC 2000*, Nashville, TN, 2000, pp. 786–791.

[46]   Ghose, K., et al., "Multimodal Localization of a Flying Bat," *Proc. IEEE ICASSP 2001,* Salt Lake City, UT, May 2001, pp. 3057–3060.

[47]   Zotkin, D., et al., "Multimodal Tracking for Smart Videoconferencing," *Proc. IEEE ICME 2001,* Tokyo, Japan, August 2001, pp. 37–40.

[48]   Raykar, V., and R. Duraiswami, "Automatic Position Calibration of Multiple Microphones," *Proc. IEEE ICASSP 2004,* Montreal, QC, Canada, Vol. 4, May 2004, pp. 69–72.

Appendix 7A   Jacobian Computations

The following are the derivatives necessary for the minimization routine. These derivatives constitute the nonzero elements of the Jacobian matrix.

$$\frac{\partial TOF_{ij}^{actual}}{\partial mx_i} = -\frac{\partial TOF_{ij}^{actual}}{\partial sx_j} = \frac{mx_i - sx_j}{c \parallel m_i - s_j \parallel}$$

$$\frac{\partial TOF_{ij}^{actual}}{\partial my_i} = -\frac{\partial TOF_{ij}^{actual}}{\partial sy_j} = \frac{my_i - sy_j}{c \parallel m_i - s_j \parallel} \qquad (7A.1)$$

$$\frac{\partial TOF_{ij}^{actual}}{\partial mz_i} = -\frac{\partial TOF_{ij}^{actual}}{\partial sz_j} = \frac{mz_i - sz_j}{c \parallel m_i - s_j \parallel}$$

Appendix 7B   Converting the Distance Matrix to a Dot Product Matrix

Assume that we chose the kth point as the origin of our coordinate system. Let d_{ij} be the distance between ith and jth point. Using the cosine law (Figure 7B.1), one can write

$$d_{ij}^2 = d_{ki}^2 + d_{kj}^2 - 2d_{ki}d_{kj}\cos(\alpha). \qquad (7B.1)$$

The dot product b_{ij} is further defined as

$$b_{ij} = d_{ki}d_{kj}\cos(\alpha). \qquad (7B.2)$$

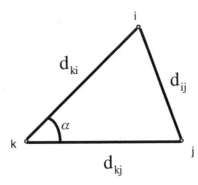

Figure 7B.1   Law of cosines.

Combining the above two equations,

$$b_{ij} = \frac{1}{2}(d_{ki}^2 + d_{kj}^2 - d_{ij}^2) \tag{7B.3}$$

However, this equation assumes that the kth point is the coordinate system origin, and we need to get the dot product matrix with the centroid as the origin. Let B and B^* be the dot product matrices with respect to the kth point and the centroid as the origin, respectively. Let X^* be the matrix of coordinates with the origin shifted to the centroid.

$$X^* = X - \frac{1}{N}I^*X \tag{7B.4}$$

where I^* is an $N \times N$ matrix with all elements equal to one (not simply an identity matrix). Now B^* can be written in terms of B as follows:

$$B^* = X^*X^{*T} = B - \frac{1}{N}BI^* - \frac{1}{N}I^*B + \frac{1}{N^2}I^*BI^*$$

Hence the ijth element in B^* is given by

$$b_{ij}^* = b_{ij} - \frac{1}{N}\sum_{l=1}^{N}b_{il} - \frac{1}{N}\sum_{m=1}^{N}b_{mj} + \frac{1}{N^2}\sum_{o=1}^{N}\sum_{p=1}^{N}b_{op} \tag{7B.5}$$

Using (7B.32), we obtain that

$$b_{ij}^* = -\frac{1}{2}\left[d_{ij}^2 - \frac{1}{N}\sum_{l=1}^{N}d_{il}^2 - \frac{1}{N}\sum_{m=1}^{N}d_{mj}^2 + \frac{1}{N^2}\sum_{o=1}^{N}\sum_{p=1}^{N}d_{op}^2 \right] \tag{7B.6}$$

This operation is also known as double centering (i.e., subtract the row and the column means from its elements, add the grand mean, and then multiply by $-\frac{1}{2}$).

CHAPTER 8

Multimodal Biometrics Involving the Human Ear

Christopher Middendorff, Kevin W. Bowyer, and Ping Yan

8.1 Introduction

Biometrics concerns itself with identifying or confirming the identity of an individual using physical or behavioral characteristics. Fingerprint and iris are generally considered to be the most accurate of the current widely used biometrics, but they are also more intrusive due to the nature of the image acquisition. A picture of the face or ear can be taken from a wide range of distances, potentially without a person's knowledge. This allows more discreet analysis of an individual, raising fewer concerns about privacy and user cooperation. Thus, ear biometrics can be a *passive, nonintrusive* method.

Face biometrics have received a great deal of attention, with only a small sampling of work suggested by [1–7]. Because people recognize each other easily from facial appearance, this method of biometric recognition seems reasonable. However, due to the changes that can occur to a face as time passes (aging) or even moment-to-moment (emotion) [8], other biometric methods have been explored. Occlusions such as glasses, lighting, and facial hair can also cause difficulty using the face as a metric [7, 9].

Ears have gained attention in biometrics due to the robustness of the ear shape and appearance over time [10–13]. The shape does not change due to emotion as the face does, and the ear is relatively constant over most of a person's life [10].

An image of the ear may be acquired either by a normal camera or by a sensor that acquires a description of 3-D shape. In the case of a normal camera, we will refer to the image as a *2-D image* or an *intensity image*. The original image may be either color or grayscale, but we will use only grayscale images in the work here. In the case of a 3-D sensor, we will refer to the image as a *3-D image*.

Biometrics may be used in either a verification scenario or an identification scenario. In a verification scenario, a person makes an identity claim, and the system acquires an image, compares the biometric to the one on file for the claimed identity, and then accepts or rejects the identity claim. Thus there is a one-to-one match of a probe biometric (the one newly acquired) to a gallery biometric (the one on file). In an identification scenario, an image of a person is acquired without the person making an identity claim, the biometric is compared to all of those on file, and if the best match to an on-file identity is good enough, then the unknown person is recognized as having that identity. Thus an identification scenario results in a one-to-many match. The term *authentication* is sometimes used to mean the same thing as *verification*, and the term *recognition* is sometimes used to mean the same thing as *identification*.

8.2   2-D and 3-D Ear Biometrics

In this section, we first survey various methods proposed for ear biometrics using 2-D intensity images and then cover methods proposed for use with 3-D shape (ear examples shown in Figure 8.1). Not surprisingly, methods initially used for face recognition have been adapted for use with ear biometrics. Just as eigenfaces is a well known approach to face recognition, several researchers have explored an eigen-ears approach. And just as iterative closest point (ICP) is a popular approach for face recognition based on 3-D shape [5], several researchers have looked at using ICP for 3-D ear biometrics [14, 15].

One of the earliest motivators for the use of the ear as a biometric dates to Iannarelli's experiment [10] and his stated ability to uniquely identify over 10,000 ears using 12 measurements of an ear normalized for size and rotation [image shown in Figure 8.2(b)], along with race and gender information. Iannarelli's

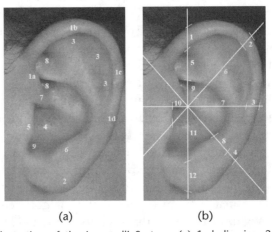

Figure 8.1   Left: an intensity image (2-D) of an ear. Right: A depth image (3-D) of an ear.

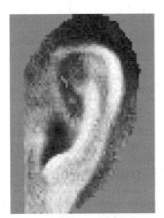 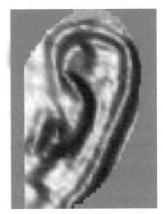

(a)   (b)

Figure 8.2   (a, b) Illustration of the Iannarelli System. (a) 1. helix rim, 2. lobule, 3. antihelix, 4. concha, 5. tragus, 6. antitragus, 7. crus of helix, 8. triangular fossa, and 9. incisure intertragica. (b) The 12 measurements for the Iannarelli System. (*From:* [16]. Reprinted from *Biometrics: Personal Identification in a Networked Society,* "Ear Biometrics," Fig. 1.1, M. Burge and W. Burger, Kluwer Academic, 1998. With kind permission of Springer Science and Business Media.)

approach was manual, with the operator making measurements on the image. Before Iannarelli, Imhofer in 1906 found that only four characteristics were needed to uniquely identify 500 ears [17].

Burge and Burger [16] proposed the ear as a biometric, citing Iannarelli. They suggested edge detection and a graph-based approach as their comparison method, to demonstrate the viability of the ear as a biometric. They showed examples of stages of processing an ear image but did not present results of any recognition experiments.

8.2.1   2-D Ear Biometrics

Principal component analysis (PCA) was the first approach used to generate experimental results for ear biometrics [1, 18]. The following explanation of PCA follows from Shapiro and Stockman [19].

This method uses a set of training images T to calculate a space composed of m orthogonal basis images $B = \{F_1, F_2, \ldots F_m\}$. These basis images are the principal components of T, which are called *eigen-ears* in the context of ear recognition. The number of principal components m has a maximum value of the minimum of the number of training images or the number of pixels in a training image. The basis images represent variance in the training images; the front vectors have the most variance between them, and the back vectors have the least. This space is often tuned to further reduce the number of basis images and reduce the dimensionality of the problem.

When this space is generated, an image I can be projected into this space and represented as $I' = a_1F_1 + a_2F_2 + \ldots a_mF_m$ with accuracy $\|I{-}I'\|/\|I\| < P\%$. If no basis images are dropped, the data can be perfectly reconstructed. When reducing the number of vectors, ideally all combinations would be tested to see what combination reduces the most while minimizing error. However, this requires 2^m evaluations, so some assumptions about the meanings of the vectors are employed to reduce the space without this exhaustive test. Vectors are usually dropped from the front (the highest variance, often representing lighting changes) and the back (the lowest variance, considered negligible).

Two steps are taken in order to ensure that the images are the same size and orientation. First, *landmark points* are chosen on the ear, and the image is transformed (by rotation, scale, and translation) to align these landmark points with the *destination landmark points*, so the feature to be used is now in the same place and orientation in every image. Second, the feature (ear) must be isolated from the surrounding elements of the image. This is done by applying a *mask* to each transformed image. The mask determines which pixels to use for PCA and which to discard. As a result, this method is sensitive to pose (as in Figure 8.6), lighting variance, and structural changes (for example, earrings, as in Figure 8.5).

Chang et al. [1] used PCA as the recognition engine in a comparison of ear and face biometrics. In these experiments, the images varied by day (88 subjects), lighting (111 subjects), and pose (101 subjects). Face and ear had comparable rank-one recognition rates in the day-variation experiment, 70.5% and 71.6%, respectively. In the lighting variation experiment, rank-one recognition was 64.9% for the face and 68.5% for the ear. In the pose variation experiment, rank-one recognition fell to just over 20% and 10%, respectively, when trained on the gallery set. This was the earliest presentation of experimental results for an ear biometrics algorithm. However, the algorithm was not fully automated because the landmark points on the ear image were marked by hand.

McFadden [18] used PCA in combination with a neural network to perform identification. In this experiment, ears were manually cropped from mugshot images. The ear edge was detected automatically and normalized before PCA was performed. The neural network was trained to determine whether or not two images matched. The function learned by the neural network improved upon the distance metric by taking inputs based on the properties of the images, as well as distances from the ear positions in a trained eigenspace, and returning a likelihood of the two images matching. Rank-one recognition using a testing size of 62 images was 58%. The rank-five recognition was 77%.

Choras [20] introduced a contour-tracing method to extract feature vectors based on the points where edge features intersected evenly spaced concentric circles, using the number of intersection points on each circle and the distances between them, as well as the locations of the contour endings and the contour bifurcations. On a set of easy images (high quality, ideal recognition conditions) chosen from a set of 240, the system obtained error-free recognition.

Choras also introduced a geometric parameter extraction method [21], decomposing the ear into two triangles and using various properties to generate feature vectors (triangle ratio method). Another method is the shape ratio, comparing the length of each meaningful contour to the distance between its endpoints. The geometrical parameters method has a perfect rank-one recognition rate using 104 probe and gallery images. This paper also introduces the angle-based contour representation model, defining a center point p_0 of the contour and concentric circles around it, and using the angle between p_0 and the intersections of the circles with the contours. The angle-based method achieved a rank-one recognition rate of 90.4% using 104 probe and gallery images.

Pun and Moon [13] gave an overview of ear biometrics. They cite the ear's smaller size and more uniform color as desirable traits for pattern recognition. Other characteristics are that it is less invasive than iris or fingerprint recognition, and more reliable than voice. They state that the principal methods of ear biometrics are PCA, force field transformation, local surface patch comparisons using range data, Voronoi diagram matching, neural networks, and genetic algorithms.

Hurley et al. [22] approach ear biometrics by modeling the image as a Gaussian force field, where the pixels exert "forces" on each other modeled after a magnetic field. The field lines (shown in Figure 8.3) created by the force field generate channels, which can be used for identification. This method is compared to PCA using manual registration of the ear images. The force field method gave a 99.2% recognition rate, using 4 samples of each of 63 subjects, taken over a period of 5 months. The PCA method gave a recognition rate of 98.4%.

Abate et al. [23] proposed a rotation-invariant descriptor. This method used a Generic Fourier Descriptor (GFD) of a polar representation of the ear. The problem they address is the positioning of the center of the Cartesian space $C(0, 0)$ around which the polar coordinates are considered, asserting that small shifts of

Figure 8.3 Force-field channels. (*From*: [22]. Reprinted from *Computer Vision and Image Understanding*, Vol. 98, No. 3, D. J. Hurley, M. S. Nixon, and J. N. Carter, "Force Field Feature Extraction for Ear Biometrics," pp. 491–512, copyright 2005, with permission from Elsevier.)

this point will cause large modifications in the coefficient values of the Fourier transform. They achieve translation invariance in this by choosing the $C(0, 0)$ as the mass center of the edge map of the input shape. In experiments, they use two datasets: set *A* containing 2 sessions with 210 images from 70 people, one looking ahead, the second looking up at a 15° angle, and the third looking up at a 30° angle; and set *B* containing 2 sessions with 72 images from 36 persons looking up with a free rotation angle. Ears from set *A* were manually extracted, and ears from set *B* were extracted using a Haar-based object detector. From testing each set, the images from the first session were used as gallery, while images from the second session were used as probe. In these experiments the GFD method outperformed the PCA method, even in the case where the probe rotation was 0°.

8.2.1.1   Challenges in Ear Biometrics Using Intensity Images

One of the primary drawbacks of appearance-based methods of ear biometrics is lighting effects. Insufficient lighting or differently placed lighting can affect the color and shadow of a feature, causing difficulties in recognition.

Accessories such as earrings can also cause problems for ear biometrics. Slight alteration in the placement of the accessory can cause a probe to differ from its gallery image. Similarly, when the subject gets his or her ear pierced (as in Figure 8.4), the probe and gallery may differ, and the recognition process may encounter difficulty between gallery acquisition and subsequent probe acquisition.

Additionally, the image content is sensitive to pose. A rotation toward or away from the camera can affect what and how much of the ear detail is captured by

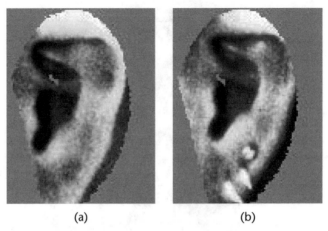

(a)   (b)

Figure 8.4 An example of a subject getting an ear pierced between subsequent sessions. Ear before piercing (a) and ear after piercing (b).

the camera. As a result, existing 2-D methods need to closely control the situations under which images are captured.

A simple experiment illustrates how strongly variation in pose can affect accuracy. Example images of the ear at several increments of 15° difference in pose are shown in Figure 8.5. The profile image is 90° off a frontal face view, or approximately straight on to the ear. The two other views in Figure 8.5 are 15° and 30° different. A set of ear images of this type were acquired for 21 subjects, using the same controlled lighting and with the images of each subject taken in the same session. An eigen-ear recognition experiment was performed, using the 90° images as the gallery, the 75° images as one set of probes, and the 60° images as a second set of probes. Under these conditions, without pose variation, we would expect nearly perfect rank-one recognition. The results are given in Table 8.1. Performance with the 15° off-angle images is 95% and with the 30° off-angle

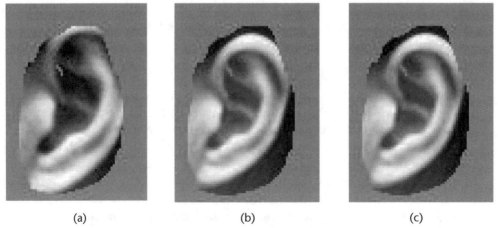

(a)   (b)   (c)

Figure 8.5 Head rotation affecting the view of the feature: (a) offset of 60°, (b) offset of 75°, and (c) offset of 90°.

Table 8.1 Optimal Settings Across Rotations

Rotation	Vectors Dropped	Energy Used	Recognition
60	0	100%	71.4%
75	0	90%	95.2%

images is 71%. These results support what Chang asserted in [1], specifically, that rotation in pose can have strong detrimental effects on performance.

8.2.2   3-D Ear Biometrics

Bhanu and Chen [24] used 3-D ear data, defining local surface patches around a feature point P, computing a surface type, and computing a 2-D histogram. Recognition was accomplished by performing the patch extraction on the probe image and using χ^2 dissimilarity to compare the histograms, filtering the results by the distance between matching patches. Verification was done by performing a transformation to align the patches and calculating a match quality, which is the ratio of corresponding points to total model points. Using 3-D ear shape from 10 subjects, they showed that the more feature patches that are used, the higher the recognition rate.

8.2.2.1   Approaches Using the Iterative Closest Point (ICP) Method

The iterative closest point (ICP) algorithm is a method of determining a rigid transformation between two clouds of points to align them as closely as possible. On each iteration, the algorithm finds closest corresponding points between the two clouds and computes a new pose for the probe to minimize the mean-square difference between the correspondences. This is done until a stopping condition is reached. In a recognition context, the closest match a probe makes to the gallery is the rank-one match. However, the ICP algorithm can be time-consuming and impractical over large galleries.

Chen and Bhanu [14] describe a two-step ICP procedure for matching 3-D ear shapes. The first step is to align the probe ear helix with a test ear helix, and the second is to perform ICP against the gallery ears, using the smallest distance as the match. In this method, ears were manually extracted from the profile face images. Using 1 gallery and 1 probe for each of 30 persons, this method achieved a rank-one recognition rate of 93.3%.

Yan and Bowyer [25] explored the accuracy of ear biometrics using 3-D shape. Automatic segmentation is used to extract the ear (described in Section 8.4). The ICP method was performed with various subsampled resolutions of the gallery and probe images, in an attempt to reduce computation time. It also used point-to-point ICP, point-to-surface (matching a point in ICP to its closest point on a calculated surface), and a combination of both. While the nonsubsampled probe and gallery gave the best performance (97.6% rank-one recognition), subsampling was found to cost about 1% degradation in recognition rate while drastically reducing computation time (from 15 to 18 seconds per match; to 2 to 3 seconds or less). The study also found that performance is better when the sampling of points on the gallery image is more dense than that of the probe image. Testing was done on a 415-subject data set with 1,386 probes.

8.3 Multibiometric Approaches to Ear Biometrics

Multibiometric methods fall into several categories [26]. One is *multialgorithm, monomodal*, which employs multiple algorithms on a single input, for example, performing color and edge algorithms on a single ear image to achieve recognition. Another is *single-algorithm, multimodal*, which uses a single method on multiple input, for example, using face and ear images as inputs to PCA as shown in Figure 8.6, or using several images from the same sensor for comparison, as recommended by Wayman [27]. The last is *multialgorithm, multimodal*, which uses different approaches on different data, for example, using PCA on 2-D images and ICP on 3-D images.

Intensity image of the ear, matched using an "eigen-ear" algorithm, rank one recognition rate = **86.5 %**

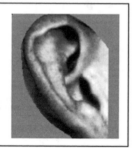

Intensity image of the face, matched using an "eigen-face" algorithm, rank one recognition rate = **83.3 %**

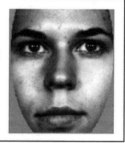

Multi-modal ear + face matched using eigen-x, rank one recognition rate = **98.7 %**

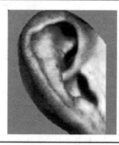 +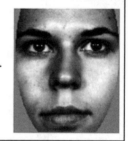

Multi-sample face + face matched using eigen-x, rank one recognition rate = **91.6 %**

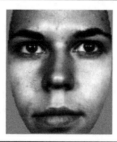 +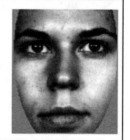

Figure 8.6 Comparing multimodal against multisample. (*Source*: [26].)

Fusion of multibiometrics is usually performed at the level of their scores, after results of the individual tests are returned but before the ranking is assigned. One example of such simple fusion is *sum*, where the score from each test is combined, and the sum is ranked. For example, a method using PCA on both 2-D and 3-D would combine the distances corresponding between the probe point and each gallery point (or, a bit more mathematically, $D_i = d_{a,i} + d_{b,i}$, where a is the projection of the 2-D probe into PCA space, b is the projection of the 3-D probe of the same person into PCA space, and the subscript indexes the gallery image). The match is the i corresponding to the smallest D. Another example is *weighted sum*, which is the same as the sum except that the metrics are given a weight besides an even split. This is useful if one metric is inherently more reliable than the other, or if it is possible to experimentally determine which metric is stronger in a given case.

If not fused at the score level (*independent* multimodal biometrics), the system may be *collaborative*, or use the results of one mode to influence another. For example, Socolinsky and Selinger [28] use 2-D information to determine the eye location and transfer that information to an infrared-driven recognition system. However, this relatively simple fusion example fails to take full advantage of the biometric data available in the 2-D information.

Moreno et al. [11] present a neural network approach to combine three different identifiers: ear feature points, ear morphology, and compression networks. Ears were manually extracted from profile images and normalized for pose and width/height ratio. In this experiment 6 photos were taken of 28 individuals, over several sessions (no time interval is given), and individuals were invited to change their expression and face orientation (angles/dimensions of rotation are not given). They use several combination methods (majority vote combination, Borda combination, Bayesian combination, and weighted Bayesian combination) and find that multialgorithm performed equivalent to their best individual method (compression network), which had a 93% identification rate.

Chang et al. [1] explored the benefits of combining face and ear biometrics, using PCA for recognition. When combined, performance improved significantly for the day-variation experiment, increasing from 70.5% rank-one recognition for face and 71.6% rank-one recognition for ear to 90% when the two metrics were combined. A statistically significant difference is also achieved in the lighting-variant experiment, with a combined rank-one score of 87.4%. These results suggest the combination of biometrics as a method of improving recognition rates.

Abate et al. [29] introduce a fractal-based technique they call BIS (face and ear: A bimodal identification system). This system uses a Haar-based object detector to detect a frontal face and profile ear. The detected features are normalized for size and segmented into regions of interest (eyes, nose, and mouth for a face; four square regions for the ear). Within these regions, a set of predefined *entry points* are used as starting points for analysis of the region content, which produces a set of domains. For each entry point, a neighbor range is optimized with respect to an affine transformation, using the 18 nearest neighbors to the entry point. Domains are organized by their centroids and the average approximation error between the domains and the entry points.

Yan and Bowyer [30] explored multimodal, multialgorithm and multi-instance Biometrics, combining 2-D intensity data with 3-D range data. Using PCA with the 2-D image and ICP with the 3-D image and weighting the modalities differently, rank-one recognition of up to 93.1% is achieved. While the ICP is weighted at 80% and the PCA is weighted only at 20%, the ICP-only method has rank-one of only 85.1%. This suggests the importance of fusing multiple methods, though the results are not statistically significant. Multialgorithm gave performance up to 90.2% (combining 3-D PCA with 3-D edge data). Multi-instance biometrics gave improvement over single-gallery, single-probe biometrics as well; 2-D PCA gave 73.4% rank one, 3-D ICP gave 81.7% rank one; combined gave up to 88.2%.

Yan [31] extended the ear and face biometric, using 3-D face and ear images. Experiments were performed with 174 subjects in the dataset, each with 2 ear shapes and 2 face shapes (example probe and gallery images shown in Figure 8.7). Rank-one recognition for face was 93.1%, and rank-one recognition for ear was 97.7%. Modalities were combined first by performing min-max normalization $(s' = (s-\text{min})/(\text{max}-\text{min})$, where s is the score), and then using sum and interval-based fusion. For both fusion rules, 100% rank-one recognition was achieved.

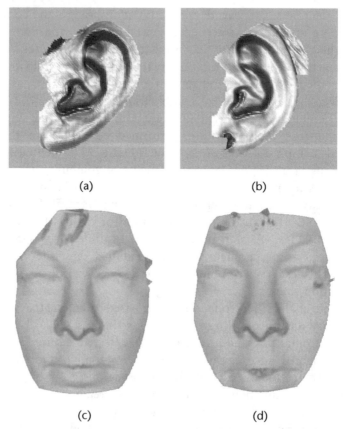

(a)   (b)

(c)   (d)

Figure 8.7 Multimodal biometrics using ear and face. The original matching score of ear alone does not return the correct match but is corrected by the face matching. (a) Gallery ear, (b) mismatched probe ear, (c) gallery face, and (d) matched gallery face. (*From:* [31]. © 2006 Pin Yang.)

Rahman and Ishikawa [32] combined 2-D PCA methods of ear and face, using profile images. In this method, features are extracted manually from images, and tests combining profile face with ear using 18 subjects had up to a 94.44% recognition rate.

8.4 Ear Segmentation

Separating the ear from the surrounding area is an important part of ear biometrics. If a biometric system is to be fully autonomous, the feature it is performing on (in this case, the ear) must be extracted automatically as well. Some methods are classifier-based (e.g., [23, 29]) while others take advantage of ear properties (concentration of edges [33], ear pit location [34]).

Chen and Bhanu [33] introduced a two-stage method of ear detection from range images. The first stage was an offline model template construction. Using manually extracted ears, the template feature vector is an averaged shape index histogram. For each ear, the shape index histogram is determined by taking the histogram of the maximum and minimum principal curvatures for each point in the image. To build the template, these histograms are averaged. To determine the candidates for ear detection, a step-edge magnitude image is generated. The image is then dilated, and connected components are labeled using 8-connectivity; features with an area $< \beta$ are removed (it is assumed that the ear will provide one of the largest edge features). The orientation of each region is determined by minimizing the sum of squares difference between an axial line and the edge region. Each region is then enclosed and the histogram calculated, and the difference between this histogram and the template histogram is calculated using χ^2 divergence. This was tested using a dataset of 30 subjects with ear images taken from two different viewpoints. These were divided into training and testing, with 30 subjects in each set. The first 20 subjects from the training set were used to generate the template. The ear detection rate is 91.5% with a 2.5% false positive rate.

A method described by Yan and Bowyer [25] uses depth information to find the nose tip from a given profile image. It then uses skin color information [9] to find the skin region. The ear pit is then detected using curvature information [35, 36], and from there the ear is segmented using an active contour algorithm that operates on the intensity image and the 3-D data [37].

McFadden [18] applies ray-based edge analysis to a manually cropped ear image, performing edge detection along rays emanating from near the center of the image instead of over the entire image. If the ray intersects a point on the outer edge of the ear, it will do so at exactly one point. The goal of the ear segmentation algorithm is to determine the outer boundary of the ear from a set of candidate points. The algorithm fuses two sets of candidates. The first set is the set of highest-priority candidates from each ray. The second is the set that best fits an ellipse determined by a small sample of the candidates. The sets were then fused based on which points provided the better contour; in this sense it is similar to a genetic algorithm, where two parents (the contours) contribute genes (the edge points) to produce a superior child (the resulting contour) based on some fitness function (each contour point's distance from its predecessor). The resulting contour is then smoothed, replacing outliers points on the contour with better-fitting points on the same ray.

8.5   Conclusions

The literature has shown that the use of multimodal biometrics can improve performance of a recognition system. However, there is no consensus on what features should be used, how they should be acquired, or even how they should be combined.

Various approaches to ear biometrics have been explored in recent years [1, 20–22]. At least one study, which used a PCA-based recognition approach on face and ear images from the same subjects, suggests that the two biometric sources have approximately equal recognition power [1]. However, other studies suggest that a PCA-based approach to ear biometrics with intensity images is not the best approach to use [20–22].

There is not yet any consensus on the best approach to use for ear biometrics with 2-D intensity images. Yan's work suggests that using 3-D shape of the ear allows better performance than using intensity images, but there is still active work on ear biometrics with 2-D images.

Studies have shown increased performance through combining face and ear biometrics [1], through combining 2-D and 3-D sensing of the ear [30], and through combining results from multiple images of the same feature taken with the same sensor [27].

The most pragmatic approach is the one put forth by Wayman [27], who suggested that although there are fascinating combinations possible, acquiring multiple images of the same modality would be the simplest since it does not require multiple sensors. While this is certainly cost-efficient, the method is still handicapped by the weaknesses of that imaging device. As we stated, taking a 2-D image means sacrificing the 3-D data, which has been demonstrated to be extremely valuable as a biometric. However, taking five traditional pictures of someone can be done in well under a second, but taking five 3-D images takes considerably longer using most 3-D sensors available today.

The ear has been demonstrated to be a valuable biometric resource. Individual modes of recognition have yielded remarkable accuracy, which is only further improved when integrating ear results from other modes. When designing a multimodal biometric system, one must consider the type of data to be acquired (e.g. 2-D or 3-D), the type of recognition algorithm performed on each data element (PCA or ICP), the output of that algorithm (the distance or error metric), the type of fusion to be performed to combine them, and the level at which it should be performed.

Acknowledgments

Biometrics research at the University of Notre Dame is supported by the National Science Foundation under grant CNS01–30839, by the Central Intelligence

Agency, by the U.S. Department of Justice/National Institute for Justice under grants 2005-DD-CX-K078 and 2006-IJ-CX-K041, by the National Geo-spatial Intelligence Agency, and by UNISYS Corp.

References

[1]   Chang, K., K. Bowyer, and V. Barnabas, "Comparison and Combination of Ear and Face Images in Appearance-Based Biometrics," *IEEE Trans. on Pattern Analysis and Machine Intelligence,* Vol. 25, 2003, pp. 1160–1165.

[2]   Victor, B., K. Bowyer, and S. Sarkar, "An Evaluation of Face and Ear Biometrics," *16th International Conference of Pattern Recognition,* 2002, pp. 429–432.

[3]   Chang, K., K. Bowyer, and P. Flynn, "Face Recognition using 2D and 3D Facial Data," *Workshop on Multimodal User Authentication,* 2003, pp. 25–32.

[4]   Phillips, P., et al., "The FERET Evaluation Methodology for Face Recognition Algorithms," *IEEE Trans. on Pattern Analysis and Machine Intelligence,* Vol. 22, No. 10, 2000, pp. 1090–1104.

[5]   Bowyer, K., K. I. Chang, and P. J. Flynn, "A Survey of Approaches and Challenges in 3D and Multi-Modal 3D+2D Face Recognition," *Computer Vision and Image Understanding,* Vol. 101, 2006, pp. 1–15.

[6]   Gordon, G., "Face Recognition Based on Depth and Curvature Features," *IEEE Conference on Computer Vision Pattern Recognition,* June 1992, pp. 108–110.

[7]   Phillips, P., et al., "Overview of the Face Recognition Grand Challenge," *IEEE Conference on Computer Vision Pattern Recognition,* Vol. 1, 2005, pp. 947–954.

[8]   Chang, K., K. Bowyer, and P. Flynn, "Effects of Facial Expression in 3D Face Recognition," *Biometric Technology for Human Identification II, Proc. of SPIE,* Vol. 5779, March 2005, pp. 132–143.

[9]   Hsu, R., M. Abdel-Mottaleb, and A. Jain, "Face Detection in Color Images," *IEEE Trans. on Pattern Analysis and Machine Intelligence,* Vol. 24, 2002, pp. 696–706.

[10]   Iannarelli, A., *Ear Identification,* Fremont, CA: Paramont Publishing Company, 1989.

[11]   Moreno, B., A. Sanchez, and J. Velez, "On the Use of Outer Ear Images for Personal Identification in Security Applications," *IEEE International Carnaham Conference on Security Technology,* 1999, pp. 469–476.

[12]   Yuizono, T., et al., "Study on Individual Recognition for Ear Images by Using Genetic Local Search," *Proc. of the 2002 Congress on Evolutionary Computation,* 2002, pp. 237–242.

[13]   Pun, K., and Y. Moon, "Recent Advances in Ear Biometrics," *Proc. of the Sixth International Conference on Automatic Face and Gesture Recognition,* May 2004, pp. 164–169.

[14]   Chen, H., and B. Bhanu, "Contour Matching for 3D Ear Recognition," *Seventh IEEE Workshop on Application of Computer Vision,* 2005, pp. 123–128.

[15]   Yan, P., and K. Bowyer, "Ear Biometrics Using 2D and 3D Images," *2005 IEEE Computer Society Conference on Computer Vision and Pattern Recognition (CVPR'05) — Workshops,* 2005, p. 121.

[16]   Burge, M., and W. Burger, "Ear Biometrics," in *Biometrics: Personal Identification in a Networked Society,* A. Jain, R. Bolle, and S. Pankanti, (eds.), Boston, MA: Kluwer Academic, 1998, pp. 273–286.

[17]   Imhofer, R., "Die Bedeutung Der Ohrmuschel Für Die Feststellung Der Identität," *Archiv für Krimi-nologie,* Vol. 26, 1906, pp. 150–163.

[18]   McFadden, F., "Personal Identification from Mugshot Ear Images," Technical Report, National Institute of Standards and Technology, 1998.

[19]   Shapiro, L., and G. Stockman, *3D Models and Matching,* Upper Saddle River, NJ: Prentice-Hall, 2001, pp. 479–526.

[20]   Choras, M., "Ear Biometrics Based on Geometrical Feature Extraction," *Electronic Letters on Computer Vision and Image Analysis,* Vol. 5, No. 3, 2005, pp. 84–95.

[21]   Choras, M., "Further Developments in Geometrical Algorithms for Ear Biometrics," *4th International Conference on Articulated Motion and Deformable Objects, AMDO 2006,* 2006, pp. 58–67.

[22]   Hurley, D., M. Nixon, and J. Carter, "Force Field Feature Extraction for Ear Biometrics," *Computer Vision and Image Understanding,* Vol. 98, 2005, pp. 491–512.

[23]   Abate, A., et al., "Ear Recognition by Means of a Rotation Invariant Descriptor," *International Conference on Pattern Recognition,* 2006, pp. 437–440.

[24]   Bhanu, B., and H. Chen, "Human Ear Recognition in 3D," *Workshop on Multimodal User Authentication,* 2003, pp. 91–98.

[25]   Yan, P., and K. Bowyer, "Biometric Recognition Using 3D Ear Shape," *IEEE Trans. on Pattern Analysis and Machine Intelligence,* Vol. 29, No. 8, 2007.

[26]   Bowyer, K.W., et al., "Multi-Modal Biometrics: An Overview," *Second Workshop on MultiModal User Authentication,* May 2006, Toulouse, France.

[27]   Wayman, J., "A Path Forward for Multi-Biometrics," *ICASSP,* 2006.

[28]   Socolinsky, D., and A. Selinger, "Thermal Face Recognition in an Operational Scenario," *IEEE Conference on Computer Vision Pattern Recognition,* Vol. 2, June 2004, pp. 1012–1019.

[29]   Abate, A., M. Nappi, and D. Riccio, "Face and Ear: A Bimodal Identification System," *Proc. Third International Conference on Image Analysis and Recognition, Part II,* 2006, pp. 297–304.

[30]   Yan, P., and K. Bowyer, "Multi-Biometrics 2D and 3D Ear Recognition," *Audio- and Video-Based Biometric Person Authentication,* 2005, pp. 503–512.

[31]   Yan, P., "Ear Biometrics in Human Identification," Ph.D. thesis, Department of Computer Science and Engineering, University of Notre Dame, Notre Dame, IN, 2006.

[32]   Rahman, M., and S. Ishikawa, "Proposing a Passive Biometric System for Robotic Vision," *Proc. of the Tenth International Symposium on Artifical Life and Robotics,* 2005.

[33]   Chen, H., and B. Bhanu, "Human Ear Detection from Side Face Range Images," *International Conference on Pattern Recognition,* 2004, pp. 574–577.

[34]   Yan, P., and K. Bowyer, "An Automatic 3D Ear Recognition System," *Proc. 3rd International Symposium on 3D Data Processing, Visualization, and Transmission, 3DPVT 2006,* 2006.

[35]   Besl, P., and R. Jain, "Invariant Surface Characteristics for 3D Object Recognition in Range Images," *Computer Vision Graphics Image Processing,* Vol. 33, 1986, pp. 30–80.

[36]   Flynn, P., and A. Jain, "Surface Classification: Hypothesis Testing and Parameter Estimation," *IEEE Conference on Computer Vision Pattern Recognition,* 1988, pp. 261–267.

[37]   Kass, M., A. Witkin, and D. Terzopoulos, "Snakes: Active Contour Models," *International Journal of Computer Vision,* Vol. 1, 1987, pp. 321–331.

CHAPTER 9

Fusion of Face and Palmprint for Personal Identification Based on Ordinal Features

Rufeng Chu, Shengcai Liao, Yufei Han, Zhenan Sun, Stan Z. Li, and Tieniu Tan

Biometric systems relying on a single technology have been deployed in many different application contexts (airports, passports, access control). However, due to the limitations of universality and accuracy, it is difficult for unimodal biometric systems to meet the accuracy requirement. By multiple modalities, enhanced performance reliability could be achieved. In this chapter, we present a face + palmprint multimodal biometric identification method and system to improve the identification performance. Effective classifiers based on ordinal features are constructed for faces and palmprints, respectively. Then, the matching scores from the two classifiers are combined using several fusion strategies to give a unique matching score. Experimental results on a middle-scale dataset, including 378 subjects and 20 pairs of images for each, have demonstrated the effectiveness and improvements over the unimodal systems.

9.1 Introduction

Biometric identification makes use of the physiological or behavioral characteristics of people, such as fingerprint, iris, face, palmprint, gait, and voice, for personal identification [1], which provides advantages over nonbiometric methods such as password, PIN, and ID cards. Its promising applications as well as the theoretical challenges have attracted much attention since the last decade.

Currently, most biometric systems deployed in real-world applications are unimodal, relying on the evidence of a single source of biometric information for authentication. Such systems cannot meet desired performance requirements [2]. For example, face image is affected by illumination, pose, and facial expression, and voiceprint by environmental noise. To overcome such difficulties, multimodal biometric systems are developed in which evidences from multiple sources of information are integrated to improve the performance. A number of studies have shown advantages of multimodal biometrics [3–12].

Brunelli and Falavigna [3] proposed a person identification system based on voice and face, and Kittler et al. [4] further evaluated such methodology using different fusion rules. Hong and Jain [5] proposed an identification system based on face and fingerprint. Wang et al. [6] proposed the fusion of face and iris. Recently, multimodal biometric systems using hand-based information are proposed. Kumar et al. [7] proposed a system combining the geometric features of the hand with palmprints. Kumar and Zhang [8] proposed an identification system based on face and palmprint, where a feed-forward neural network is used to integrate individual matching scores and generate a combined decision score. Alternatively, Feng et al. [9] presented a face and palmprint multimodal biometric system by fusion of features extracted by PCA or ICA. Ribaric and Fratric [10] described a biometric identification system based on eigen-palm and eigen-finger features with fusion applied at the matching score level. Recently we have also seen fusion of ear and face [11] and fusion of face, fingerprint, and hand geometry [12].

Face and palmprint multimodal biometrics are advantageous due to the use of noninvasive and low-cost image acquisition. We can easily acquire face and palmprint images using two touchless sensors simultaneously. Existing studies in this approach [8, 9] employ holistic features for face representation, such as PCA and ICA, and results are shown with small datasets (less than 100 subjects) reported. Note that PCA and ICA are sensitive to global variation of faces [13, 14], such as illumination and inaccurate alignment. On the other hand, although various palmprint representations have been proposed, such as line features [15], feature points [16], Fourier spectrum [17], eigen-palm features [18], Sobel and morphological features [19], texture energy [20], wavelet signatures [21], Gabor phase [22], fusion code [23], competitive code [24], how to model the palmprint pattern effectively and efficiently has so far not been well addressed.

In contrast to the holistic features, local appearance features, more stable to the global variation, have been widely found to be useful and powerful for face recognition. These include local features analysis (LFA) [25], Gabor wavelet-based features [26–28] and local binary pattern (LBP) [29]. For palmprint recognition, competitive performances are also reported by using local features [22–24].

It is believed that the human vision system uses a series of levels of representation, with increasing complexity. A recent study on local appearance or fragment (or local region) based object recognition [30] shows that features of intermediate complexity are optimal for basic visual task of classification, and mutual information for classification is maximized in a middle range of fragment size. Existing approaches suggest a trade-off between the complexity of features and the complexity of the classification scheme. Using fragment features is therefore advantageous [31] in that the number of features used for classification is reduced from richer information content of the individual features, and that a linear classifier may suffice when proper fragment features are selected, while for simple generic features the classifier has to use higher-order properties of their distributions.

We consider a class of simple local features: that of ordinal relationship. Ordinal features are defined based on the qualitative relationship between two image regions and are robust against various intraclass variations [32–34]. For example, they are invariant to monotonic transformations on images and are flexible enough to represent different local structures of different complexity. Sinha [33] shows that several ordinal measures on facial images, such as those between eye and forehead and between mouth and cheek, are invariant with different persons and imaging conditions, and thereby develops a ratio-template for face detection. In addition, Schneiderman [35] also uses an ordinal representation for face detection.

Sinha exploited ordinal information of several attribute dimensions, such as intensity, color, and shape, to construct a unique face signature of a scene [33]. Lipson et al. applied an ordinal technique to image database indexing [36]. Bhat and Nayar employed the relative intensity values in image windows for stereo correspondence [37]. By combining ordinal measures and co-occurrence, Partio et al. obtained better texture retrieval results than traditional gray-level co-occurrence matrices [38]. Ordinal features have been used for recognition of palmprints [39] and faces [40].

In this chapter, we present a new multimodal biometric system for fusion of face and palmprint based on ordinal features. Effective classifiers are constructed for each of the modalities. Different strategies are employed for fusing palmprint and face classifiers on a middle-scale dataset. While Thoresz [34] believed that ordinal features may be only suited for simple detection and categorization but too weak for fine discrimination tasks, such as personal identification, our work presented here shows the power of ordinal features for biometric identification. Experimental results have demonstrated the effectiveness of the proposed system.

The rest of this chapter is organized as follows. In Section 9.2 we introduce ordinal features. In Section 9.3 we describe the details of the proposed multimodal biometric system. Experimental results and conclusions are presented in Section 9.4 and Section 9.5, respectively.

9.2   Ordinal Features

Ordinal features come from a simple and straightforward concept that we often use. For example, we could easily rank or order the heights or weights of two persons, but it is hard to answer their precise differences. For computer vision, the absolute intensity information associated with a face can vary because it can change under various illumination settings. However, ordinal relationships among neighborhood image pixels or regions present some stability with such changes and reflect the intrinsic natures of the object.

An ordinal feature encodes an ordinal relationship between two concepts. Figure 9.1 gives an example in which the average intensities between regions A and B are compared to give the ordinal code of 1 or 0. Ordinal features are efficient to compute. Moreover, the information entropy of the measure is maximized because the ordinal code has nearly equal probability of being 1 or 0 for arbitrary patterns.

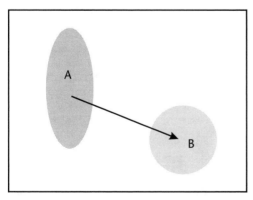
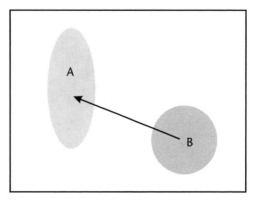

Figure 9.1   Ordinal measure of relationship between two regions. An arrow points from the darker region to the brighter one. Left: Region A is darker than Region B, in other words, $A < B$. Right: Region A is brighter than Region B, or $A > B$.

According to the spatial relationship between image regions, ordinal measure can be classified into two categories: local ordinal measure and nonlocal ordinal measure.

9.2.1 Local Ordinal Features

Local ordinal measures depict the comparison of adjacent image regions, which is well-suited to represent the object possessing a rich source of local sharp variations, such as iris [41] and palmprint [39]. A common practice to compare adjacent image regions is based on differential filters. After filtering, the local region covered by the operator is coded as 1 or 0 based on the sign of the filtering result.

In fact, many existing palmprint recognition methods [22–24] used this information implicitly [39]. For example, Gabor-based encoding filters used in palm code [22] are essentially local ordinal operators (see Figure 9.2). For odd Gabor filtering of local palmprint region, the image regions covered by two excitatory lobes are compared with the image regions covered by two inhibitory lobes [Figure 9.2(b)]. The filtered result is qualitatively encoded as 1 or 0 based on the sign of this inequality. Similarly, even Gabor-generated palm code is mainly determined by the ordinal relationship between one excitatory lobe-covered region and two small inhibitory lobe-covered regions [Figure 9.2(d)]. Because the sum of

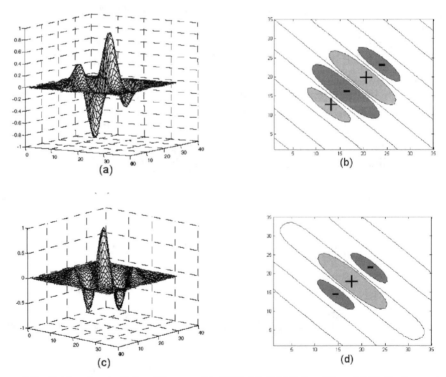

Figure 9.2 Odd and even Gabor filters used in [22]. (a) Odd Gabor filter. (b) Ordinal comparison of image regions using odd Gabor filter: + denotes excitatory lobe covered image region, and − represents inhibitory lobe covered image region. (c) Even Gabor filter. (d) Ordinal comparison of image regions using even Gabor filter.

original even Gabor filters coefficients is not equal to 0, the average coefficient value is reduced from the filter to maximize the information content of the corresponding palm code.

In addition, ordinal relationship is not restricted to intensity measurement. As a byproduct of Gabor phase measure (i.e., ordinal intensity measure), the orientation energy or magnitude was also obtained by orthogonal Gabor filtering. Thus it is possible to combine ordinal intensity measures and ordinal energy measures together. In [23], the local energy along four different orientations was compared to each other to obtain the maximum. Then the palmprint is represented using the Gabor-filtered ordinal intensity measures whose basic lobes are along the maximum energy orientation. In [24], orientation of the dominant line segment is regarded as the palmprint feature. Even Gabor filter is used to filter the local image region along six different orientations, obtaining the corresponding contrast magnitudes. Based on the winner-take-all competitive rule, the index (ranging from 0 to 5) of the minimum contrast magnitude was represented by three bits, namely, competitive code. Due to the success of these methods, we conclude that the ordinal measures are perhaps the most suitable representation for palmprint-based identification systems.

9.2.2 Nonlocal Ordinal Features

For face image, due to the similar facial shape of different persons, it is difficult to effectively represent different faces by simple local ordinal comparison. Balas and Sinha [42] extend differential filters to "dissociated dipoles" for nonlocal comparison, which can compare small regions across large distances, shown in Figure 9.3. Like differential filters, a dissociated dipole also consists of an excitatory and an inhibitory lobe, but the limitation on the relative position between the two lobes is removed. There are three parameters in dissociated dipoles:

- *The scale parameter σ*: On one hand, the noise suppression requires a coarse scale representation of the image structure. On the other hand,

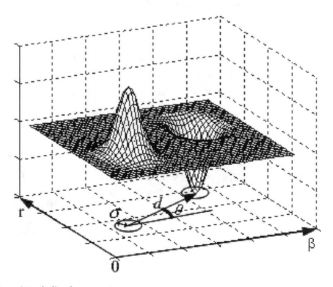

Figure 9.3 Dissociated dipole operator.

the discriminating power derives from fine details. The result of this trade-off is that an intermediate scale should be carefully chosen or information at large scale and fine scale would be fused. For dipoles with a Gaussian filter, the standard deviation σ is an indicator of the scale;

- *The interlobe distance d*: This is defined as the distance between the centers of the two lobes. When d is equal to the size of the lobes in a dipole, the operator is essentially a local filter. So the differential filters can be seen as a special case of dipoles;

- *The orientation θ*: This is the angle between the line joining the centers of the two lobes and the horizontal line. It is in the range from 0 to 2π.

We extend dissociated dipoles to dissociated multipoles, as shown Figure 9.4. While a dipole tells us the orientation of a slope edge, a multipole can represent more complex image microstructures. A multipole filter can be designed for a specific macrostructure, by using appropriate lobe shape configuration. This gives much flexibility for filter design.

To be effective for face recognition or image representation, there are three rules in the development of dissociated multipoles (DMPs):

- Each lobe of a DMP should be a low-pass filter. On one hand, the intensity information within the region of the lobe should be statistically estimated; on the other hand, the image noise should attenuated by low-pass filtering;

- To obtain the locality of the operator, the coefficients of each lobe should be arranged in such a way that the weight of a pixel is inverse proportional to its distance from the lobe center. Gaussian mask satisfies this; there are other choices as well;

- The sum of all lobes' coefficients should be zero, so that the ordinal code of a nonlocal comparison has equal probability of being 1 or 0. Thus the entropy of a single ordinal code is maximized. In the examples shown in Figure 9.4, the sum of two excitatory lobes' weights is equal to the inhibitory lobes' total absolute weights.

In our experiments, we designed 24 disassociated multipole ordinal filters as shown in Figure 9.5. The filter sizes are all 41×41 pixels. The Gaussian parameter is uniformly $\sigma = \pi/2$. The interpole distances are $d = 8, 12, 16, 20$ for the 2-poles

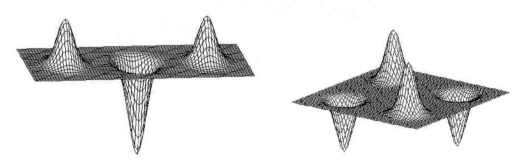

Figure 9.4   Dissociated multipole filters: tri- and quad-pole filters.

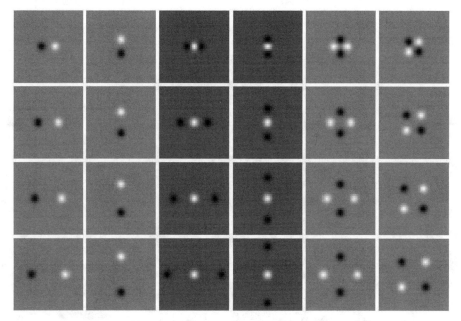

Figure 9.5 The 24 ordinal filters used in the experiments.

and 4-poles, and $d = 4, 8, 12, 16$ for the 3-poles. For 2-poles and 3-poles, the directions are 0 and $\pi/2$; for the 4-poles, the directions are 0 and $\pi/4$. Figure 9.6 shows the correspondingly filtered images of a face.

9.3 Multimodal Biometric System Using Ordinal Features

Figure 9.7 shows the block diagram of the proposed multimodal biometric system based on the fusion of face and palmprint at the matching score level. Firstly, effective face and palmprint ordinal features are extracted for matching. By comparing with the templates stored in the database, the matching scores of each classifier are generated. Then the scores output from the two classifiers are combined using several fusion strategies to give a unique matching score. Finally, a decision about whether to accept or reject a user is made.

9.3.1 Face Recognition

There are a large number of ordinal features generated by various ordinal types and pixel locations; therefore, the initial ordinal feature set is of high dimensionality. However, the intrinsic dimension of the face pattern may not be so high. A further processing is needed to remove the redundancy and build effective classifier. This is done in this work by using the following AdaBoost algorithm [43]:

Input: Sequence of N weighted examples
$$\{(x_1, y_1, w_1), (x_2, y_2, w_2), \ldots, (x_n, y_n, w_n)\};$$
Initial distribution P over the n examples;
Weak learning algorithm WeakLearn;
Integer T specifying number of iterations;

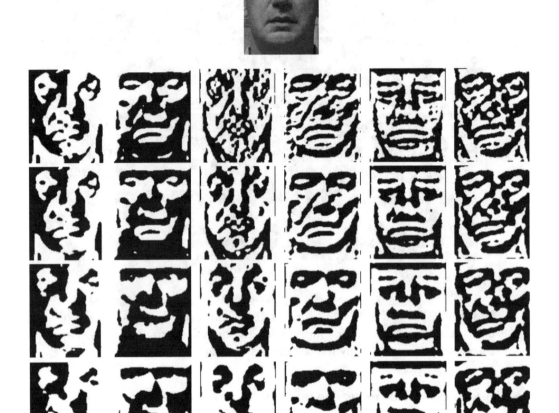

Figure 9.6   The filtered images of a face for the 24 ordinal filters.

Initialize $w_i^1 = P(i)$ for $i = 1, \ldots, n$;
For $t = 1, \ldots, T$:
    1. Set $p_i^t = w_i^t / \sum_i w_i^t$;
    2. Call WeakLearn, providing it with the distribution p;
        get back hypothesis $h_t(x_i) \in \{0, 1\}$ for each x_i;
    3. Calculate the error of $h_t : \epsilon_t = \sum_{i=1}^{N} p_i^t |h_t(x_i) - y_i|$;
    4. Set $\beta_t = \frac{\epsilon_t}{(1-\epsilon_t)}$;
    5. Set the new weights to $w_i^{t+1} = \beta_i^{1-|h_t(x_i)-y_i|}$;
Output the hypothesis

$$H(x) = \begin{cases} 1 & \text{if } \sum_{t=1}^{T} \left(\log \frac{1}{\beta_t}\right) h_t(x) \geq \sum_{t=1}^{T} \left(\log \frac{1}{\beta_t}\right) \\ 0 & \text{otherwise} \end{cases}$$

AdaBoost iteratively learns a sequence of weak hypotheses $h_t(x)$ and linearly combines them with the corresponding learned weights $\log(1/\beta_t)$. Given a data distribution p, AdaBoost assumes that a WeakLearn procedure is available for learning a sequence of most effective weak classifiers $h_t(x)$.

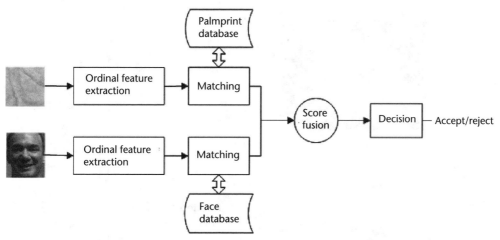

Figure 9.7   Block diagram of face and palmprint multimodal biometric system.

The simplest weak classifier can be constructed for each pixel and each filter type, called single bit weak classifier ($SBWC$). Considering that the $SBWC$ may be unstable to image noise and alignment error, a more involved weak classifier can be designed based on a spatial local subwindow instead of a single bit. The advantage is that some statistic over a local subwindow can be more stable than that at a bit. In this scheme, the Hamming distance can be calculated between the ordinal values in the two corresponding subwindows. The Hamming distance as a weak classifier can be used to make a weak decision for the classification. The use of subwindows gives one more dimension of freedom. A different size leads to a different weak classifier. In our experiments, 20 subwindow sizes are used: 6×6, 12×12, ..., 120×120, and the length of the side is incremented by 6.

9.3.2   Palmprint Recognition

Due to different illumination settings, stretching, and misalignment, the acquired palmprint signal varies significantly. A robust palmprint image representation method is need to ensure the high discriminant power. As analyzed in Section 9.2.1, a possible improvement could be made by choosing well-designed ordinal measures as the palmprint representation. We propose a novel palmprint representation, namely, *orthogonal line ordinal features* (*OLOF*), as illustrated in Figure 9.8, where normalized subimage is referenced by finger gaps using an algorithm similar to Zhang et al. [22]. *OLOF* is so called because the two regions involved in ordinal comparison are elongated or linelike, and the two are geometrically orthogonal.

The ideas are motivated by the most stable and robust ordinal measures available in palmprint pattern, in other words, randomly distributed negative line segments versus their orthogonal regions. In low resolution palmprint images, the line patterns are mainly constituted by principal lines and wrinkles, whose intensity is much lower than their orthogonal regions. Of course detection of all line segments in a palmprint is impossible in real-time applications. Nevertheless, if we apply thousands of ordinal operators onto a palmprint image, most of them correspond to robust ordinal measures.

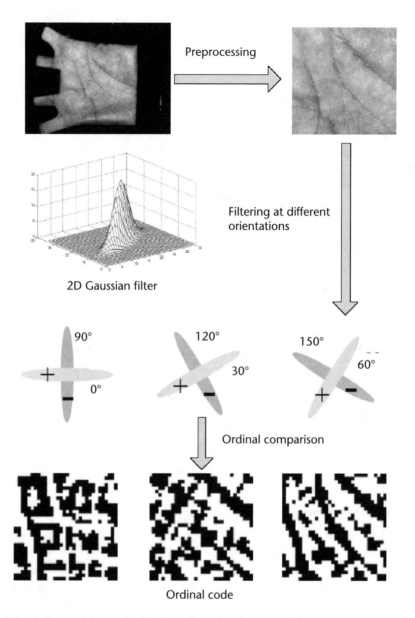

Figure 9.8 Orthogonal line ordinal features for palmprint recognition.

Here we use 2-D Gaussian filter to obtain the weighted average intensity of a linelike region. Its expression is as follows:

$$f(x, y, \theta) = exp\left\{-\frac{1}{2} \times \left[\left(\frac{x\cos\theta + y\sin\theta}{\delta_x}\right)^2 - \left(\frac{-x\sin\theta + y\cos\theta}{\delta_y}\right)^2\right]\right\} \qquad (9.1)$$

where θ denotes the orientation of 2-D Gaussian filter, δ_x denotes the filter's horizontal scale, and δ_y denotes the filter's vertical scale. We control the scale ratio δ_x/δ_y higher than 3 to make its shape like a line (see Figure 9.8).

The orthogonal line ordinal filter, comparing two orthogonal line-like palm-print image regions, is specially designed as follows:

$$OF(\theta) = f(x, y, \theta) - f\left(x, y, \theta + \frac{\pi}{2}\right) \qquad (9.2)$$

For each local region in normalized palmprint image, three ordinal filters, $OF(0)$, $OF(\pi/6)$, and $OF(\pi/3)$, are performed on it to obtain three bit ordinal codes based on the sign of filtering results. Finally, three ordinal templates named as ordinal code are obtained as the feature of the input palmprint image (Figure 9.8). The matching metric is also based on Hamming distance.

9.3.3   Fusion of Face and Palmprint

In the context of biometrics, there are mainly three levels of fusion methods for combining two (or more) biometric systems [12]: (1) fusion at the feature extraction level, where the feature vectors of multiple biometric modalities are concatenated to create a new feature vector to represent the individual, (2) fusion at the matching score level, where the matching scores of multiple classifiers are combined to generate a single scalar score, (3) fusion at the decision level, where each matcher outputs its own Boolean result and the fusion process fuses them together by a combination algorithm such as AND, OR, and so forth.

In our multimodal biometric system, the fusion is performed at the matching score level. Kittler et al. [44] developed a common theoretical framework for combining classifiers and showed that many existing schemes could be considered as special cases of compound classification where all the pattern representations are used jointly to make a decision. The fusion procedure can be formulated as follows [44]. Given s_1 and s_2 (matching score of each classifier), a tester T is assigned to one of the two possible classes ω_1 (genuine) and ω_2 (imposter), or

$$\begin{aligned} assign \quad T \to \omega_j, \quad if \\ P(\omega_j|s_1, s_2) = \max_k P(\omega_k|s_1, s_2), \quad j = 1, 2 \end{aligned} \qquad (9.3)$$

where $P(\omega_j\backslash \text{verts}_1, s_2)$ are the posteriori of classes ω_j.

There are two approaches for consolidating the scores obtained from different matchers. One is to formulate it as a combination problem and use simple fusion rules, such as sum, product, max, and min rule [44], to combine the two matching scores and compare the result to a threshold. The other is to formulate it as a classification problem and treat the matching scores of different biometrics as a feature vector and use linear discriminant analysis (LDA) to classify the vector as being genuine or an impostor.

9.3.3.1   Sum, Product, Max, and Min Rules

In the fusion approach, the individual matching scores are combined to generate a single scalar score, which is then used to make the final decision. Let S_1 and S_2 be the matching scores generated by face and palmprint classifier, respectively.

To ensure a meaningful combination of the scores from different modalities, the scores must be first transformed to a common domain prior to combining them. This is known as score normalization. In our experiments, the following normalization techniques are used: simple min-max, Z-score, and hyperbolic tangent (tanh) [45]. Note that all the following score normalization methods are only for the combination approaches. The definitions are as follows:

- Min-max normalization is used for normalizing each of the matching scores to a scale of 0 to 1,

$$S'_i = \frac{S_i - \min(S_i)}{\max(S_i) - \min(S_i)}, \quad i = 1, 2 \tag{9.4}$$

  where $\min(S_i)$ and $\max(S_i)$ denote the overall minimum and maximum value of S_i;
- Z-Score normalization method transforms the scores to a distribution with mean of 0 and standard deviation of 1.

$$S'_i = \frac{S_i - \mu_i}{\sigma_i}, \quad i = 1, 2 \tag{9.5}$$

  where μ_i and σ_i denote the arithmetic mean and standard deviation, respectively;
- Tanh normalization method is a stable statistical technique [46]. It maps the raw scores to the (0, 1) range:

$$S'_i = \frac{1}{2}\left[\tanh\left(0.01 \frac{S_i - \mu_i}{\sigma_i}\right) + 1\right], \quad i = 1, 2 \tag{9.6}$$

After score normalization, the final fused matching scores can be obtained using the following strategies:

- Sum rule:

$$S = (S'_1 + S'_2)/2 \tag{9.7}$$

- Product rule:

$$S = S'_1 \cdot S'_2 \tag{9.8}$$

- Max rule:

$$S = \max\{S'_1, S'_2\} \tag{9.9}$$

- Min rule:

$$S = \min\{S'_1, S'_2\} \tag{9.10}$$

9.3.3.2 Linear Discriminant Analysis

Linear discriminant analysis (LDA) helps transform the two-dimensional score vectors $\vec{s} = (s_1, s_2)$ into a new subspace that maximizes the between-class distance while minimizing the within-class distance. Consider a set of feature vectors $\vec{s}$ for each sample of an object with known class labels ω_j ($j = 1, 2$). The classification problem is then to find a good predictor of the class ω_j for any observation $\vec{s}$.

LDA is known to be Bayes optimal if the probability density functions $p(\vec{s}|\omega_1)$ and $p(\vec{s}|\omega_2)$ are both normally distributed with equal covariance matrix. Using LDA, one can measure the probability $p(\omega_j|\vec{s})$ by calculating the distance between the observation and the class prototype in a discriminant subspace induced by the following projection [47].

$$\vec{W} = \Sigma^{-1}(m_1 - m_2) \tag{9.11}$$

$$\vec{m}_i = \frac{1}{n_i} \sum_{\vec{s} \in D_i} \vec{s} \tag{9.12}$$

$$\Sigma = \sum_{i=1,2} \sum_{\vec{s} \in D_i} (\vec{s} - \vec{m}_i)(\vec{s} - \vec{m}_i)^T \tag{9.13}$$

where Σ and $\vec{m}_i$ denote the covariance and mean of the input feature vector, respectively; and D_1 and D_2 denote the set of samples from the genuine and impostor classes, respectively.

9.4 Experiments

9.4.1 Data Description

We evaluate the proposed multimodal system on a data set including 7,560 pairs of images from 378 subjects. The face images are collected from FRGC 2.0 Experiment 1 face database [48], which contains 535 subjects and about 28,804 still images. The number of images for each subject varies from 4 to 152, coming from several sessions. We selected 7,560 images from 378 different subjects with 20 samples for each. All these face images are normalized to 142×120 and preprocessed using the method [49]. These face images were mainly subject to illumination, poses, and facial expressions. Examples of typical face images are shown in Figure 9.9.

The palmprint images are collected from PolyU Palmprint Database [50], which contains 7,752 images corresponding to 386 different palms. There are around 20 samples in each of these palms. We selected 7,560 images from 378

Figure 9.9 Sample face images from the FRGC2.0 database.

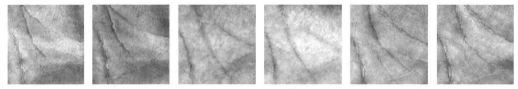

Figure 9.10   Sample palmprint images from PolyU Palmprint Database.

different palms, where 20 samples for each were collected in two sessions, in other words, 10 samples were captured in each session respectively. The average interval between those two collections was two months. After preprocessing, the input palmprint image is normalized to 128×128. The variations in the palmprint images are mainly due to illumination, stretching, and misalignment. Some palmprint images are shown in Figure 9.10.

Though face images in FRGC 2.0 database do not correspond to the palmprint images, Ross et al. [12] showed that biometric modalities were mutually independent. This allows us to randomly pair face and palmprint to obtain a multimodal image set for each of these 378 subjects. Each pair contains 20 face images and 20 palmprint images. In our experiments, we divide the images into two partitions. The first 3,780 pairs of images, including 189 subjects and 20 pairs of images for each, are used for training, and the rest are used for testing.

9.4.2   Experimental Results and Evaluation

As for palmprint matching, a training procedure is actually not required. Each palmprint image was directly used to obtain a characteristic, for example, ordinal feature vector of size being 384 bytes. For face matching, each of the images was filtered by 24 ordinal filters. Local subwindows of ordinal features are used to construct weak classifier based on Hamming distance for AdaBoost Learning. Finally, a strong classifier, consisting of 2,014 weak classifiers, is obtained. For illustration, the first five learned weak classifiers are shown in Figure 9.11.

In both training and testing set, there are totally 35,910 intraclass (genuine) samples and 7,106,400 extra-class (impostor) samples for each. The genuine and impostor matching scores from the training set were used to learn the LDA classifier. Then we test the performance of face, palmprint, and different types of fusion classifiers on the testing set. For comparison, the holistic feature based (PCA-based here) method is also evaluated using the same set of training and test images, where eigenface and eigen-palm features are used for face and palmprint representation,

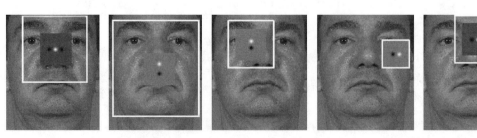

Figure 9.11   The first five features and associated subwindow sizes selected by AdaBoost learning.

respectively. Figure 9.12 and Figure 9.13 show the ROC curves derived from the scores for the intra- and extra-class pairs.

The ROC curves suggest that the multimodal biometrics can offer substantial performance gain. In addition, as we analyzed, local features, for instance ordinal features here, are experimentally illustrated to be superior to the holistic features, such as PCA here. Moreover, they are not only good for improving

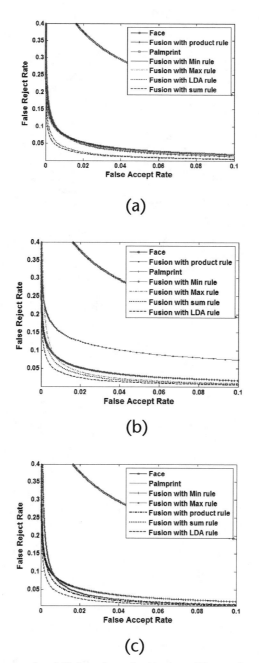

Figure 9.12 ROC curves for PCA-based method using different fusion strategies and score normalization methods ((a) min-max; (b) Z-score; and (c) tanh).

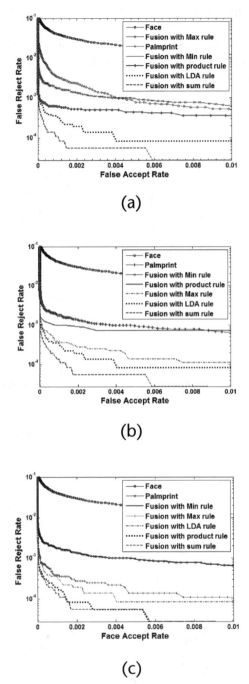

(a)

(b)

(c)

Figure 9.13   ROC curves for the proposed method in this chapter using different fusion strategies and score normalization methods ((a) min-max; (b) Z-score; and (c) tanh).

the performance of face and palmprint classifiers, but they are also good for the fusion schemes.

As shown in Figure 9.13, only the max rule performs slightly worse than a single palmprint classifier when using min-max score normalization method.

In particular, it is found that fusion with sum rule method has the highest accuracy for all score normalization techniques. The performance of LDA classifier is similar to that of the sum rule-based one. The performance of product rule is not stable if different score normalization methods are utilized.

For further comparison with different fusion methods, two measurements are employed. One is the equal error rate (EER) described by a point in ROC when false accept rate is equal to false reject rate; and another is the d' (d-prime) [51], which is shown in Table 9.1, where d' is a statistical measure of how well a biometric system can discriminate between different individuals. The definition is

$$d' = \frac{|m_1 - m_2|}{\sqrt{(\delta_1^2 + \delta_2^2)/2}} \tag{9.14}$$

where m_1 and δ_1 denote the mean and variance of intraclass feature vector, respectively, while m_2 and δ_2 denote the mean and variance of extra-class feature vector, respectively. The larger the d' value is, the better a biometric system performs at discriminating between individuals.

As shown in Table 9.1, the EER of PCA-based methods are all significantly larger than that of our proposed method, while the d' values of PCA-based method are all significantly smaller than that of our proposed method, which also indicates the superiority of the proposed method.

For the proposed method, we can see from Table 9.1 that sum rule–based method achieves better performance (in bold font). A similar conclusion can be drawn for LDA. For other fusion rules, we see that the product rule is good for improvement of EER, which is significant, but the discriminating index drops a little when min-max or Z-score normalization is employed, while for the max rule all indices are better than that of a single face or palmprint classifier when Z-score or tanh normalization is utilized.

Table 9.1   Comparison of Accuracy Measures for Different Classifiers

Algorithm	Score Normalization Method	Proposed Method		PCA-Based Method	
		EER (%)	d'	EER (%)	d'
Face	—	1.160	4.4033	14.88	1.9719
Palmprint	—	0.160	6.0813	3.66	3.9704
Fusion with sum rule	Min-max	**0.028**	**7.3916**	**2.03**	4.1151
	Z-score	0.032	7.3645	2.65	3.6728
	Tanh	0.032	7.3723	2.70	3.6741
Fusion with product rule	Min-max	0.067	5.7714	3.27	3.1131
	Z-score	0.091	4.0360	8.19	1.9563
	Tanh	0.032	7.2822	2.72	3.6484
Fusion with max rule	Min-max	0.220	5.4496	3.45	**4.1460**
	Z-score	0.056	6.2805	2.98	3.5837
	Tanh	0.056	6.2903	2.96	3.5860
Fusion with min rule	Min-max	0.150	6.7518	3.66	2.0724
	Z-score	0.160	5.5558	3.66	2.7978
	Tanh	0.150	5.5598	3.67	2.7981
Fusion with LDA	—	0.044	7.1244	2.18	3.9260

9.5 Conclusions

We have presented a multimodal biometric identification system in fusion of palmprint and face, which takes advantage of ordinal measure for palmprint and face representations. Our work has shown that well-designed ordinal features can be powerful enough for complex tasks such as personal identification, and this has countered the existing view that ordinal features are believed to be only suited for face detection and too weak for fine discrimination tasks [34].

In addition, we have investigated the multimodal fusion methods at matching score level, and examined the multimodal biometric system on a middle-scale population (378 subjects, 20 pairs each). The experimental results show that the proposed system significantly improves the performance of identification systems. Furthermore, our score-level fusion experiments show that sum rule and LDA generally perform better than min rule and max rule. Thanks to the high accuracy and low computation cost, the best combination is a simple sum rule with min-max score normalization.

Acknowledgments

This work was supported by the following funds: Chinese National Natural Science Foundation Project #60518002, Chinese National 863 Program Projects #2006AA01Z192 and #2006AA01Z193, Chinese National Science and Technology Supporting Platform Project #2006BAK08B06, and Chinese Academy of Sciences 100 People Project.

References

[1] Jain, A., R. Bolle, and S. Pankanti, *Biometrics: Personal Identification in Networked Society*, Boston, MA: Kluwer, 1999.

[2] Ross, A., and A. Jain, "Multimodal Biometrics: An Overview," *Proc. of 12th European Signal Processing Conference (EUSIPCO)*, 2004, pp. 1221–1224.

[3] Brunelli, R., and D. Falavigna, "Person Identification Using Multiple Cues," *IEEE Trans. on Pattern Analysis and Machine Intelligence*, Vol. 17, 1995, pp. 955–966.

[4] Kittler, J., et al., "Combining Evidence in Personal Identity Verification Systems," *Pattern Recognition Letters*, Vol. 18, 1997, pp. 845–852.

[5] Hong, L., and A. Jain, "Integrating Faces and Fingerprints for Personal Identification," *IEEE Trans. on Pattern Analysis and Machine Intelligence*, Vol. 20, 1998, pp. 1295–1307.

[6] Wang, Y., T. Tan, and A. Jain, "Combining Face and Iris Biometrics for Identity Verification," *Proc. of Intl. Conf. on Audio- and Video-Based Person Authentication*, 2003, pp. 805–813.

[7] Kumar, A., et al., "Personal Verification Using Palmprint and Hand Geometry Biometric," *Proc. of Intl. Conf. on Audio- and Video-Based Person Authentication*, 2003, pp. 668–678.

[8] Kumar, A., and D. Zhang, "Integrating Palmprint with Face for User Authentication," *Proc. Multi Modal User Authentication Workshop*, 2003, pp. 107–112.

[9] Feng, G., et al., "When Faces Are Combined with Palmprints: A Novel Biometric Fusion Strategy," *Proc. of International Conference on Biometric Authentication*, 2004, pp. 701–707.

[10]   Ribaric, S., and I. Fratric., "A Biometric Identification System Based on Eigenpalm and Eigenfinger Features," *IEEE Trans. on Pattern Analysis and Machine Intelligence*, Vol. 27, 2005, pp. 1698–1709.

[11]   Chang, K., et al., "Comparison and Combination of Ear and Face Images in Appearance Based Biometrics," *IEEE Trans. on Pattern Analysis and Machine Intelligence*, Vol. 25, 2003, pp. 1160–1165.

[12]   Ross, A., A. Jain, and J. Qian, "Information Fusion in Biometrics," *Pattern Recognition Letters*, Vol. 24, 2003, pp. 2115–2125.

[13]   Turk, M., and A. Pentland, "Eigenfaces for Recognition," *Journal of Cognitive Neuroscience*, Vol. 3, No. 1, March 1991, pp. 71–86.

[14]   Bartlett, M., H. Lades, and T. Sejnowski, "Independent Component Representations for Face Recognition," *Proc. of the SPIE, Conference on Human Vision and Electronic Imaging III*, Vol. 3299, 1998, pp. 528–539.

[15]   Zhang, D., and W. Shu, "Two Novel Characteristics in Palmprint Verification: Datum Point Invariance and Line Feature Matching," *Pattern Recognition*, Vol. 32, 1999, pp. 691–702.

[16]   Duta, N., A. Jain, and K. Mardia, "Matching of Palmprint," *Pattern Recognition Letters*, Vol. 23, 2001, pp. 477–485.

[17]   Li, W., D. Zhang, and Z. Xu, "Palmprint Identification by Fourier Transform," *International Journal of Pattern Recognition and Artificial Intelligence*, Vol. 16, 2002, pp. 417–432.

[18]   Lu, G., D. Zhang, and K. Wang, "Palmprint Recognition Using Eigenpalms Features," *Pattern Recognition Letters*, Vol. 24, 2003, pp. 1463–1467.

[19]   Han, C., et al., "Personal Authentication Using Palmprint Features," *Pattern Recognition*, Vol. 36, 2003, pp. 371–381.

[20]   You, J., et al., "On Hierarchical Palmprint Coding with Multi-Features for Personal Identification in Large Databases," *IEEE Trans. on Circuits and Systems for Video Technology*, Vol. 14, 2004, pp. 234–243.

[21]   Zhang, L., and D. Zhang, "Characterization of Palmprints by Wavelet Signatures via Directional Context Modeling," *IEEE Trans. on Systems, Man, and Cybernetics, Part B*, Vol. 34, 2004, pp. 1335–1347.

[22]   Zhang, D., et al., "On-Line Palmprint Identification," *IEEE Trans. on Pattern Analysis and Machine Intelligence*, Vol. 25, 2003, pp. 1041–1050.

[23]   Kong, W., and D. Zhang, "Feature-Level Fusion for Effective Palmprint Authentication," *Proc. of International Conference on Biometric Authentication*, 2004, pp. 761–767.

[24]   Kong, W., and D. Zhang, "Competitive Coding Scheme for Palmprint Verification," *Proc. of International Conference Pattern Recognition*, Vol. 1, 2004, pp. 520–523.

[25]   Penev, P., and J. Atick, "Local Feature Analysis: A General Statistical Theory for Object Representation," *Neural Systems*, Vol. 7, No. 3, 1996, pp. 477–500.

[26]   Lades, M., et al., "Distortion Invariant Object Recognition in the Dynamic Link Architecture," *IEEE Trans. on Computers*, Vol. 42, 1993, pp. 300–311.

[27]   Wiskott, L., et al., "Face Recognition by Elastic Bunch Graph Matching," *IEEE Trans. on Pattern Analysis and Machine Intelligence*, Vol. 19, No. 7, 1997, pp. 775–779.

[28]   Liu, C., and H. Wechsler. "Gabor Feature Based Classification Using the Enhanced Fisher Linear Discriminant Model for Face Recognition," *IEEE Trans. on Image Processing*, Vol. 11, No. 4, 2002, pp. 467–476.

[29]   Ahonen, T., A. Hadid, and M. Pietikainen, "Face Recognition with Local Binary Patterns," *Proc. of the European Conference on Computer Vision*, Prague, Czechoslovakia, 2004, pp. 469–481.

[30]   Ullman, S., M. Vidal-Naquet, and E. Sali, "Visual Features of Intermediate Complexity and Their Use in Classification," *Nature Neuroscience*, Vol. 5, No. 7, 2002.

[31]   Naquet, M., and S. Ullman, "Object Recognition with Informative Features and Linear Classification," *Proc. of IEEE International Conference on Computer Vision*, Nice, France, 2003.

[32]   Sadr, J., et al., "Toward the Fidelity of Local Ordinal Encoding," *Proc. of the Fifteenth Annual Conference on Neural Information Processing Systems*, Vancouver, British Columbia, Canada, December 3–8, 2001.

[33]   Sinha, P., "Toward Qualitative Representations for Recognition," *Proc. of the Second International Workshop on Biologically Motivated Computer Vision*, Tubingen, Germany, November 22–24, 2002, pp. 249–262.

[34]   Thoresz, K., "On Qualitative Representations for Recognition," Master's thesis, Massachusetts Institute of Technology, Cambridge, MA, July 2002.

[35]   Schneiderman, H., "Toward Feature-Centric Evaluation for Efficient Cascaded Object Detection," *Proc. of IEEE Computer Society Conference on Computer Vision and Pattern Recognition*, Washington, D.C., June 27–July 2, 2004, pp. 1007–1013.

[36]   Lipson, P., E. Grimson, and P. Sinha, "Toward Configuration Based Scene Classification and Image Indexing," *Proc. of IEEE Computer Society Conference on Computer Vision and Pattern Recognition*, San Juan, Puerto Rico, June 17–19, 1997, pp. 1007–1013.

[37]   Bhat, D., and S. Nayar, "On Ordinal Measures for Image Correspondence," *IEEE Trans. on Pattern Analysis and Machine Intelligence*, Vol. 20, No. 4, April 1998, pp. 415–423.

[38]   Partio, M., B. Cramariuc, and M. Gabbouj, "Toward Texture Similarity Evaluation Using Ordinal Co-Occurrence," *Proc. of IEEE International Conference on Image Processing*, Singapore, October 24–27, 2004, pp. 1537–1540.

[39]   Sun, Z., et al., "Ordinal Palmprint Represention for Personal Identification," *Proc. of IEEE Computer Society Conference on Computer Vision and Pattern Recognition*, June 20–25, 2005, pp. 279–284.

[40]   Liao, S., et al., "Face Recognition Using Ordinal Features," *Proc. of IAPR International Conference on Biometric*, Hong Kong, January 2006, pp. 40–46.

[41]   Sun, Z., T. Tan, and Y. Wang, "Toward Robust Encoding of Local Ordinal Measures: A General Framework of Iris Recognition," *Proc. of International Workshop on Biometric Authentication*, Prague, Czech Republic, May 2004, pp. 270–282.

[42]   Balas, B., and P. Sinha, "Toward Dissociated Dipoles: Image Representation Via Nonlocal Comparisons," CBCL Paper #229/AI Memo #2003–018, Massachusetts Institute of Technology Computer Science and Artificial Intelligence Laboratory, Cambridge, MA, August 2003.

[43]   Freund, Y., and R. Schapire, "A Decision-Theoretic Generalization of On-Line Learning and an Application to Boosting," *Journal of Computer and System Sciences*, Vol. 55, No. 1, August 1997, pp. 119–139.

[44]   Kittler, J., et al., "On Combining Classifiers," *IEEE Trans. on Pattern Analysis and Machine Intelligence*, Vol. 20, No. 3, January 1998, pp. 226–239.

[45]   Samoska, N., "Evaluation and Performance Prediction of Multimodal Biometric Systems," Master's thesis, West Virginia University, 2006.

[46]   Huber, P., *Robust Statistics*, New York: John Wiley and Sons, 1981.

[47]   Duda, R., P. Hart, and D. Stork, *Pattern Classification*, 2nd ed., New York: John Wiley and Sons, 2000.

[48]   Phillips, P., et al., "Overview of the Face Recognition Grand Challenge," *Proc. of IEEE Computer Society Conference on Computer Vision and Pattern Recognition*, 2005.

[49]   Gross, R., and V. Brajovic, "An Image Preprocessing Algorithm for Illumination Invariant Face Recognition," *Proc. 4th International Conference on Audio- and Video-Based Biometric Person Authentication*, Guildford, U.K., June 9–11, 2003, pp. 10–18.

[50]   PolyU Palmprint Database, http://www.comp.polyu.edu.hk/biometrics/.

[51]   Daugman, J., and G. Williams, "A Proposed Standard for Biometric Decidability," *Proc. CardTech/SecureTech Conference*, 1996, pp. 223–234.

CHAPTER 10

Human Identification Using Gait and Face

Amit Kale, Amit K. Roy Chowdhury, and Rama Chellappa

Recognition of humans from arbitrary viewpoints is an important requirement for different applications such as intelligent environments, surveillance, and access control. For optimal performance, the system must use as many cues as possible from appropriate vantage points and fuse them in meaningful ways. In this chapter we discuss fusion of face and gait cues from a monocular video. We use a view invariant gait recognition algorithm for gait recognition and a face recognition algorithm using particle filters. We employ decision fusion to combine the results of our gait and face recognition algorithms. We consider two fusion scenarios: hierarchical and holistic. The first employs the gait recognition algorithm when the person is far away from the camera and passes the top few candidates to the face recognition algorithm. The second approach involves combining the similarity scores obtained individually from the face and gait recognition algorithms. Simple rules like the Sum, Min, and Product are used for combining the scores. The results of fusion experiments are demonstrated on the NIST database, which has outdoor gait and face data of 30 subjects.

10.1 Introduction

Identification of humans from arbitrary viewpoints is an important requirement for different tasks, including perceptual interfaces for intelligent environments, covert security, and access control. Different modalities can be used for identification based on the number of pixels on the individual. If the person is far away from the camera, it is hard to get face information at a high enough resolution for recognition tasks. However, when available, it yields a very powerful cue for recognition. A modality that can be detected and measured when the subject is far away from the camera is human gait, or the style of walking. For optimal performance, the system must use as many cues as possible and combine them in meaningful ways. Information may be fused in two ways. The data available may be fused and a decision can be made based on the fused data (data fusion), or each signal/feature can be matched separately, using possibly different techniques, and the decisions made may be fused (decision fusion).

The gait of a person is best reflected when he or she presents a side view (referred to in this chapter as a canonical view) to the camera. Hence, most gait recognition algorithms rely on the availability of the side view of the subject. For face recognition, on the other hand, it is desirable to have frontal views of the person's face. The most general solution to perform integrated face and gait recognition from arbitrary views would be to estimate 3-D models for face and gait. While there has been some progress in building 3-D models for faces, a recent work being [1], the problem of building reliable 3-D models for articulating objects like the human body still remains a hard problem. One way to exploit current recognition algorithms for frontal face and side gait without resorting to 3-D models is to synthesize canonical views, given arbitrary views of the person. In [2], Shakhnarovich et al. compute an image-based visual hull from a set of monocular views, which is then used to render virtual canonical views for tracking and recognition. Gait recognition is achieved by matching a set of image features based on moments extracted from the silhouettes of the synthesized probe video to the gallery. The visual hull is also used to render frontal face images. Eigenfaces [3] are used for face recognition. In a later work, Shakhnarovich and Darrell [4] studied the fusion of face and gait cues for this multicamera indoor environment. Zhou et al. [5] present an approach to fusion of gait and face that uses side views of noncooperating subjects, and that rely on side views of face, making use of PCA and MDA for gait and face features and combining them at the score level.

In general, the visual-hull approach for performing integrated face and gait recognition requires at least two cameras. In this chapter we present experimental results for fusion of face and gait for the single camera case. We considered the NIST database, which contains outdoor face and gait data for 30 subjects. In the NIST database, subjects walk along an inverted Σ pattern (see Figure 10.1). In one segment of the NIST database, the subjects walk at an angle to the exact side view [in Figure 10.1(b)]. In the last segment the person provides a nearly frontal view of his face to the camera [in Figure 10.1(c)]. This final segment can be used for face recognition. In [6], we presented a view-invariant gait recognition algorithm for the single camera case, along with some experimental evaluations. In this chapter we present the results of our view-invariant gait recognition algorithm in [6] on the

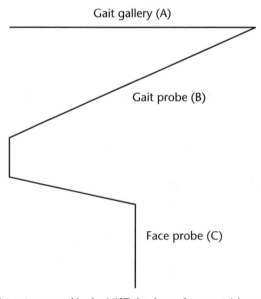

Figure 10.1 Inverted Σ pattern used in the NIST database. Segment A is used as the gallery for gait recognition, while segment B (which is at a 330° angle to the horizontal part) is used as a probe. The last part of the sequence, where the person presents a front view to the camera (segment C), was used as the probe for still-to-video face recognition using [7], the gallery consisting of static faces for the 30 subjects.

NIST database. The algorithm is based on the planar approximation of the person, which is valid when the person walks far away from the camera. In [7], an algorithm for probabilistic recognition of human faces from video was proposed, and the results were demonstrated on the NIST database. We employ decision fusion, which is a special case of data fusion (see [8]) to combine the results of our gait recognition algorithm and the face recognition algorithm. We consider two fusion scenarios: hierarchical and holistic. The first involves using the gait recognition algorithm as a filter to pass on a smaller set of candidates to the face recognition algorithm. The second involves combining the similarity scores obtained individually from the face and gait recognition algorithms. Simple rules like the Sum and Product are used for combinining the scores.

10.2   Framework for View-Invariant Gait Recognition

The imaging setup is shown in Figure 10.2. We assume that the person walks with a translational velocity $\mathbf{V} = [v_X, 0, v_Z]^T$ along the line AC at an angle θ to the canonical direction AB. Assuming that we can find the location (x_{ref}, y_{ref}) of the person's head at the start of such a segment, a sequential Monte Carlo particle filter [9] is used to track the head of the person to get $\{(x^i(t), y^i(t)), w^i(t)\}$ where the superscript denotes the index of the particle and $\omega^i(t)$ denotes the probability weight for the estimate $(x^i(t), y^i(t))$. Assuming constant velocity models and small motion between successive frames, we can show using the optical flow–based SfM equations that the angle traced by centroid of the person's head α is related to θ by:

$$\cot(\alpha)(x_{ref}, y_{ref}) = \frac{x_{ref} - f\cot(\theta)}{y_{ref}} \tag{10.1}$$

Thus, given f and (x_0, y_0), we can compute θ. Knowing (x_0, y_0), $\cot(\alpha)$ and θ, f can be computed as part of a calibration procedure. For a derivation of (10.1), see Appendix 10A.

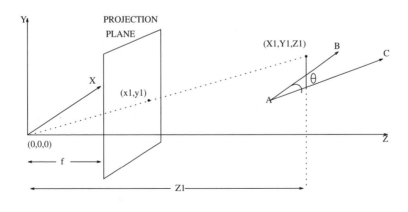

Figure 10.2   Imaging geometry.

Having obtained the angle θ, we need to synthesize the canonical view. Let Z denote the distance of the object from the image plane. If the dimensions of the object are small compared to Z, then the variation in θ, $d\theta \approx 0$. This essentially corresponds to assuming a planar approximation to the object. Let $[X_\theta, Y_\theta, Z_\theta]'$ denote the coordinates of any point on the person (as shown in Figure 10.2) who is walking at an angle $\theta \geq 0$ to the plane passing through the starting point $[X_{ref} Y_{ref} Z_{ref}]'$ and parallel to the image plane, which we shall refer to, hereafter, as the canonical plane. Computing the 3-D coordinates of the synthesized point involve a rotation around the line passing through the starting point and taking the perspective projection. We can show (see Appendix 10A) that

$$x_0 = f \frac{x_\theta \cos(\theta) + x_{ref}(1 - \cos(\theta))}{-\sin(\theta)(x_\theta + x_{ref}) + f}$$

$$y_0 = f \frac{y_\theta}{-\sin(\theta)(x_\theta + x_{ref}) + f} \qquad (10.2)$$

where $x = f\frac{X}{z}$ and $y = f\frac{Y}{z}$. Equation (10.2) is attractive since it does not involve the 3-D depth; rather, it is a direct transformation of the 2-D image plane coordinates in the noncanonical view to get the image plane coordinates in the canonical one. Thus, knowing the azimuth angle θ, we can obtain a synthetic canonical view using (10.2). Novel view synthesis as described above corrects for the distortion of appearance-based features, including height and leg-swing [6].

Given $\{\theta^i(t), w^i(t)\}$ derived using the particle filter to obtain α and (10.1), we use the MAP estimate of θ for synthesizing the corresponding probe sequence. Keeping in mind the limited training data available and dealing with a different number of frames in the gallery and probe, a template-matching technique based on dynamic time warping [10] is used for recognition. In order to assess the utility of our method without being affected by the choice of a particular image feature, we choose to use the entire image as the feature with binary correlation as a local distance measure. The gait recognition algorithm yields a score that is the cumulative binary correlation distance between the probe video sequence and the gallery.

10.3   Face Recognition from Video

Face recognition has been an active area of research for a long time. Most of the work in face recognition has focused on still-to-still situations. Often the probe consists of a video sequence of the unknown subject while the gallery contains static images of the subjects. Different strategies have been developed for this still-to-video scenario. Most approaches involve detecting the face and tracking it over time; when the frame becomes large enough, recognition is performed using still-to-still recognition approaches. Zhou et al. [7] argue against such an approach since it has unresolved issues; such as criteria for selecting good frames, estimation of parameters for registration, and they present a tracking-and-recognition approach that attempts to resolve uncertainties in tracking and recognition simultaneously in a unified probabilistic framework. A time series model is used to fuse temporal information in a probe video, which simultaneously characterizes the kinematics and identity using a motion vector and an identity variable. The joint posterior density of the motion vector and the identity variable is estimated at every time instant and then propagated to the next time instant. Marginalization over the motion vector yields a robust estimate of the posterior distribution of the identity variable. A computationally efficient sequential importance sampling algorithm is used to estimate the posterior distribution. It was shown that a degeneracy in the posterior probability of the identity variable occurs, leading to improved recognition. The algorithm yields the score in terms of a posterior probability $P(i/X)$ where i denotes a gallery person and X denotes the probe video.

10.4 Fusion Strategies

The ultimate goal of designing pattern recognition systems is to achieve the best possible classification performance for the task at hand. As discussed in [11], fusion of multiple sources of evidence is likely to yield tangible benefits in terms of improved efficiency and accuracy of the identification system. Two different approaches exist for fusion. The first employs multiple experts to provide opinions of the same biometric data. The second involves using different modalities from the input. We focus on this approach for fusion.

To improve efficiency of a multimodal biometric system, one can adopt multistage combination rules whereby subjects may be coarsely classified by a less accurate classifier, passing a smaller set of likely candidates to a more accurate classifier. The results of the gait classifier, for example, can be used to pass a smaller number of candidates to the more accurate face recognition unit. Alternatively, decisions from the different classifiers can be combined directly using simple rules such as sum, product and so forth. In this case it is first necessary to transform the scores obtained from the different classifiers in order to make them comptible. The transformation should be such that the relative ordering of the scores is not altered. In other words, the transformation function should be monotone. Some of the commonly used transformations include linear, logarithmic, exponential, and logistic. The purpose of these transformations is, first, to map the scores to the same range of values; and, second, to change the distribution of the scores. For example, the logarithmic transformation puts strong emphasis on the top ranks, whereas the lower ranked scores that are transformed to very high values have a quickly decreasing influence. A detailed discussion of score transformation is given in [12] in the context of combining classifiers for face recognition. The face recognition algorithm yields a match score, which is a probability, while the gait recognition algorithm yields a distance measurement. In order to make the scores comparable before fusing them, we apply an appropriate transformation to the gait scores discussed in the experimental section. Note that the score transformation is necessary only when the scores of the face and gait recognition algorithms are to be directly combined. In our experiments we describe the results of the two fusion strategies for face and gait cues.

10.5   Experimental Results

The NIST database consists of 30 people walking along an inverted Σ-shaped walking pattern as shown in Figure 10.1. The segment A is used as the gallery for gait recognition, while the segment B (which is at an angle 33° to the horizontal part) is used as a probe. As explained in Section 10.2, the person's head is tracked using a particle filter. Using (10.1), $\theta \sim (\theta^i(t), w^i(t))$ is obtained. Using $\tilde{\theta}(t) = $ arg $\max_{w^j}(\theta^i(t), w^j(t))$, the image of the unknown person $X(t)$ is synthesized using 10.2. A few images from the NIST database are shown in Figure 10.3. The gait recognition result is shown in Figure 10.4(a) and 10.4(d). The last part of the sequence, where the person presents a front view to the camera (segment C), was used as the probe for still-to-video face recognition using [7], the gallery consisting

(a)

(b)

(c)

Figure 10.3 Examples for the NIST database: (a) gallery images of person walking parallel to the camera; (b) unnormalized images of person walking at 33° to the camera; and (c) synthesized images for (b).

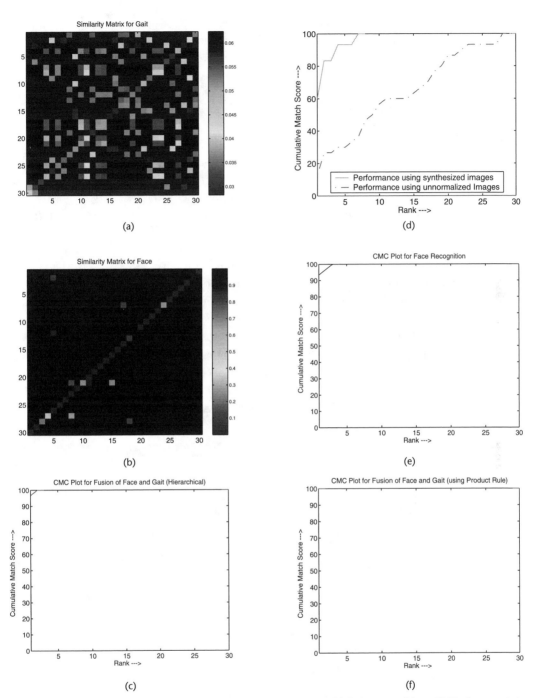

Figure 10.4   Similarity matrices for (a) gait recognition, (b) face recognition; CMC characteristics for (d) gait and (e) face; and CMC curves for (c) hierarchical and (f) holistic fusion.

of static faces for the 30 subjects. The results for face recognition are shown in Figures 10.4(b) and 10.4(e).

We now present the results of the fusion of face and gait cues. As mentioned before, in order to combine the scores from the face and gait classifiers directly, it is necessary to make them comparable. We used the exponential transformation for converting the scores obtained from the gait recognition viz. given that the match score for a probe X from the gallery gaits is given by $S_{X1}, \ldots, S_{XN}$, we obtain the transformed scores $\exp(-S_{X1}), \ldots, \exp(-S_{XN})$. Finally, we normalize the transformed scores to sum to unity. We also tried logistic and logarithmic score transformation methods. The results obtained using these were comparable to the exponential case.

Hierarchical Fusion: For a given unknown person X, we rank order the scores in descending order $RS_1, \ldots, RS_N$. Following the normalization, we note that scores that are very low get mapped to a constant equal value. This value may be different for different people. Hence we consider the difference between successive RS_is in order to come up with a threshold. We compute the differences $D_1, \ldots, D_{N-1}$ where $D_i = RS_i - RS_{i-1}$. Finally, we consider the sum of D_is over a sliding window of size 3, $SD_i = \sum_{j=i}^{\max(i-3,1)} D_j$. Hierarchical fusion involves choosing a threshold T and passing only the candidates i, for which $SD_i > T$ to the face recognition unit. Although it is tempting to choose T to be as high as possible, it should be noted that due to the lower accuracy of gait recognition, this may lead to the true person not being in the set of individuals passed to the face recognition algorithm. The plot of SD as a function of rank is shown in Figure 10.5 when X is chosen as each of the 30 individuals. Note that for a given T the number of candidates passed on to the face recognition algorithm varies depending upon the identity of X. For the NIST database we chose $T = 0.015$, and the number of candidates passed on to the face recognition algorithm varied from four to eight.

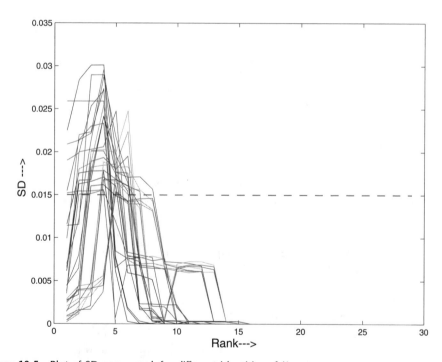

Figure 10.5 Plot of SD versus rank for different identities of X.

On an average, the top six matches from the gait recognition algorithm were passed to the face recognition algorithm. The CMC plot for the resulting hierarchical fusion is shown in Figure 10.4 (c). Note that the top match performance has gone up to 97% from 93% for this case. The more important gain, however, is in terms of the number of computations required. This number drops to one-fifth of its previous value. This demonstrates the value of gait as a filter.

Holistic Fusion: If the requirement from the fusion is that of accuracy as against computational speed, alternate fusion strategies can be employed. Assuming that gait and face can be considered to be independent cues, a simple way of combining the scores is to use the sum or product rule [11]. Both the strategies were tried. The CMC curve for either case is as shown in Figure 10.4(f). In both cases the recognition rate is 100%.

10.6   Conclusion

In this chapter, we presented experimental results for fusion of face and gait for the single camera case. We considered the NIST database, which contains outdoor face and gait data for 30 subjects. We used the method described in [6] for gait recognition and the method described in [7] for face recognition. Decision fusion, which is a special case of data fusion, was used to combine the results of the face and gait recognition algorithms. We demonstrated the use of gait as a filter in building more efficient multimodal biometric systems. We also showed the results of directly combining the scores obtained by the individual algorithms to improve the overall recognition rates.

Acknowledgments

The work reported in this chapter was supported by the DARPA/ONR grant N00014-00-1-0908.

References

[1]   Chowdhury, A., and R. Chellappa, "Face Reconstruction from Video Using Uncertainty Analysis and a Generic Model," *CVIU*, Vol. 91, No. 1–2, July–August 2003, pp. 188–213.

[2]   Shakhnarovich, G., L. Lee, and T. Darrell, "Integrated Face and Gait Recognition from Multiple Views," *Proc. of IEEE Conference on Computer Vision and Pattern Recognition*, December 2001.

[3]   Turk, M., and A. Pentland, "Eigenfaces for Recognition," *Journal of Cognitive Neuroscience*, Vol. 3, No. 1, 1991, pp. 71–86.

[4]   Shakhnarovich, G., and T. Darrell, "On Probabilistic Combination of Face and Gait Cues for Identification," *Proc. of the IEEE International Conference on Face and Gesture*, 2002.

[5]   Zhou, X., B. Bhanu, and J. Han, "Human Recognition at a Distance in Video by Integrating Face Profile and Gait," *Proc. of AVBPA*, 2005, pp. 533–543.

[6]   Kale, A., A. Chowdhury, and R.Chellappa, "Towards a View Invariant Gait Recognition Algorithm," *Proc. of IEEE AVSS*, 2003, pp. 143–150.

[7]   Zhou, S., V. Krueger, and R. Chellappa, "Probabilistic Recognition of Human Faces from Video," *CVIU*, Vol. 91, 2003, pp. 214–245.

[8]   Kokar, M., and J. Tomasik, "Data vs. Decision Fusion in the Category Theory Framework," *FUSION 2001*, 2001.

[9]   Isard, M., and A. Blake, "Contour Tracking by Stochastic Propagation of Conditional Density," *Proc. of ECCV*, No. 1, 1996, pp. 343–356.

[10]   Kale, A., et al., "Gait Analysis for Human Identification," *Proc. of AVBPA*, June 2003.

[11]   Kittler, J., et al., "On Combining Classifiers," *IEEE Trans. on Pattern Analysis and Machine Intelligence*, March 1998, pp. 226–239.

[12]   Achermann, B., and H. Bunke, "Combination of Classifiers on the Decision Level for Face Recognition," Technical Report, Institut fur Informatik und angewandte, Mathematik, Universitat Bern, 1996.

Appendix 10A   Mathematical Details

10A.1 Proof of (10.1)

Assuming a constant velocity model

$$X(t) = X_{ref} + v_X t \tag{10A.1}$$

$$Y(t) = Y_{ref} \tag{10A.2}$$

$$Z(t) = Z_{ref} + v_Z t \tag{10A.3}$$

Under perspective projection,

$$x(t) = f\frac{X(t)}{Z(t)} = f\frac{X_{ref} + v_X t}{Z_{ref} + v_Z t} \tag{10A.4}$$

$$y(t) = f\frac{Y(t)}{Z(t)} = f\frac{Y_{ref}}{Z_{ref} + v_Z t} \tag{10A.5}$$

Taking the time derivatives and simplifying

$$\dot{x}(t) = f\frac{Z_{ref} v_X - X_{ref} v_Z}{(Z_{ref} + v_Z t)^2} \tag{10A.6}$$

$$\dot{y}(t) = f\frac{Y_{ref} v_Z}{(Z_{ref} + v_Z t)^2} \tag{10A.7}$$

Dividing (10A.6) by (10A.7), we get

$$\cot(\alpha) = \frac{\dot{x}(t)}{\dot{y}(t)} \tag{10A.8}$$

$$= \frac{Z_{ref}v_X - X_{ref}v_Z}{-Y_{ref}v_Z} \tag{10A.9}$$

Noting that $\cot(\theta) = \dfrac{v_X}{v_Z}$ and multiplying and dividing (10A.9) by $\dfrac{f}{Z_{ref}v_Z}$, we get

$$\cot(\alpha) = \frac{f\cot(\theta) - f\frac{X_{ref}}{Z_{ref}}}{-f\frac{Y_{ref}}{Z_{ref}}} \tag{10A.10}$$

$$= \frac{x_{ref} - f\cot(\theta)}{y_{ref}} \tag{10A.11}$$

where $x_{ref} = f\dfrac{X_{ref}}{Z_{ref}}$ and $y_{ref} = f\dfrac{Y_{ref}}{Z_{ref}}$

10A.2   Proof of (10.2)

Let $[X_\theta, Y_\theta, Z_\theta]'$ denote the coordinates of any point on the person who is walking at an angle $\theta \geq 0$ to the plane passing through the starting point $[X_{ref}\, Y_{ref}\, Z_{ref}]'$ and parallel to the image plane, which we shall refer to, hereafter, as the canonical plane. Rotation of this point on to the canonical plane by θ yields

$$
\begin{bmatrix} X_{int} \\ Y_{int} \\ Z_{int} \end{bmatrix} =
\begin{bmatrix} \cos(\theta) & 0 & \sin(\theta) \\ 0 & 1 & 0 \\ -\sin(\theta) & 0 & \cos(\theta) \end{bmatrix}
\begin{bmatrix} X_\theta - X_{ref} \\ Y_\theta - Y_{ref} \\ Z_\theta - Z_{ref} \end{bmatrix}
$$
$$
= \begin{bmatrix} (X_\theta - X_{ref})\cos(\theta) + (Z_\theta - Z_{ref})\sin(\theta) \\ (Y_\theta - Y_{ref}) \\ -(X_\theta - X_{ref})\sin(\theta) + (Z_\theta - Z_{ref})\cos(\theta) \end{bmatrix} \tag{10A.12}
$$

The coordinates of the synthesized point with respect to the coordinate frame attached to the camera are given as

$$
\begin{bmatrix} X_0 \\ Y_0 \\ Z_0 \end{bmatrix} =
\begin{bmatrix} X_{int} \\ Y_{int} \\ Z_{int} \end{bmatrix} +
\begin{bmatrix} X_{ref} \\ Y_{ref} \\ Z_{ref} \end{bmatrix} \tag{10A.13}
$$

which simplifies to

$$\begin{bmatrix} X_0 \\ Y_0 \\ Z_0 \end{bmatrix} = \begin{bmatrix} X_\theta\cos(\theta) + Z_\theta\sin(\theta) - X_{ref}\cos(\theta) - Z_{ref}\sin(\theta) + X_{ref} \\ Y_\theta \\ -X_\theta\sin(\theta) + Z_\theta\cos(\theta) - X_{ref}\sin(\theta) - Z_{ref}\cos(\theta) + Z_{ref} \end{bmatrix} \qquad (10A.14)$$

Under perspective projection,

$$x_0 = f\frac{X_0}{Z_0}, y_0 = f\frac{Y_0}{Z_0} \qquad (10A.15)$$

$$x_\theta = f\frac{X_\theta}{Z_\theta}, y_\theta = f\frac{Y_\theta}{Z_\theta} \qquad (10A.16)$$

Substituting from (10.16) in (10.17), we get

$$x_0 = f\frac{X_\theta\cos(\theta) + Z_\theta\sin(\theta) - X_{ref}\cos(\theta) - Z_{ref}\sin(\theta) + X_{ref}}{-X_\theta\sin(\theta) + Z_\theta\cos(\theta) - X_{ref}\sin(\theta) - Z_{ref}\cos(\theta) + Z_{ref}} \qquad (10A.17)$$

$$y_0 = f\frac{Y_\theta}{-X_\theta\sin(\theta) + Z_\theta\cos(\theta) - X_{ref}\sin(\theta) - Z_{ref}\cos(\theta) + Z_{ref}} \qquad (10A.18)$$

Multiplying and dividing the numerator and denominator of (10A.17) and (10A.18) by $\frac{f}{Z_\theta}$, we get

$$x_0 = f\frac{x_\theta\cos(\theta) + f\sin(\theta) + f\frac{X_{ref}}{Z_\theta}(1 - \cos(\theta)) - f\frac{Z_{ref}}{Z_\theta}\sin(\theta)}{-x_\theta\sin(\theta) + f\cos(\theta) - f\frac{X_{ref}}{Z_\theta}\sin(\theta) + f\frac{Z_{ref}}{Z_\theta}(1 - \cos(\theta))} \qquad (10A.19)$$

$$y_0 = f\frac{y_\theta}{-x_\theta\sin(\theta) + f\cos(\theta) - f\frac{X_{ref}}{Z_\theta}\sin(\theta) + f\frac{Z_{ref}}{Z_\theta}(1 - \cos(\theta))} \qquad (10A.20)$$

Now for Z_{ref}, $Z_\theta \gg 0$, $f\frac{X_{ref}}{Z_\theta} \approx f\frac{X_{ref}}{Z_{ref}} = x_{ref}$ and $\frac{Z_{ref}}{Z_\theta} \approx 1$, this means that we consider the person far from the camera and for a few cycles only when he does not move too far from the starting point.

Thus, (10A.19) and (10A.20) simplify to

$$x_0 = f\frac{x_\theta\cos(\theta) + x_{ref}(1 - \cos(\theta))}{-\sin(\theta)(x_\theta + x_{ref}) + f} \qquad (10A.21)$$

$$y_0 = f\frac{y_\theta}{-\sin(\theta)(x_\theta + x_{ref}) + f} \qquad (10A.22)$$

PART III

Multimodal Systems and Issues

CHAPTER 11

Sensor Fusion and Environmental Modeling for Multimodal Sentient Computing

Christopher Town

This chapter describes an approach to multisensory and multimodal fusion problems in which computer vision information obtained from calibrated cameras is integrated with a sentient computing system known as *SPIRIT*. The SPIRIT system employs an ultrasonic location infrastructure to track people and devices in an office building and model their state. It is shown how the resulting location and context data can be fused with a range of computer vision modules to augment the system's perceptual and representational capabilities.

Vision techniques include background and object appearance modeling, face detection, segmentation, and tracking modules. Integration is achieved at the system level through the metaphor of shared perceptions, in the sense that the different modalities are guided by and provide updates to a shared world model. This model incorporates aspects of both the static (e.g., positions of office walls and doors) and the dynamic (e.g., location and appearance of devices and people) environment. The shared world model serves both as an ontology of prior information and as a language of context for applications. Fusion and inference are performed by Bayesian networks that model the probabilistic dependencies and reliabilities of different sources of information over time. This chapter shows that the fusion process significantly enhances the capabilities and robustness of both sensory modalities, thus enabling the system to maintain a richer and more accurate world model.

Section 11.1 discusses the concept of sentient computing and presents an outline of the SPIRIT system. Related work is discussed in Section 11.2. In Section 11.3.1 the problem of integrating visual information with the SPIRIT system is introduced, and the computer vision methods employed are presented in Section 11.3.2. Modeling of visual appearance of tracked entities is described in detail in Section 11.3.3. The fusion of several vision algorithms with information from the SPIRIT system is then presented in Section 11.3.4 and algorithms for combined multi-object tracking are described. Section 11.4 shows how these techniques can be applied to augment the representation of dynamic aspects of the SPIRIT world model. Section 11.4.2 shows how multimodal fusion yields higher accuracy and additional sensory data for tracking of people, while Section 11.4.3 presents results for enhanced modeling and visualization of the office environment.

11.1 Sentient Computing—Systems and Sensors

11.1.1 Overview

Efforts in ubiquitous computing [1] are increasingly focused on providing a model of computing in which the proliferation of relatively cheap communications, sensors, and processing devices is leveraged in such a way as to make the resulting systems aware of aspects of their environment and the interactions that take place within it. The goal of what is termed *sentient*[1] [2] or *context-aware* [3] computing is to enable systems to perceive the world and relate to it in much the same way as people do, thereby creating the illusion of a shared perception that carries with it an implicit understanding of *context*. Indeed it can be argued that deriving an accurate representation of context is a holy grail of human computer interaction [4], as it would allow people to interact much more naturally with computer systems in a way that is pervasive and largely transparent to the user.

Sentient computing thus aims to model aspects of the context within which human-computer interactions take place in order to better infer and anticipate user intentions and requirements. This is achieved by integrating information from a range of networked sensors and processors distributed throughout a (typically indoor) space in order to maintain an internal representation, or *world model*, of that environment. Applications utilize the world model in order to obtain implicit knowledge of user context. To realize the goal of shared perception, the robustness and accuracy of sensory data and its interpretation must approximate that of human beings in the chosen domain [5], and the world model must maintain an accurate up-to-date representation of context.

As Section 11.2 will briefly describe, there are a number of sensor technologies and modalities that have been employed in building sentient computing systems. Since vision is our primary sensory modality, it, too, has attracted interest from researchers wishing to build sentient computing systems. However, each technology exhibits its own drawbacks, which intrinsically limit the capabilities of the overall system. Much recent work has therefore focused on combining different sensory modalities, often to the effect that one sensor system is found to be most reliable with additional complementary modalities augmenting its performance.

The sentient computing system considered in this chapter (see Section 11.1.2) uses ultrasound to track tagged devices such as computers and phones within an office. It is currently the most accurate large-scale wireless tracking system of its kind. Nevertheless, systems of this kind have a number of limitations arising from the fact that they are largely restricted to tracking the 3-D location of ultrasonic tags, which must be attached to objects of interest. On the other hand, visual information holds the promise of delivering richer representations of the world without an inherent need to tag salient entities. Computer vision offers a range of capabilities such as detection, classification, and tracking, which are important prerequisites of a context-aware computing system [6]. However, apart from the need to deploy sufficient numbers of cameras to ensure adequate coverage, machine vision remains hampered by problems of *generality* and *robustness*, which

---

1. From *sentient:* having the ability to perceive via the senses.

reduce its suitability as a primary (or sole) sensory modality. There is clearly much scope for work that integrates information from these disparate sources.

11.1.2 The SPIRIT System

A sentient computing environment uses sensor and resource status data to maintain a model of the world that is shared between users and applications. Sensors and telemetry are used to keep the model accurate and up to date, while applications see the world via the model. A richer and more accurate model enables applications to better perceive context and thereby interact with users in a more natural way. For example, a call-routing application could use location information to forward a call to whichever phone is closest to the intended recipient, but if that person appears to be in a meeting, it may instead notify the recipient using a vibrating pager or forward the call to a voice mail facility or the office receptionist.

The SPIRIT[2] [7, 8] system was originally developed at AT&T Laboratories Cambridge, where it was in continuous operation by 50 staff members. The system is currently deployed throughout large parts of the Computer Science Department at Cambridge University (http://www.cl.cam.ac.uk). As shown in Figure 11.1, the system uses mobile ultrasonic sensor devices known as *bats* and a receiver infrastructure to gather high-resolution location information for tagged objects, such as people and machines. Such information is used to maintain a sophisticated world model of the office environment where it has been deployed. Applications can register with the system to receive notifications of relevant events to provide them with an awareness of the spatial context of user interactions. The achieved spatial granularity is better than 3 cm for over 95% of bat observations (assuming only small motion), and bats may be polled using radio base stations with a variable quality of service to give update frequencies of up to 25 Hz (shared among all bats assigned to a given radio base station) while remaining scalable to hundreds of tagged people and devices in a large office. The bats are equipped with two buttons, two LEDs, and a sound chip to allow them to be used as portable input-output devices.

The driving paradigm is that of "computing with space" [2, 9] — physical location and spatial context (typically expressed in terms of containment and proximity) together with the attributes and capabilities of entities and devices present at a given time drive the behavior of applications built upon the system. Some applications such as LabView (shown in Figure 11.2) allow users to navigate and browse the world model itself, while others respond to particular configurations of interest. Colocation and spatial composition can be used to infer aspects of context (e.g., "User A has entered office O," "User B is using his bat as a 3-D mouse to control the scanner in corridor C," "User B has picked up PDA P"), which can influence or trigger application behavior, hence space itself becomes part of the user interface. Current SPIRIT applications include follow-me event notification, personnel and resource localization, office visualization, user authentication, desktop teleporting, virtual 3-D interfaces, and location support for augmented reality.

---

2. Originally an acronym for spatially indexed resource identification and tracking.

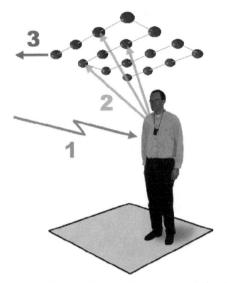
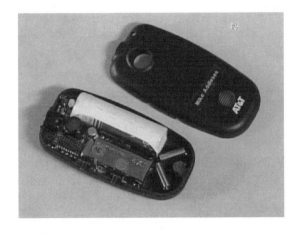

Figure 11.1 Left: Function of the SPIRIT location system. A bat sensor is triggered by radio (1), emits an ultrasonic pulse (2), and time-of-flight of the pulse is measured by receivers in the ceiling (3) to compute 3-D position. Right: One of the wireless tracking tags known as a bat. The device is about the size of a matchbox and features two buttons for input and two LEDs and a sound chip for output. It contains a radio receiver, ultrasonic transmitter, microchip with a unique 48-bit ID, and an AA lithium battery supplying enough power for up to 12 months of operation under normal conditions.

11.1.3 Motivation and Challenges

Although the SPIRIT system has proven effective in providing fairly fine-grained spatial context upon which sentient computing applications can be built, difficulties remain [10]. Bat system spatial observations are limited to the location of the bat sensor, which is polled sporadically by a central base station. Each bat has an associated identity (e.g., Digital camera 1, or User J. Smith), which may carry associated semantics (e.g., digital cameras must be operated by a person, people can exhibit certain patterns of movement). However, only objects tagged with bats can be tracked, and the model of the environment is static unless other sensors (e.g., light switches and temperature dials) provide information on it.

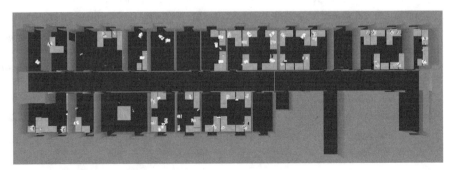

Figure 11.2 The LabView application displays a 3-D real-time map representing the state of the world model. The bird's-eye view shown provides an overview of the sentient office (at the Laboratory of Communications Engineering) and objects such as furniture, devices, and people.

Computer vision methods can provide multimodal human-computer interfaces with transparent detection, recognition, and surveillance capabilities, but on their own suffer from a lack of robustness and autonomy in real-world interaction scenarios. The integration of distinct sources of information about the world in light of application-specific constraints holds great promise for building systems that can optimally leverage different sensory capabilities and failure characteristics. Vision offers the possibility of acquiring much richer representations of entities in terms of their orientation, posture, and movements. It can also detect and to some extent classify and track additional features of the static (e.g., furniture) and dynamic (e.g., people and portable devices not equipped with bats) environment. It may also be used to smooth over some of the difficulties inherent in an ultrasonic location infrastructure, thereby making it more robust. Information from the SPIRIT world model can in turn be used to provide constraints to the fusion process, to (re)initialize computer vision modules, and to act as a focus of attention mechanism.

11.1.4 Sentient Computing World Model

The SPIRIT system maintains an internal dynamic representation of the office environment, including objects and events within it. This world model [11] comprises a static part consisting of those aspects of the environment that are not monitored using the ultrasonic location system, and a dynamic part consisting of those objects and devices that are. The former includes the location and spatial extent of rooms, walls, windows, doors, and items of furniture such as desks and shelves that were manually added to the model. The latter tracks personnel, visitors, portable devices, and other assets that have been tagged with one or more of the bats.

The role of the sensor systems is to keep the model consistent and accurate. As shown in Figure 11.3, applications see a description of the environment that is abstracted away from the sensor level. The interpretation of such information is

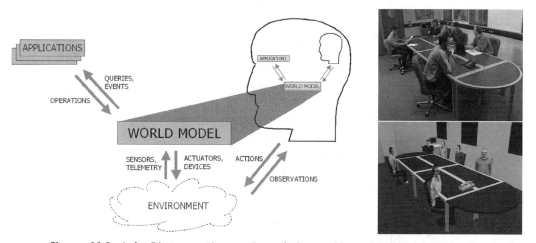

Figure 11.3 Left: Diagrammatic overview of the world model maintained by the sentient computing system. Right: The world perceived by (top) users and (bottom) the sentient computing system (LabView visualization).

application-dependent; for example, routing a phone call may require finding the phone that is closest to a given person (provided the person is in the same room), whereas a follow-me desktop session would need to know something about the user's orientation relative to available screens in order to select the best one for display.

While the evolution of context and context-dependent interpretation is dynamic, a prior notion of what comprises models of context and means of inferring it is required. In the approach presented here, this prerequisite is expressed through an ontology consisting of the world model maintained by the SPIRIT system, which is augmented with information gained through a number of visual cues. Computer vision has for some time recognized the importance of integrating top-down and bottom-up information using knowledge hierarchies and domain constraints. Furthermore, the circular problem of defining context in terms of perceptions and interpreting these perceptions in terms of the given context naturally gives rise to a solution framework based on feedback.

11.2 Related Work

Efforts in ubiquitous and context-aware computing have employed a range of sensor systems [12, 13] such as accelerometers; touch-sensitive surfaces; and, more commonly, sound (microphones, ultrasound) or light (infrared, radio) at different frequencies, in order to detect or track people and devices.

Most indoor location systems such as the infrared-based Active Badge system [14] only allow positioning at room accuracy. Exceptions include the RADAR [15] project, which uses a series of base stations transmitting in the ISM radio band to provide an indoor location system with a median resolution in the range of 2m to 3m. The Cricket system [16] uses a series of ultrasonic beacons placed throughout a building at known locations, which periodically emit both ultrasonic pulses and a radio signal. Devices can use the time difference of the two pulses to determine the closest beacon and estimate their distance from it to within 5 cm to 25 cm. Much current research focuses on the use of ultrawideband (UWB) radio positioning [17] and other wireless technologies such as Bluetooth and 802.11b. The requirements for low-cost sensor systems in applications such as retail stock control and supply chain management have fostered interest in cheap, passive tagging technologies such as radio frequency identification (RFID) tags [18].

Outdoor location is a more challenging problem due to the lack of a controlled environment and suitable infrastructure. Most systems derive positioning information from the global positioning system (GPS) [19], which permits spatial resolutions of about 10m, although greater accuracy is achievable through the use of precalibrated reference points. The use of information from differential timing and directional signal strength variation with respect to multiple base stations allows handset localization to within a few hundred meters of the widely used GSM mobile phone system [20]. Given the intrinsic drawbacks of any given positioning technology, sensor fusion approaches are gaining popularity [21]. A number of architectures have been proposed for this purpose, for example, [22].

The role of computer vision in practical office-scale sentient systems has largely been restricted to the detection of tagged objects [23], although vision-based systems for gesture recognition and motion analysis have been under development for some time [6, 24–26]. Approaches relying on probabilistic integration of different sources of visual evidence such as face detectors and models for motion, shape, and color have shown particular promise (e.g., [27, 28]). However, the difficulties of deploying perceptual user interface mechanisms on the basis of vision alone are exacerbated by problems such as brittleness and lack of real-time performance and generality. Many vision-based systems have consequently been developed for fairly circumscribed control tasks where a limited vocabulary of pointing and selection gestures is appropriate. A truly pervasive system, however, requires a richer basis for interactions and a notion of context strong enough to recognize when no intended interaction with the system occurs.

These problems have led researchers to investigate fusion of vision with other sensory modalities [25]. Most such systems rely on integration with sound in the audible range via microphone assemblies, which has proven particularly popular for videoconferencing applications [29, 30]. There are also systems that have integrated vision as a secondary modality to an existing system, for example, [31], which deploys visual gait recognition as an additional identifying cue to a system based on a pressure-sensitive active floor. Multimodal user localization is also an important topic in robotics research [32], where information from stereo or omnidirectional cameras mounted on a moving robot is often combined with sensors, such as laser range finders.

Bayesian approaches to multimodal fusion have been gaining prominence in the vision and other research communities. Tracking algorithms that perform concurrent probabilistic integration of multiple complementary and redundant cues has been shown to be much more robust than tracking those that utilize only a single cue [33, 34]. Tracking on the basis of multiple sources of information is also demonstrated by Choudhury et al. [35], who present a system that fuses auditory and visual cues for speaker detection, while Torralba et al. [36] describe work that highlights the importance of context for object and place recognition in situated vision systems.

11.3   Sensor Fusion

11.3.1   Sensory Modalities and Correspondences

In order to fuse data from the visual and SPIRIT modalities, one must translate between their underlying representations. This requires translation between the 3-D SPIRIT and 2-D image reference frames and synchronization of SPIRIT events with corresponding video frames acquired by a particular camera. Further details are available in [37, 38].

11.3.1.1   Frame of Reference

Visual input is acquired from cameras placed at known locations within the SPIRIT world frame. Both the positions of the cameras and their intrinsic and extrinsic parameters were calibrated carefully. Intrinsic parameters were estimated using a chessboard calibration pattern and the Matlab toolbox developed by Jean-Yves Bouguet [39]. Several images (typically 20 to 30) of the chessboard were analyzed, and camera parameters were estimated through correspondence analysis of positions of the corner points to establish planar homographies. The camera model consists of 8 parameters consisting of the coordinates of the 2×1 effective focal length and optic center vectors, the skew coefficients accounting for nonorthogonality of the x-y axes, and 4 distortion coefficients representing radial and tangential distortions of the lens.

Camera position, view area, and extrinsic parameters were determined by means of a surveying application running on top of the SPIRIT system. This allowed feature points such as the position of the camera, the corners of its field of view, and calibration points visible by it to be localized very accurately in the 3-D SPIRIT coordinate system. In order to determine the translation between 3-D SPIRIT points and their 2-D pixel coordinates when projected onto the camera's image plane, the image coordinates of the calibration points were also measured. After performing such measurements for a set of coplanar points, the extrinsic parameters consisting of a rotation matrix R and a translation vector T in 3-D space can be determined numerically. These steps make it possible to determine which objects should be visible (in the absence of occlusions) from a given camera, and to calculate the projection of 3-D bat system coordinates onto the image plane of that camera with a mean error of a few pixels.

11.3.1.2   Synchronization

SPIRIT location events need to be precisely synchronized with associated video frames. The synchronization can be initialized manually using the buttons on the bat device. Arrival events for people entering the view area of a camera can be used to perform automatic resynchronization by using the visual tracking method and a motion history window to interpolate locations and correlate these to bat sensor sightings. Together with the camera calibration process described above, this enables data in both the spatial and temporal domain to be translated between the SPIRIT system and the visual information captured by the cameras.

11.3.1.3   Quality of Service

Location information captured by a sensor infrastructure is a limited resource, and variable rates of quality of service are therefore imposed by the SPIRIT scheduler to determine the rate at which location events are generated for a given bat. The frequency at which bats are polled is reduced when the device is stationary. An internal accelerometer allows sampling rates to be increased when the bat is in motion, or if it is to be used as a "3-D mouse" to drive particular applications. However, there is some latency before an increased polling rate comes into effect.

11.3.1.4   Accuracy

The accuracy of the SPIRIT location events is primarily affected by the properties of the sensor technology and the quality of the camera calibration and frame synchronization. Ultrasound imposes intrinsic limits on update frequency and resolution due to its propagation characteristics and the filtering that is applied to dampen echoes and remove spurious (e.g., multipath) observations.

11.3.1.5   Visibility

The SPIRIT world model contains the locations of walls, doors, and windows, thereby making it possible to determine constraints on the environment viewed by each camera to deduce which objects and entities known to the system are likely to be visible by a given camera. A certain amount of occlusion reasoning may also be performed on this basis by computing a spatial ordering and predicting likely regions of overlap between tagged objects. However, there are many aspects of the world model such as furniture and the state of doors (open versus closed) that are not fully modeled and must therefore be inferred by other means.

11.3.2   Vision Algorithms

This section provides an overview of the vision techniques that have been implemented to provide additional surveillance information on the sentient computing environment and objects within it.

11.3.2.1   Skin and Face Detection

Human skin color is modeled as a region in HSV space [40]. Histogram equalization is applied to the entire image (or a target region predicted by other means), and candidate pixels that lie within the HSV subspace are clustered into regions using morphological operators to remove noise. Then face detection methods are applied to candidate regions identified by means of skin color classification across the whole image or selectively to regions likely to contain faces as determined by the other modalities (i.e., head regions predicted by SPIRIT person observations or blob tracker–based appearance models). In the former case, ellipse fitting is applied to the skin clusters, and clusters may be split based on how elongated they are. Face detection is applied to PCA-transformed subwindows of the candidate region at multiple scales.

Two face detection methods were trained: the first uses a generative mixture of Gaussians model trained using expectation maximization, and the second consists of polynomial kernel SVM classifiers. In both cases, the classifiers are arranged in a two-level cascade with the first classifier acting as a fast rejection filter for the second classifier, which was trained by incorporating test set misclassifications into the training set for the second stage. The two classification schemes are combined using simple disjunction of their binary classification decisions. This may increase false positive rates but ensures that fewer faces are missed.

11.3.2.2   Background Modeling and Foreground Detection

As described [41], the system maintains a background model and foreground motion history that are adapted over time. A motion history matrix M_t is used to identify a background image bim_t of pixels undergoing sufficiently slow change, which can then be used to reliably update the background model B_t and estimate its variance. Pixels are deemed to be part of the dynamic foreground if they exceed a difference threshold that is a multiple of the background variance σ_t^B, and if they are not deemed to be part of a shadow as determined by the DNM1 algorithm (see [37, 41]). Following these steps, candidate foreground pixels are subjected to morphological operations (dilation and erosion) to reduce noise in the final estimate.

11.3.2.3   Blob Analysis and Tracking

Foreground pixels are clustered using connected component analysis to identify moving regions ("blobs"). These are then parameterized using shape (bounding box, center of gravity, major axis orientation) and appearance measures as described in Section 11.3.3. Blobs are tracked using a Kalman filter or Condensation tracker with a second order motion model. Tracked objects are matched to detected blobs using a weighted dissimilarity metric that takes into account differences in predicted object location vs. blob location and changes in shape and appearance.

11.3.2.4   Occlusion Reasoning

To make tracking more robust, the object-to-blob assignment stage features a Bayesian network for reasoning about occlusions and object interactions based on observed or predicted overlap of object bounding boxes and failures of object assignment. Dynamic occlusions can also be disambiguated by using 3-D SPIRIT data to predict spatial ordering of tracked objects, while static occlusions, object arrivals, and object departures can often be resolved with reference to the world model.

11.3.2.5   Static Scene Segmentation and Region Classification

The region segmentation facilitates correspondence analysis between the world model and the scene viewed by a camera for environmental analysis and constrained tracking. The segmentation method due to [42] is applied to video frames at infrequent intervals. This method segments images into nonoverlapping regions by computing a Canny-style color edge detector and generating Voronoi seed points from the peaks in the distance transform of the edge image. Regions are grown agglomeratively from seed points with gates on color difference with respect to the boundary color and mean color across the region. A texture model based on discrete ridge features is also used to describe regions in terms of texture feature orientation and density. Sets of properties for size, color, shape, and texture are computed for each region. These properties are fed into artificial neural network classifiers that have been trained to classify regions into wood, cloth, carpet, and internal walls. As explained in Section 11.4.3, the classifiers were trained on images taken in the SPIRIT office and were found to perform well in identifying areas that would otherwise have been mislabeled (e.g., skin rather than wood) and in identifying furniture and wooden doors.

11.3.3   Fusion and Adaptation of Visual Appearance Models

This section describes how fusion of three different appearance models enables robust tracking of multiple objects on the basis of color information and by using the visual tracking framework described in Section 11.3.2. In this chapter, short-term variation in object color is modeled nonparametrically using adaptive binning histograms. Appearance changes at intermediate time scales are represented by semiparametric (Gaussian mixture) models, while a parametric subspace method (robust principal component analysis, RPCA [43]) is employed to model long-term stable appearance. Fusion of the three models is achieved through particle filtering and the democratic integration method. Further details are available in [37, 38].

11.3.3.1   Adaptive Binning Color Histogram

The optimal number and width of histogram bins is determined by means of k-means clustering with color differences computed in the CIELAB space using the CIE94 distance d_{kp}. The clustering is repeated n times or until no pixels are left unclustered. Matching of tracked objects with candidate blobs is performed using weighted correlation. The similarity between two histogram bins is calculated by using a weighted product of the bin counts $H[i]$ and $H[j]$, where the weight w_{ij} is determined from the volume of intersection V_s between the two bins.

In order to incorporate some longer-term appearance variation and smooth over fluctuations, the histograms are adapted using exponential averaging. Given a color histogram H_t calculated for a blob at frame t and a smoothed object color histogram S_{t-1} from frame $t-1$, the new smoothed object color histogram S_t for frame t is given by $S_t = \alpha H_t + (1-\alpha) S_{t-1}$ where $\alpha = 1 - e^{-\frac{1}{\lambda}}$ determines the rate of adaptation. This is set to increase with increasing object speed in order to keep track of rapidly moving objects.

11.3.3.2   Gaussian Mixture Model

The conditional density for a pixel ψ belonging to an object O can be represented by a mixture of M Gaussians. In this chapter, mixture modeling is performed in hue-saturation (HS) space to gain a degree of illumination invariance. Model estimation is performed on blob pixels using subsampling for efficiency and discarding samples whose intensity value is very low or close to saturation. Components are estimated using k-means with priors computed from the proportion of samples in each cluster. The parameters of the Gaussians (mean and covariance) are calculated from the clusters. Model order selection is performed using cross-validation on a training and validation set randomly selected from the pixel samples. The training set is used to train a number of models of a different order M by means of the expectation maximization algorithm. Adaptation of the GMM over time is performed using the approach suggested in [44].

11.3.3.3   Robust Principal Component Analysis

In order to acquire a stable model of object appearance over longer time scales, an extension of the robust principal component analysis (RPCA) method proposed by De la Torre and Black [45] is applied. RPCA enhances standard PCA by means of a pixel outlier process using M-estimators. To ensure adequate performance for tracking, RPCA has been extended in this work using a robust incremental subspace learning technique to efficiently recompute the eigenspace. In addition, rather than computing RPCA over image intensity alone, RPCA was applied to 1-D color statistics histograms derived from the color distribution of each object in HSV space. Following Hanbury [46], the saturation-weighted hue mean histogram $H_{S\ell}$ (where hue H is measured as an angle in the range {$0°$, $1°$, ..., $360°$}) is calculated at each sample luminance level $\ell \in \{0, 1, 2, ..., N\}$. Re-estimation of the RPCA coefficients is performed incrementally by adapting the method proposed in [47].

11.3.3.4   Adaptative Fusion by Democratic Integration

The observation density is modeled by a function that contains Gaussian peaks where the observation density is assumed to be high, that is, where an object could have generated a set of blobs with high probability. Each Gaussian peak corresponds to the position of a blob, and the peak is scaled by the object-blob distance. The likelihood $\mathcal{L}$ for a particle is computed as $\mathcal{L}(\mathbf{Z}_t|\mathbf{X}_t) \propto e^{-k \times \mathrm{dist}^2}$, where dist is a distance under one of the appearance models of the local image patch at a given particle and the object under consideration, and k is a constant. Likelihoods are calculated using the condensation (conditional density propagation) algorithm for each of the three appearance modeling schemes above and combined using:

$$\mathcal{L}(\mathbf{Z}_t|\mathbf{X}_t) \propto [\mathcal{L}_{rpca}(\mathbf{Z}_t|\mathbf{X}_t)]^{\alpha_1}[\mathcal{L}_{chist}(\mathbf{Z}_t|\mathbf{X}_t)]^{\alpha_2}[\mathcal{L}_{gmm}(\mathbf{Z}_t|\mathbf{X}_t)]^{\alpha_3} \qquad (11.1)$$

where $0 \leq \alpha_1, \alpha_2, \alpha_3 \leq 1$ are the reliability weights for each appearance model, initialized to $\frac{1}{3}$.

Adaptation of the weights in (11.1) is performed dynamically during tracking by extending the idea of democratic integration [28] to the condensation framework. Four separate observation likelihoods are computed: one for the joint appearance model, and three for each of the RPCA, adaptive histogram and GMM appearance cues. Condensation is performed separately for each of the observation functions, resulting in four hypotheses, R_{fused}, R_{rpca}, R_{chist}, and R_{gmm}, which are regions where the object is thought to be in the current frame. Each region centroid is obtained by computing the expectation of the respective particle sets for each cue.

The Euclidean distances $E_{k,t}$ between the centroid of R_{fused} and the centroids of R_{rpca}, R_{chist}, R_{gmm} at time t are then calculated. Since the joint observation function is assumed to exhibit the best performance, appearance cues that result in relatively large values of $E_{k,t}$ are considered less reliable in the current frame, and their reliability weight is lowered accordingly. A score $\gamma_{k,t}$ is computed for each cue k using the equation $\gamma_{k,t} = \frac{\tanh(-aE_{k,t}+b)+1}{2}$ where a, b are constants (set to 2 and 5, respectively). Given $\gamma_{k,t}$, the weights $\alpha_{k,t}$ for each cue k are then adapted using first order exponential averaging, that is, $\alpha_{k,t+1} = \beta\gamma_{k,t} + (1 - \beta)\alpha_{k,t}$, where β controls the rate of adaptation (setting $\beta = 0.75$ was found to give good results in most sequences). Performing the condensation algorithm four times during each frame was found not to be a bottleneck since most of the computation time is required for the particle distances (which need only be computed once per frame).

11.3.4   Multihypothesis Bayesian Modality Fusion

A viable multimodal fusion method must generate reliable results that improve upon the individual modalities while maintaining their fidelity and uncertainty information for higher-level processing and adaptation of the fusion strategy. The approach taken here is essentially a development of Bayesian modality fusion [27, 48] for multiobject tracking. It uses a Bayesian graphical network (shown in Figure 11.4) to integrate information from the different sources. Discrete reliability indicator variables (R_S, R_F, R_D, R_C, and R_B) are used to model how reliable each modality is at the current time. At present each variable may take on one of the values low, normal, or high. The network serves as a shared template from which individual tracking hypotheses are derived. Hypotheses are instantiated by SPIRIT observations or the blob tracking framework, thus allowing tracking of people who are not tagged with a functioning bat device or who are not currently visible by a given camera. Other visual cues such as skin color and face detection serve as supporting modalities. Spatial and object-specific ontologies from the world model or the region segmentation and classification methods provide contextual constraints and guide the generation of hypotheses.

Reliabilities are adapted on the basis of manually specified rules over reliability indicators, such as motion and appearance variation, and performance feedback measures, such as consistency and log-likelihood of the observations under each

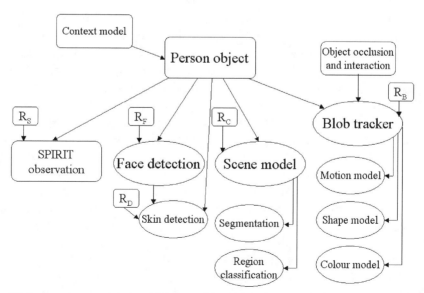

Figure 11.4 Bayesian graphical model for multimodal fusion. Reliability variables allow adaptive integration of different sources of information.

modality. Posterior probabilities for each hypothesis can then be computed by integrating all available information using the fusion network. The position and spatial extent of tracked people is computed by reliability-weighted interpolation of the object bounding box deduced from the SPIRIT observation (if available for the current observation) and blob tracker appearance model.

Each hypothesis maintains its own set of reliability variables and competes for observations with other hypotheses. The conditional probabilities (including the dependency on reliability variables) of the underlying network structure were initially set by hand, but have since been re-estimated by means of the EM algorithm on statistics gathered from manually labeled training sequences consisting of over 3,000 frames. Temporal evolution of the model occurs via a Kalman or particle filter applied to the color blob tracker and through the modification of reliability variables in light of current observations. This update stage introduces a coupling of the observation models for the individual modalities. Some results from a meeting room sequence will be shown in Section 11.4.2.

SPIRIT observations consist of projected 3-D positions of bat sensor locations together with information on the type and identity of the observed object as available from the SPIRIT world model. The world model contains information about people's height and the probable position of their bat on their body, and hence the projective mapping of the person's likely silhouette onto the camera's image plane can be calculated. Location events are generally quite accurate but are assigned a reduced reliability if they are not well-synchronized with the current frame or if the bat has been undergoing rapid motion.

The reliability of tracked blobs depends on the correspondence between predicted and actual position and appearance dissimilarity. Face detection can be a powerful cue for head position but becomes unreliable when there is too much variation in appearance due to movement, occlusions, or changes in posture. Evidence for false positives consists of detections in regions of high motion energy or areas where there is no expectation of faces being observed, in other words, where the other modalities fail to hypothesize the appearance of a person. This is particularly the case for areas of skin color (such as a wooden door or table) where one or more faces have been detected, but which are unlikely to coincide with the appearance of a human head due to their position or shape. Conversely, face detections in regions where head presence is predicted by other modalities lead to an increase in reliability of the face cue for the given hypothesis. Skin detections may be used to model other body parts such as hands and legs. The scene model serves to disambiguate skin detections by dictating low reliability in regions that are likely to lead to false detection (e.g., wood). The scene model consists of the static region segmentation of each frame and the neural network classifications of each region. Areas of high motion energy lead to blur, which degrades the reliability of the segmentation and classification. Segmentation is also unreliable when similarly colored objects overlap.

The integration process computes a probability for each tracked object given the current observations and reliabilities. Person position is computed by weighted interpolation of the object bounding box deduced from the SPIRIT observation and blob tracker object currently associated with a given hypothesis, taking into account their respective reliabilities. Skin color, face detections, and the scene model serve as supporting modalities, whereas the output of the SPIRIT and blob tracker maintain object identity and can serve as instantiating modalities, that is, a new object hypothesis must be supported by either a tracked color blob or a SPIRIT observation (see Figure 11.5). In the latter case, both location events for people and devices assigned to a particular person can be used. People generally wear their Bat sensor at a particular calibrated position on their body, and together with the known distance and projective properties of the camera, this can be used to instantiate an expected 2-D occupancy region for the person, even if no useful blob tracker can be assigned to the hypothesis at present. Face detections contribute to the combined probability if they occur within the upper third of the bounding box, and skin color contributes if it is found anywhere within this region. Objects that are tracked only on the basis of SPIRIT information but don't appear to be visible in the current frame continue (for a while) to be represented by a hypothesis whose likelihood is adjusted according to the occlusion reasoning described above.

Hypotheses can be viewed as competing for observations, since each visual cue and SPIRIT location event may only be used to support one tracked object in a given frame. Hypotheses are removed after their probability has dropped below a threshold for a certain number of frames but may be revived if a new SPIRIT location event tagged with the same object ID occurs. New hypotheses are instantiated in response to blob or bat observations. Due to the relative brittleness of the visual cues alone, new hypotheses are given a low initial probability until they have been confirmed by a SPIRIT observation or have proven stable over several frames. This allows people who are not wearing a functioning bat device to be tracked. SPIRIT data is also particularly valuable in maintaining object identity across occlusions

Figure 11.5   Example images from the test sequences. Rectangles indicate bounding boxes of blob tracker objects. Solid ellipses are skin clusters; those with a cross were deemed to contain a face. Light-dotted ellipses and numbers indicate hypotheses resulting from the fusion process described later.

(although in some cases this is also possible on the basis of blob appearance and predicted motion), and to generate expectations (expressed as hypotheses) for people who are about to enter the visual field of a camera. Hence, the bat system and the federation of visual cues may each serve to guide the combined system's focus of attention by instantiating hypotheses and generating expectations.

11.4 Environmental Modeling Using Sensor Fusion

Having visual information as an additional sensory modality is useful when the SPIRIT system has trouble detecting a person (e.g., the person is not wearing a Bat or it is temporarily concealed), or when an application requires additional information about a person's posture, direction of gaze, gestures, interactions with devices and other people, or facial expression to enhance visually mediated human computer interaction and provide a richer model of the context in which such interactions take place.

11.4.1 Experimental Setup

To facilitate the integration of visual information into the world model, surveillance cameras were deployed in various parts of the sentient office (see map in Figure 11.2), namely, one at the western entrance, one facing east along the corridor, and two in the meeting room (third room from the left at the bottom part of the map). Additional experimental sequences were also taken in some of the other rooms of the office. The cameras used were two standard Philips Webcams yielding a picture resolution and frame rate of (320×240 pixels, 12 fps) and (640×480 pixels, 15 fps), respectively.

A number of sequences (S1, …, S10) featuring a total of seven individuals were acquired to experiment with different scenarios and activities, such as one or more people entering or leaving a room, holding a meeting, drawing on the whiteboard, walking in front of each other, and so forth. Several thousand frames from ten sequences were manually annotated[3] by marking the bounding boxes of people's bodies and heads visible within each frame. A brief description of each follows:

- *S1*: One person enters the meeting room, walks around, and draws on the whiteboard. In addition to body outline and head region, the position of the person's hands was also annotated to allow for further research into gesture recognition and related applications.
- *S2*: Two people enter the meeting room one after the other, greet each other, and walk around the room at different speeds, frequently occluding one another.
- *S3*: Two people hold a meeting, entering and leaving the room one after the other. At various points they sit around the meeting room table, get up, draw on the whiteboard, and pass objects to one another.

---

3. For reasons of time, not every sequence was exhaustively annotated. In most cases the annotations are limited to a subset of the actual footage and/or only labeled every fifth frame, which is the also the rate at which the visual analysis is usually performed.

- *S4*: The same scene depicted in sequence S3, but viewed from a different camera mounted on a different wall and at a different height from that in S3.
- *S5*: A similar scenario as in S2 and S3, but involving up to five people simultaneously. The participants enter and eventually leave the room one after the other. They perform actions such as greeting one another, trading places, drawing and pointing on the whiteboard, and manipulating objects such as an hourglass, pens, and papers.
- *S6*: A group of five people engaging in a more casual social interaction in an office room filmed by a camera mounted on a tripod. At different times they are standing, walking around, shaking hands, sitting down, and leaving and re-entering the camera's view from different directions.
- *S7*: Scene shot from a raised tripod such that the camera was facing along a long corridor. Up to four people can be seen entering and leaving rooms at different distances from the camera, meeting each other, walking past or next to one another, walking toward and past the camera, and so forth. The camera focus and resolution were particularly poor, and the types of movement exhibited are of a kind especially challenging to many visual tracking and appearance-modeling methods.
- *S8*: Sequence filmed from a camera about 1.8m above the ground facing the main entrance to the LCE. Five different people can be seen entering and leaving the laboratory, sometimes individually and sometimes in pairs. The scene includes people opening the door for others and one person tail-gating behind another once the door is open, which are situations of interest in terms of access security. The entrance door also marks the outer boundary of the space in which the SPIRIT system was deployed, which means that there are far fewer available SPIRIT observations.
- *S9*: A similar scenario to that depicted in sequence S8, except that the camera is filming people entering a particular office (with the camera positioned just inside the office facing the door).
- *S10*: This sequence shows up to five people in the small reception area of the LCE. They are at various points walking around, sitting down, and interacting with the wall-mounted plasma screen (which runs a version of the LabView application that may be controlled by means of the bat).

11.4.2 Enhanced Tracking and Dynamic State Estimation

11.4.2.1 Evaluation of Vision-Based Adaptive Appearance Modeling

To evaluate the adaptive appearance models and the fusion mechanism discussed above, testing was carried out on a number of indoor surveillance sequences that were acquired in the same office in which the SPIRIT system has been deployed (see Section 11.4.1). The tracking conditions are especially demanding due to the presence of intermittent bright lighting, skin-colored walls, motion blur, and occlusions as the people interact. Figure 11.6 shows how the fusion framework makes tracking robust with respect to occlusions and movements of people (the results shown are for sequence *S7* discussed in Section 11.4.1). In Figure 11.7 it is

Figure 11.6 Indoor tracking results using vision cues only. The rectangles indicate bounding boxes of tracked objects, and object identity assignments throughout the sequence. Top: tracking using only blob features and distances. Bottom: tracking using the robust fusion of adaptive appearance models as described in Section 11.3.3. Note how this allows identity of tracked entities (indicated by bounding box intensities) to be maintained during and across occlusions.

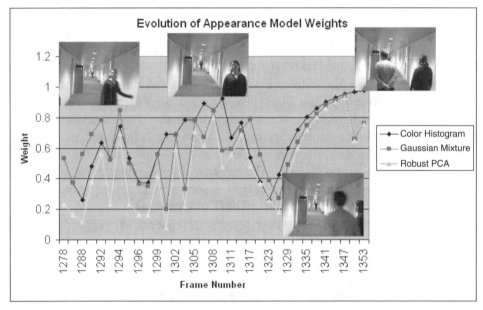

Figure 11.7 Evolution of reliability weights and object color appearance models. Left: Graph plotting the reliabilities of the appearance model cues for the woman shown in the test sequence. There is an initial rise in the reliability of all models due to the clear visibility of the woman. The large fall in reliability at frame 1,320 onwards is due to occlusion by the man entering the scene. After the occlusion, the appearance models successfully recover, and their reliability increases very rapidly. Note the lag of the RPCA (and in some cases the Gaussian mixture) model behind the color histogram model due to their slower adaptation.

shown how the appearance modeling improves accuracy in light of erroneous blob hypotheses generated by the background differencing and blob detection framework.

11.4.2.2   Evaluation of Multisensory Fusion

Using the combined tracking framework, position estimates can be made more robust and accurate. As described in Section 11.3.4, this can be achieved through Bayesian multimodal fusion. Figure 11.8 shows sample results for a meeting

Figure 11.8 Examples of tracking results obtained for sequence S5: Rectangles denote object hypotheses derived from the two modalities dark gray and the resulting fused hypothesis light gray. Ellipses indicate face detections.

scenario with multiple participants. As indicated below, additional information apart from a person's location can be inferred through the joint integration of the various perceptual cues.

As described in Section 11.4.1, several sequences were manually annotated with ground truth information in order to analyze tracking performance. Figures 11.9, 11.10, 11.11, and 11.12 show performance data for sequences S2, S3, S5, and S6, respectively. For each sequence, results are shown that compare performance when tracking is performed using the two modalities on their own (i.e., only vision or only SPIRIT information) and for the fusion method described above. The performance measures plotted are the mean distance-from-track TD, the detection rate DR, and the false positive rate FR, computed for each frame in the sequence. Consequently, a value of DR close to 1 indicates that all objects are being tracked in a given frame while FR close to 0 means that there are few false positives (spurious instances of objects that do not correspond to any objects marked in the ground truth for that frame). The measure TD characterizes the mean accuracy of object tracks in terms of the distance between the centers of

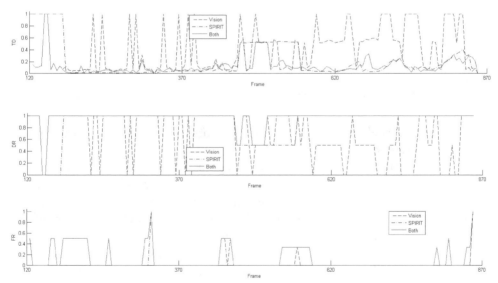

Figure 11.9 Comparative tracking results for test sequence S2 when using the two modalities, both in isolation and combined by means of the fusion process. Top: Mean distance-from-track. TD; Middle: detection rate. DR; Bottom: False positive rate FR. In each case, the solid line shows the value of the given performance measure for the outcome of the fusion method, while the dashed and dotted lines indicate results when using the vision and SPIRIT modalities in isolation, respectively.

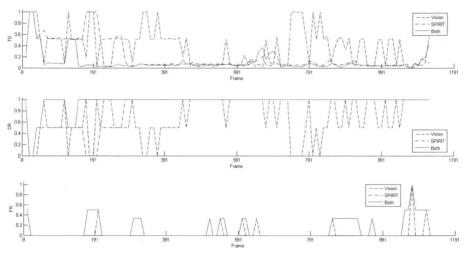

Figure 11.10   Comparative tracking results for test sequence S3.

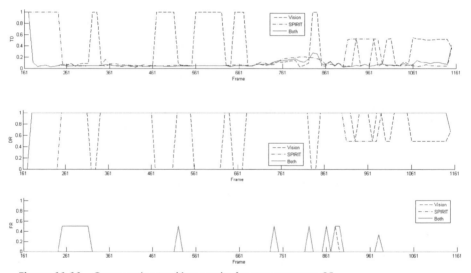

Figure 11.11   Comparative tracking results for test sequence S5.

gravity of tracked objects and ground truth objects (which takes a value of 1 if no such correspondence can be established).

In order to summarize these results, figures for overall recall and precision are shown in Table 11.1, where $Recall = \text{mean(DR)}$ and $Precision = \text{mean}(N_{tp}/(N_{tp} + N_{fp}))$ (N_{tp} is the number of true positives and N_{fp} the number of false positives for each processed frame in the sequence). As can be seen from the results, tracker accuracy and performance are generally enhanced by the combined fusion process compared to the outcome using vision modules or only SPIRIT observations, especially in difficult situations such as object-object occlusions. The system can exploit the multimodal redundancies to successfully track objects that are only detected by one of the tracking components. Reliability indicators allow the system to discount modalities that fail in particular circumstances and rely on those that are likely to give accurate results, thus ensuring that the fusion process

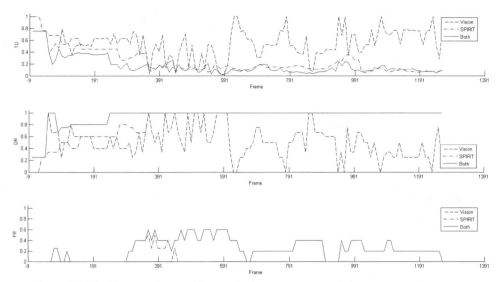

Figure 11.12   Comparative tracking results for test sequence S6.

delivers results that are as good as and sometimes better than those of the modality that performs best at a given time.

However, these results show that the fusion method sometimes incurs in a slightly increased false positive rate compared to either of the component modalities, which may lead to a reduction in precision alongside significant increases in recall. It is possible to optimize the precision-recall trade-off to best suit the requirements of particular applications of the sentient computing environment. For example, security applications are likely to require high recall, whereas tracking for human-computer interaction may need high precision.

To analyze how this can be done, some of the parameters affecting the fusion process were optimized with the goal of assessing the level of recall achievable by the fusion method at the same level of precision as that of the SPIRIT system as shown in Table 11.1. No modifications were made to the Bayesian fusion network or any other aspects of the system, but 20 different combinations of values were evaluated for the 3 internal software variables p_thresh, p_retention, and fobj_retention. In the case of results shown in Table 11.1, these variables had been set to 0.40, 0.30, and 10, respectively. The parameter p_thresh specifies the minimum probability (as calculated by the Bayesian network in Figure 11.4) that a hypothesis must satisfy in order to be regarded as a tracked object. In order to

Table 11.1   Overall Mean Recall and Precision for Test Sequences S2, S3, S5, and S6

Modality	Vision		SPIRIT		Fusion	
Sequence	Recall	Precision	Recall	Precision	Recall	Precision
S2	0.674	0.907	0.857	0.976	0.963	0.852
S3	0.673	0.931	0.845	0.933	0.960	0.868
S5	0.729	0.948	0.875	1.000	0.987	0.906
S6	0.501	0.996	0.860	0.731	0.943	0.747

For each sequence, recall and precision were computed by comparing tracking results obtained using vision and SPIRIT information (both in isolation and combined through Bayesian fusion) with manually labeled ground truth annotations.

Table 11.2   Recall Achieved by the Fusion Method at a Comparable Level of Precision as That Yielded by the SPIRIT Modality on Its Own, as Shown in Table 11.1

Sequence	Recall	Precision	p_thresh	p_retention	fobj_retention
S2	0.979	0.974	0.75	0.20	10
S3	0.956	0.940	0.70	0.20	5
S5	0.882	0.992	0.75	0.20	10
S6	0.952	0.730	0.40	0.30	5

The values of the three variables affecting the fusion process that were optimized to achieve these results are also shown.

enable tracking of objects whose hypotheses have temporarily dropped below this threshold, p_retention specifies the minimum probability that must be associated with a tracked object hypothesis in order for it to continue to be tracked. fobj_retention specifies the maximum number of frames during which a hypothesis may be retained in this way before it is discarded, unless its probability once again rises above p_thresh.

Table 11.2 shows resulting recall and precision values together with the values of the three aforementioned variables that were selected to bring the fusion system's precision as close as possible to that of the SPIRIT modality for each of the four test sequences. It can be seen that the fusion method exhibits greater accuracy (as measured by recall) than the bat system for comparable levels of precision.

11.4.2.3   Appearance Modeling and Visualization

Appearance models acquired during visual tracking are used to improve the system's visualization capabilities. Apart from parameters such as people's height, the world model also stores a template image of each person's appearance (full frontal and back), which was taken using a digital camera. This template is used as a texture map applied to a generic 3-D model to render a more faithful representation of each person in the 3-D view created by the LabView application. Using the multimodal framework, it is possible to update the color distribution of the person's upper- and lower-body clothing to better represent their appearance on a given day.

This is done by performing a weighted combination of the hue, saturation, and luminance values of each of the pixels of the stored template image with the mean color of the corresponding region on the person's current color histogram appearance model. The process is performed separately for the upper body and leg regions. In the case of the template image, the rough boundaries between these two regions and the head are known a priori and skin detection is applied to prevent pixels belonging to exposed skin regions such as the hands and face from being updated. For the appearance model acquired dynamically during tracking, face and skin detection combined with the estimated height of the person (as computed from the Bat coordinates) are used to identify nonskin areas in the leg and upper body regions to determine their mean color. Depending on the precise settings of the visual appearance modeling process, the resulting color estimate already features robustness to noise and short-term variation. The results of the occlusion reasoning process and SPIRIT z-value are taken into account to determine whether the estimate is reliable, that is whether one of the two body regions is occluded dynamically (i.e., by another person) or statically (e.g., if the person is sitting down at a table).

The visual appearance of clothing is notoriously difficult to model realistically, let alone acquire automatically from image data. The simple mean color-based scheme used here gave good results for its intended application—the acquisition of updated texture maps of a person's appearance through visual tracking at run time. Different weights for the combination of hue, saturation, and luminance values were tested. In order to retain the characteristic texture of the person's clothing while updating its color, the luminance values from the template image are weighted more strongly while the weights associated with the new mean hue and saturation are set to higher values. Pixels with very high or very low luminance values must be weighted differently to ensure adequate results. The results are necessarily better for reasonably uniformly colored garments, although the region segmentation could in principle be used to provide finer granularity of the acquired models. Figure 11.13 shows results obtained for two people tracked in sequence S5. The process is complicated by the fact that the cameras used have poor color balance and that the lighting conditions tend to be poor. To compensate for these effects, global histogram equalization was applied to the frames prior to estimation of the appearance models, and a room-specific color bias was used to improve the perceptual quality and representativeness of the obtained color mean values used during updating.

11.4.2.4 Pose Estimation

A basic estimate of people's orientation is calculated by the SPIRIT system based on the fact that the human body shields some of the ultrasound emitted from the bat device. However, this has proven to be inaccurate and unreliable, particularly in some locations such as the corners of rooms or in the presence of surfaces that may reflect some of the ultrasonic pulse. For example, SPIRIT applications that dynamically project information onto a display screen or activate a device (such as a phone) in response to some event (such as bat button being pressed or a phone call being rerouted) need information about both the target person's position and the direction they are facing in order to select the most appropriate resource accessible and available to them.

Better pose estimates can be computed by combining visual information with the more precise measurements of a person's location obtainable through fusion with the SPIRIT system. Two relatively simple methods were implemented for this purpose. They both rely on shoulder detection and integration of cues from visual tracking, face detection, and the bat system observations (see Figure 11.14). Detection of the shoulders is greatly simplified due to the constraints that can be inferred from the various modalities. The bounding boxes derived from the visual- and SPIRIT-based tracking hypotheses allow a good initial estimate of shoulder position. Additional inferences regarding pose can be made on the basis of face and skin detections and other contextual information. For example, whether the person is sitting down or standing up can be inferred on the basis of the z-coordinate of the person's bat, and through occlusion reasoning combined with an analysis

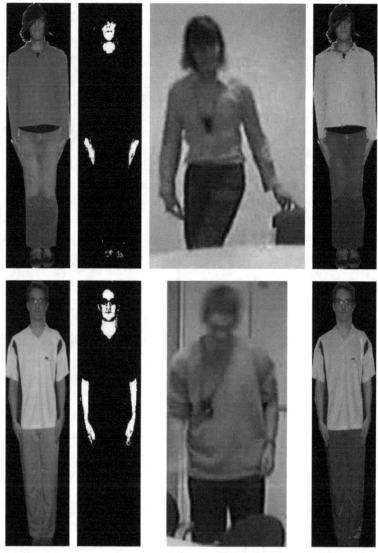

Figure 11.13 Example of dynamically updated visual appearance texture maps for two people. From left to right: original template image used for texture mapping; skin mask (detected skin pixels indicated in white); sample frame showing the same person being detected by the tracking framework; and updated template resulting from a combination of the original image with color information acquired by the color appearance modeling.

(by means of the neural network region classifiers) of any occluding regions (such as a wooden table or desk).

Shoulder detection is then performed by analyzing the edges determined by the segmentation method within the predicted regions. Curve fitting using a second order polynomial is applied to pixels on candidate edges as shown in Figure 11.15. The edge that matches (and produces the best match of any other edges) the expected shape of a shoulder contour as determined by the root mean squared error criterion is then selected within each of the two predicted shoulder regions for each person. In this way, 0, 1, or both shoulders may be identified. The midpoints of

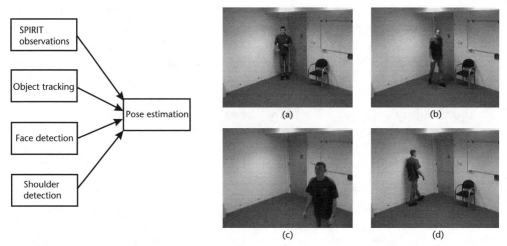

Figure 11.14 Left: Overview of the pose estimation process. Right: Examples of successful shoulder detections. Images (a), (b), (c), and (d) are four screenshots (in that order) from a video sequence.

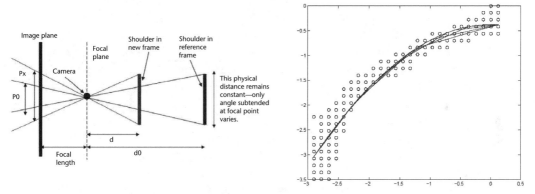

Figure 11.15 Perspective two-point algorithm for pose estimation. Left: Basis of the two-point algorithm. Right: Fitting of shoulder contour to edge points.

the detected shoulder curve together with the 2-D position of the Bat can then serve as input to pose estimation algorithms.

There are closed form solutions to the pose estimation problem given n points, although the solutions are not generally unique for 3 points or fewer. In the present case one has some knowledge of the position of the Bat on the person (assuming it is always worn in roughly the same location, such as hanging from a cord around the neck) and the actual separation of the person's shoulders (either by prior measurement or by assuming a standard value). Thus, given the 3-D position of the Bat (point P_3) and knowledge of the distances between the shoulder points (P_1 and P_2) and between the shoulder points and the Bat, one can infer the orientation of the plane spanned by (P_1, P_2, P_3) from the projections of these points (Q_1, Q_2, Q_3) onto the image plane. However, the three-point algorithm was found to be sensitive to errors in the measurements of the shoulder points and Bat location. One can often make the simplifying assumption that the three points (two shoulder points and one bat position) lie on a plane that is orthogonal to the ground plane, in other words, the person is upright. In that case the 3-D bat position can be used to estimate the distance from the shoulder points to the camera and the pose can now be estimated from the shoulder points alone. The two-point algorithm thus simplifies to computing the pose angle from $\Theta = \cos^{-1}(\frac{di_{12}}{dn})$, where di_{12} is the current distance between the shoulder points in the image and dn is their projected separation if the person were facing the camera (see Figure 11.15).

11.4.2.5   Identification and Verification

Visual information can be of use in detecting potential intruders (people not wearing functioning bats) and providing vision-based biometric verification of identity. By comparing the number and identity of people tracked visually and by the SPIRIT system, people who are not wearing a functioning bat can be detected. For security purposes, one can easily envisage an application that raises an event if such sightings are not expected to occur and forwards an image of the person(s) in question to an administrator. Another application currently under development concerns user authentication for security critical applications, for example, those that allow users to automatically unlock office doors or automatically log in to a computer in their vicinity. The SPIRIT system currently provides security only on the basis of something you have (the user's bat device), which may be augmented by applications also requiring something you know such as a password. Using computer vision, this could be enhanced by something you are, that is, by integrating visual biometric verification [49] of identity through analysis and recognition of facial characteristics, gait, iris patterns, and so forth.

By utilizing the combined fusion mechanisms, information from the SPIRIT system could be integrated to pose the problem in terms of verification (i.e., whether the person is who their bat indicates they are) rather than the much harder problem of unconstrained recognition. In addition, the SPIRIT system allows such prior information about the person's purported identity and actual biometric signature to be acquired in a nonintrusive and transparent way. The system could use the bat sensor information to detect that a person is present and select one or more camera views to verify their identity as indicated by the identity tag of their bat. The location of the bat can then be used to constrain the search window for a head and face detector, which forwards an image of the detected face to a face recognizer. Rather than solving the extremely difficult problem of general face recognition, visual authentication can then be approached as a verification problem. Moreover, it can be greatly constrained by fusing other kinds of information about assumed identity, face location, lighting conditions, and local office geometry.

11.4.3   Modeling of the Office Environment

The world model's static component can also be augmented using vision-based environmental modeling techniques. Solutions to several tasks have been implemented. Further details and results are given in [37, 50, 51].

11.4.3.1 Enhanced Visualization

The visualization of rooms can be made more compelling by acquiring texture maps of actual features such as decorations, carpet coloration, and white board surfaces. Figure 11.16 presents an overview of the steps involved in modeling textured surfaces and inferring dynamic state (door open or closed, see below) by integrating visual and SPIRIT information.

Given that the view-angle and position of a camera can be calibrated and correlated with the reference frame and office metadata of the world model, it is possible to reverse the projective mapping introduced by a camera and acquire viewpoint-normalized images of objects and surfaces of interest that can be used for texture mapping. Such texture maps are then used to provide enhanced and up-to-date visual models of the environment that can be viewed using applications such as LabView as discussed above. Figure 11.16 illustrates the geometric relationships between the SPIRIT and image plane coordinate systems. Once the relationship between camera and world (i.e., SPIRIT) coordinate frames has been established, a normalized (full frontal) view of the object or surface in question can be recovered numerically by inverting the perspective relationship. Figure 11.17 shows an example from a sequence captured by one of the cameras. An object of interest (plasma screen) whose 3-D bounding points is back-projected in order to obtain a view-normalized texture map of the given object in the original scene.

Such texture maps can then be rendered as part of the LabView application to augment its 3-D representation of the environment. Figure 11.18 shows an example of such enhanced room visualization. This would be of use in a range of applications such as video conferencing and seamless shared virtual environments. Participants located in different physical offices equipped with the sentient computing system could thus access and browse a unified view onto the world model, which includes dynamic features of interest such as the location and

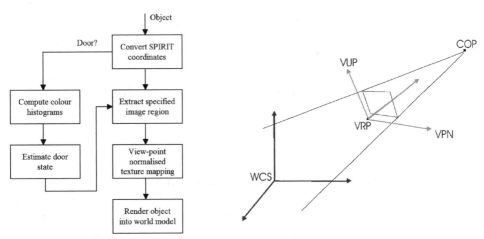

Figure 11.16 Left: Overview of the object extraction and state estimation process. Right: Perspective correction method for view-normalized texture mapping of real world objects and surfaces. The camera reference frame is defined in the world coordinate system (WCS) through the view reference point (VRP), the view plane normal (VPN), the view up vector (VUP), and the center of projection (COP).

Figure 11.17   Perspective correction and texture-mapping example.

Figure 11.18   Visually augmented representation of the world model. Left: LabView-generated view of an office. Right: The rendered scene has been augmented with texture maps acquired from a calibrated camera.

identity of people and the visual content of resources such as certain displays, whiteboards, and projector screens.

11.4.3.2   State Estimation

Aspects of environmental state such as that of doors (open, closed, ajar) and lights (on, off) can be determined. Door and light state can be estimated quite robustly using simple color histogram thresholding techniques. The classifier is a simple support vector machine that was trained on PCA-transformed images of the door region. Although some types of occlusions cause misclassifications, the method currently correctly identifies door state in over 80% of cases on a challenging test set.

11.4.3.3   Environmental Discovery

Semistatic parts of the environment such as furniture and computer screens can be recognized using vision techniques. As shown in Figure 11.19, neural networks (multilayer perceptrons and radial basis function networks) were trained to classify segmented images according to categories of manmade material (wall, carpet, cloth, and wood). The classifiers achieve correct classification of 92.8% to 96.5% on a large (about 500 images) test set of images. Classified image regions can then serve as an intermediate representation for object detection and recognition. Figure 11.19 shows two examples of Bayesian networks that were trained to detect two

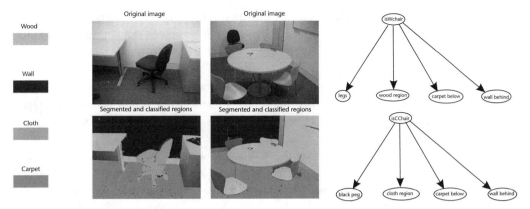

Figure 11.19 Left: Examples of neural network–based classification of man-made material categories in segmented images of the sentient office. Right: Bayesian networks for recognition of wooden (top) and cloth-backed (bottom) chairs based on image segmentation and classification information.

different categories of chairs commonly found in the office where the SPIRIT system has been deployed. The detectors are fairly robust with respect to scale, position, and orientation of the target objects, although occlusions remain a challenge (see also [37, 51]).

11.5   Summary

As computer vision continues to mature, it is likely to play an increasingly important role in the rapidly growing field of ubiquitous computing and indoor surveillance. This chapter presents a novel approach to harnessing the benefits of computer vision within the context of a sentient computing system deployed throughout an office space. It is shown how different computer vision methods such as tracking algorithms and appearance models can be fused with information from an ultrasonic surveillance system to significantly augment the capabilities and robustness of the system's world model.

The sentient computing system provides a variable granularity spatial model of the environment and a reliable device-tracking facility that can be used to automatically (re)initialize and refocus vision modules whenever an event or scene context of interest is observed by a camera. Changing reliabilities of different sources of information are handled robustly in a unified inference framework based on Bayesian networks. A hierarchy of object and environmental properties is used to integrate different hypotheses about the perceived context. The resulting world model serves as a shared representation of context that is made available to users and applications.

Unlike related approaches to Bayesian modality fusion [27, 48] for multiobject tracking, this approach is capable of incorporating both 2-D and 3-D visual and nonvisual data and does not rely on an expensive space partitioning. The number of tracked objects and the size of the modeled environment are not fixed, object arrivals and departures are handled robustly, and the approach scales smoothly as additional sensor resources (cameras or bat tags) are added or become unavailable.

Unlike fusion methods such as Kalman filtering or condensation, high-dimensional inputs can be modeled and both symbolic structural information and continuous data can be integrated into the Bayesian network.

Nevertheless, as explained in Section 11.3.3, the condensation algorithm proved effective for fusion of the visual appearance models and to integrate such information with target positions estimated by the vision-based tracker. Democratic integration is an efficient yet simple method for adapting the weights used in unimodal fusion of the visual cues. However, these methods are only used to perform vision-based fused to maintain hypotheses associated with particular objects as required by particular applications such as a security camera near the entrance of the office. Unlike the Bayesian fusion framework, they cannot easily handle differences, reliabilities, and accuracies associated with different sensor modalities and are too slow to be suitable for constrained real-time tracking in a large-scale distributed multisensory environment.

A number of applications of the sentient computing technology can in turn benefit from the video interpretation framework through the fusion of the ultrasonic and visual modalities. As further described in [37], such applications include pose and orientation estimation, biometric security, recognition of static elements of the environment such as whiteboards and furniture, and enhanced visualization for applications such as teleconferencing and shared augmented reality environments. Having visual information as an additional sensory modality is useful when the SPIRIT system has trouble detecting a person (e.g., the person is not wearing a Bat or it is temporarily concealed), or when an application requires additional information about a person's posture, direction of gaze, gestures, interactions with devices and other people, or facial expression.

To ensure sufficient performance to enable real-time processing, the fusion of individual perceptual modalities can be set up as a hierarchy where inexpensive detectors (e.g., finding the rough outline of a person) narrow down the search space to which more specific modules (e.g., a face spotter or gesture recognizer) are applied. The sensor fusion and information integration adds value to both the visual and ultrasonic modality by complementing their capabilities and adapting to error characteristics exhibited by the different sources of information at different times.

Acknowledgments

The author gratefully acknowledges financial support from the Royal Commission for the Exhibition of 1851.

References

[1]   Weiser, M., "The Computer for the 21st Century," *Scientific American*, Vol. 265, No. 3, September 1991, pp. 66–75.

[2]   Hopper, A., "Sentient Computing—The Royal Society Clifford Paterson Lecture," *Philosophical Transactions of the Royal Society of London*, Vol. 358, No. 1773, 2000, pp. 2349–2358.

[3]   Schilit, B., N. Adams, and R. Want, "Context-Aware Computing Applications," *Proc. Workshop on Mobile Computing Systems and Applications*, 1994.

[4]   Dey, A., "Understanding and Using Context," *Personal and Ubiquitous Computing*, Vol. 5, No. 1, 2001, pp. 4–7.

[5]   Erickson, T., "Some Problems with the Notion of Context-Aware Computing," *Communications of the ACM*, Vol. 45, No. 2, 2002, pp. 102–104.

[6]   Crowley, J., J. Coutaz, and F. Berard, "Things That See: Machine Perception for Human Computer Interaction," *Communications of the ACM*, Vol. 43, No. 3, 2000, pp. 54–64.

[7]   Addlesee, M., et al., "Implementing a Sentient Computing System," *IEEE Computer*, Vol. 34, No. 8, 2001, pp. 50–56.

[8]   Harle, R., and A. Hopper, "Deploying and Evaluating a Location-Aware System," *Proc. MobiSys 2005*, 2005.

[9]   Hazas, M., J. Scott, and J. Krumm, "Location-Aware Computing Comes of Age," *IEEE Computer*, February 2004, pp. 95–97.

[10]   Mansley, K., A. Beresford, and D. Scott, "The Carrot Approach: Encouraging Use of Location Systems," *Proc. of UbiComp.*, Springer, 2004, pp. 366–383.

[11]   Harle, R., "Maintaining World Models in Context-Aware Environments," Ph.D. thesis, University of Cambridge Engineering Department, Cambridge, U.K. 2004.

[12]   Smailagic, A., D. Siewiorek, and J. Anhalt, "Towards Context Aware Computing: Experiences and Lessons Learned," *IEEE Journal on Intelligent Systems*, Vol. 16, No. 3, 2001, pp. 38–46.

[13]   Hightower, J., and G. Borriello, "Location Systems for Ubiquitous Computing," *IEEE Computer*, Vol. 34, No. 8, 2001, pp. 57–66.

[14]   Want, R., et al., "The Active Badge Location System," Technical Report 92.1, AT&T Laboratories Cambridge, 1992.

[15]   Bahl, P., V. Padmanabhan, and A. Balachandran, "Enhancements to the RADAR User Location and Tracking System," Technical Report, Microsoft Research, 2000.

[16]   Priyantha, N., et al., "The Cricket Compass for Context-Aware Mobile Applications," *Mobile Computing and Networking*, 2001, pp. 1–14.

[17]   Lee, J., and R. Scholtz, "Ranging in a Dense Multipath Environment Using an UWB Radio Link," *IEEE Journal on Selected Areas in Communications*, Vol. 20, No. 9, 2002.

[18]   Want, R., et al., "Bridging Physical and Virtual Worlds with Electronic Tags," *Proc. CHI'99*, 1999, pp. 370–379.

[19]   Getting, I., "The Global Positioning System," *IEEE Spectrum*, Vol. 30, No. 12, 1993, pp. 36–47.

[20]   Duffett-Smith, P., and G. Woan, "GSM CURSOR," International Patent Specification No. 9519087.2.

[21]   Hightower, J., B. Brumitt, and G. Borriello, "The Location Stack: A Layered Model for Location in Ubiquitous Computing," *Workshop on Mobile Computing Systems and Applications*, 2002.

[22]   Nord, J., K. Synnes, and P. Parnes, "An Architecture for Location Aware Applications," *Proc. of the Hawaii Int. Conference on System Sciences*, 2002.

[23]   Ipina, D., and A. Hopper, "TRIP: A Low-Cost Vision-Based Location System for Ubiquitous Computing," *Personal and Ubiquitous Computing*, Vol. 6, No. 3, 2002, pp. 206–219.

[24]   Gavrila, D., "The Visual Analysis of Human Movement: A Survey," *Computer Vision and Image Understanding: CVIU*, Vol. 73, No. 1, 1999, pp. 82–98.

[25]   Turk, M., "Computer Vision in the Interface," *Communications of the ACM*, Vol. 47, No. 1, 2004.

[26] Cerney, M., and J. Vance, "Gesture Recognition in Virtual Environments: A Review and Framework for Future Development," Technical Report, Human Computer Interaction Center, Iowa State University, 2005.

[27] Sherrah, J., and S. Gong, "Continuous Global Evidence-Based Bayesian Modality Fusion for Simultaneous Tracking of Multiple Objects," *Proc. International Conference on Computer Vision*, 2001.

[28] Spengler, M., and B. Schiele, "Towards Robust Multi-Cue Integration for Visual Tracking," *Lecture Notes in Computer Science*, Vol. 2095, 2001, pp. 93–106.

[29] Stillman, S., and I. Essa, "Towards Reliable Multimodal Sensing in Aware Environments," *Proc. of the 2001 Workshop on Perceptual User Interfaces*, 2001.

[30] Crowley, J., et al., "Perceptual Components for Context Aware Computing," *Proc. Ubicomp. 2002*, 2002.

[31] Cattin, P., D. Zlatnik, and R. Borer, "Biometric System Using Human Gait," *Proc. on Mechatronics and Machine Vision in Practice (M2VIP)*, 2001.

[32] Fritsch, J., et al., "Multimodal Anchoring for Human-Robot Interaction," *Robotics and Autonomous Systems*, Vol. 43, No. 2, 2003.

[33] Nummiaro, K., E. Koller-Meier, and L. Gool, "An Adaptive Color-Based Particle Filter," *Image and Vision Computing*, Vol. 21, 2003, pp. 99–110.

[34] Perez, P., J. Vermaak, and A. Blake, "Data Fusion for Visual Tracking with Particles," *Proc. of IEEE*, Vol. 92, No. 3, March 2004, pp. 495–513.

[35] Choudhury, T., et al., "Boosting and Structure Learning in Dynamic Bayesian Networks for Audio-Visual Speaker Detection," *Proc. Intl. Conference on Pattern Recognition*, 2002.

[36] Torralba, A., et al., "Context-Based Vision System for Place and Object Recognition," *Proc. International Conference on Computer Vision*, 2003.

[37] Town, C., "Ontology Based Visual Information Processing," Ph.D. thesis, University of Cambridge, Cambridge, U.K. 2004.

[38] Town, C., "Multi-Sensory and Multimodal Fusion for Sentient Computing," *Int. Journal of Computer Vision*, Vol. 71, No. 2, 2007, pp. 235–253.

[39] Bouguet, J., MATLAB Calibration Toolbox, http://www.vision.caltech.edu/bouguetj.

[40] Garcia, C., and G. Tziritas, "Face Detection Using Quantized Skin Color Regions Merging and Wavelet Packet Analysis," *IEEE Trans. on Multimedia*, Vol. 1, No. 3, 1999, pp. 264–277.

[41] Town, C., "Ontology-Driven Bayesian Networks for Dynamic Scene Understanding," *Proc. Intl. Workshop on Detection and Recognition of Events in Video*, 2004.

[42] Sinclair, D., "Smooth Region Structure: Folds, Domes, Bowls, Ridges, Valleys, and Slopes," *Proc. Conference on Computer Vision and Pattern Recognition*, 2000, pp. 389–394.

[43] De la Torre, F., and M. Black, "Robust Principal Component Analysis for Computer Vision," *Proc. International Conference on Computer Vision*, 2001.

[44] McKenna, S., Y. Raja, and S. Gong, "Object Tracking Using Adaptive Color Mixture Models," *Proc. Asian Conference on Computer Vision*, 1998, pp. 615–622.

[45] De la Torre, F., and M. Black, "Robust Parameterized Component Analysis: Theory and Applications to 2D Facial Appearance Models," *Computer Vision and Image Understanding*, 2003.

[46] Hanbury, A., "Circular Statistics Applied to Color Images," *8th Computer Vision Winter Workshop*, Vol. 91, No. 1–2, July 2003, pp. 53–71.

[47] Skocaj, D., and A. Leonardis, "Robust Continuous Subspace Learning and Recognition," *Proc. Int. Electrotechnical and Computer Science Conference*, 2002.

[48]   Toyama, K., and E. Horvitz, "Bayesian Modality Fusion: Probabilistic Integration of Multiple Vision Algorithms for Head Tracking," *Proc. Asian Conference on Computer Vision*, 2000.

[49]   Bengio, S., et al., "Confidence Measures for Multimodal Identity Verification," *Information Fusion*, Vol. 3, 2002, pp. 267–276.

[50]   Town, C., "Vision-Based Augmentation of a Sentient Computing World Model," *Proc. Int. Conference on Pattern Recognition*, 2004.

[51]   Song, Y., and C. Town, "Visual Recognition of Man-Made Materials and Structures in an Office Environment," *Proc. Intl. Conference on Vision, Video and Graphics*, 2005.

CHAPTER 12

An End-to-End eChronicling System for Mobile Human Surveillance

Gopal Pingali, Ying-Li Tian, Shahram Ebadollahi, Jason Pelecanos, Mark Podlaseck, and Harry Stavropoulos

Rapid advances in mobile computing devices and sensor technologies are enabling the capture of unprecedented volumes of data by individuals involved in field operations in a variety of applications. As capture becomes ever more rich and pervasive, the biggest challenge is in developing information processing and representation tools that maximize the utility of the captured multisensory data. The right tools hold the promise of converting captured data into actionable intelligence resulting in improved memory, enhanced situational understanding, and more efficient execution of operations. These tools need to be at least as rich and diverse as the sensors used for capture, and they need to be unified within an effective system architecture. This chapter presents our initial attempt at such a system and architecture that combines several emerging sensor technologies, state-of-the-art analytic engines, and multidimensional navigation tools into an end-to-end electronic chronicling [1, 2] solution for mobile surveillance by humans.

12.1 Introduction: Mobile Human Surveillance

There are a number of applications today in which information is collected in a mobile and pervasive manner by numerous people in the field going about their jobs, businesses, lives, and activities. For instance, law enforcement personnel patrol certain areas, look for interesting or suspicious activity, and report and take action on such activities for security, surveillance, and intelligence-gathering purposes. In this kind of an application there is a lot of value in being able to access, review, and analyze the information gathered by people in the field both by the individual gathering the information as well as their peers and higher officers who would like to compare notes, integrate information, form a more complete picture of what is happening, and discover new patterns and insights. Similarly, emergency personnel such as firefighters and emergency medical teams can record their activities and review operations for effectiveness, failure points, people involved, operational insights, and so forth. Another example is the army, in which soldiers are involved in battlefield, peacekeeping, and anti-insurgency operations.

Our systems goal is to develop an end-to-end electronic chronicling system for mobile workers to much more effectively capture, relive, analyze, report, and reuse their experiences from field operations, enabling:

- *Auto-diary creation:* As a worker operates on a mission, her wearable system should generate a richly annotated multimedia diary.
- *Chronicle navigation and reporting tools:* The worker should be able to produce effective after-action mission and intelligence reports by using a unified chronicle management and navigation system that enables them to find events and content of interest, drill down to the desired level of detail, and find interesting correlations.
- *Theater-level navigation and search:* The system should enable appropriate people (such as a higher officer) to combine and navigate the chronicles from multiple workers to obtain the bigger picture. The system should also allow individual workers to subscribe to events of interest and receive automatic notifications.

To this end we are developing a system and architecture (Figure 12.1) that supports multimodal and multisensory capture, provides a variety of sensor and data analytics to detect and extract events of interest, and provides multi dimensional navigation and search capabilities.

There are several technical challenges in developing such an electronic chronicling system.

- *Wearable capture system:* This should be easily carried by a mobile worker, integrate appropriate sensors to create a rich record of the user's activities, ensure synchronized real-time capture of all data, support real-time annotation(both automatic and manual), and work uninterrupted to last the length of the worker's mission.

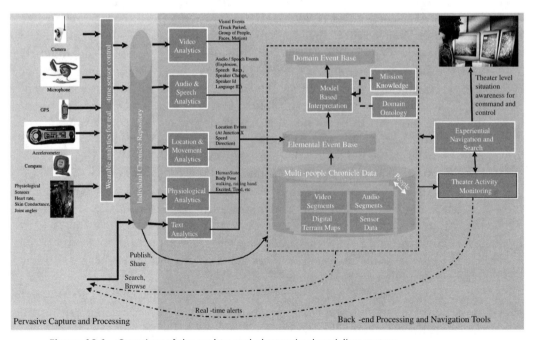

Figure 12.1 Overview of the end-to-end electronic chronicling system.

- *Analytics for event detection:* These should analyze the significant volume of multisensory chronicle data to extract events of interest, enabling the users to access the data by such events. Clearly, the challenges here span the gamut of individual sensory processing technologies such as image analysis, computer vision, speech recognition, movement analysis, and location understanding. The challenge is also to combine these analytics effectively into an integrated system that presents the users with the ability to review episodes and derive situational understanding.
- *Event and data management:* The system should have appropriate representations and assimilation mechanisms to link raw data, events derived from sensory processing, and domain-level events of interest to end users.
- *Experiential navigation techniques:* The utility of such a chronicling system is heavily dependent on the kind of mechanisms provided for browsing, filtering, and searching. End users should be able to relive their experience or the experiences of others, look for events and episodes of interest, or perform deeper analysis to derive valuable insights.

Clearly, our goals are ambitious and demand an ongoing research effort, both on individual technologies and also on a platform that effectively unifies these technologies into an end-to-end electronic chronicling solution. In this chapter, we provide an overview of our approach and present a version of such an integrated eChronicling system.

12.2 Related Work

The most popular related research theme to eChronicling has been personal information management and retrieval of personal information [3–7]. The advent of wearable devices and wearable computing [8] has had significant influence on the recent emergence of such efforts (e.g., [9]) around personal chronicles that enable rich capture of an individual's life and retrieval based on context [10]. Other efforts [11–14] have addressed the difficulties in organizing the regular information stored on computers. The area of group- and enterprise-level chronicles [15–17] remains largely unexploited, especially relative to the ongoing efforts on personal chronicles. A promising area for research is mining of captured data [18] and generation of associated alerts and notifications. Another related area is distributed event-based systems [19–23], although these have not yet addressed event extraction from multisensory captured data as discussed here.

Video surveillance is the use of computer vision and pattern recognition technologies to analyze information from situated sensors [24–27]. The key technologies are video-based detection and tracking, video-based person identification, and large-scale surveillance systems. A significant percentage of basic technologies for video-based detection and tracking were developed under a U.S. government-funded program called Video Surveillance and Monitoring (VSAM) [24]. This program looked at several fundamental issues in detection, tracking, autocalibration, and multicamera systems [28–30]. There has also been research on real-world surveillance systems in several leading universities and research labs [31]. The next generation of research in surveillance is addressing not only issues in detection and tracking but also issues of event detection and automatic system calibration [32]. The second key challenge of surveillance—namely, video-based person identification—has also been a subject of intense research. Face recognition has been a leading modality with both ongoing research and industrial systems [33, 34]. A U.S. government research program called Human ID at a Distance addressed the challenge of identifying humans at a distance using techniques such as face-at-a-distance and gait-based recognition [35]. One of the most advanced systems research efforts in large-scale surveillance systems is the ongoing U.S. government program titled Combat Zones That See [36]. This program explores rapidly deployable smart camera tracking systems that communicate over ad hoc wireless networks, transmitting track information to a central station for the purposes of activity monitoring and long-term movement pattern analysis. There are several technical challenges that need to be addressed to enable the widespread deployment of smart surveillance systems. The video surveillance systems, which run 24 hours per day and 7 days per week, create a large amount of data, including videos, extracted features, alerts, and so forth. How to manage this data and make it easily accessible for query and search are other challenges.

Audio analytics, also known as audio scene analysis [37], provide the ability to label acoustic events in an audio recording. Some examples include performing speech transcription, speaker identification, and identifying machine sounds. Audio surveillance has traditionally taken second position after video monitoring in terms of its widespread use. However, it is recognized that in many applications, audio analytics provide complementary information to video analytics. Research outcomes in audiovisual speech recognition [38] exemplify this. A significant benefit of audio analytics is the in-depth detail that is potentially available. One example includes a person being recorded while discussing a complex situation in his or her environment. In essence, the human in the environment analyzed the context of the situation and then provided a verbal description of the events that occurred. The audio recording may then be readily transcribed, indexed, and searched. Past research in the audio scene analysis area has investigated many categories of audio events, from speech recognition [39] to snore and sleep apnea detection to finding explosions in videos [40].

12.3   System Architecture and Overview

Figure 12.1 presents a conceptual overview of our system. The left part of the figure depicts support for a variety of wearable capture sensors and personal devices to enable pervasive capture of information by individuals. Supported sensors include digital cameras, microphones, GPS receiver, accelerometer, compass, skin conductance sensors, heart rate monitors, and so forth. The user captures data through these sensors, which are either worn or carried. On-board processing on a wearable computer provides real-time control for data capture, and, to some extent, local analysis of captured data. In addition to data captured through wearable sensors, the system also allows input of event logs and corresponding data from personal devices such as personal computers and PDAs (for more on this, refer to [15, 17, 41]). The user can also enter textual annotations both during wearable capture and while working on a PC/PDA. The data thus captured is stored with appropriate time stamps in a local individual chronicle repository. This electronic chronicle [1, 2], in short, represents a rich record of activity of the individual obtained both from wearable sensors while in the field and event loggers on the PC while at his or her desk.

The right portion of Figure 12.1 depicts the back-end processing and navigation tools to analyze, extract, manage, unify, and retrieve the information present in electronic chronicles. First, a variety of analytic engines process the chronicled data, including those that process image/video, speech/audio, text, location and movement, and physiological data. The results of this analysis are elemental events detected at the sensory data level. These represent metadata, or machine-derived annotations representing events in the original captured data. Examples of such events include detection of a face, a moving vehicle, or somebody speaking a particular word. The system also includes a data management component that stores the raw chronicle data, the elemental events, and domain-level events obtained from further analysis of the elemental events. The latter are events that are expressed in terminology specific to a domain based, for example, on mission knowledge, and specific ontology. For example, detection of an explosive sound or visible fire is an elemental event in our terminology, while detection of a fuel gas incident of type propane involving toxic release with high severity is a domain-specific event.

The rightmost portion of Figure 12.1 indicates tools for navigation and retrieval of the chronicled data and associated events and metadata. These tools communicate with the data management component and allow the user to search, explore, and experience the underlying data. These tools apply at an individual level for browsing personal data and also across data shared by multiple individuals. In the latter case, they enable overall situation awareness based on data from multiple people and enable theater-level search and planning.

An important challenge in a plug-in architecture system is the flexibility to adapt the base architecture to different domains with differing sensing, analytics, and navigational needs. In order to provide such flexibility, we provide a clean separation between the sensors, the analytics, the data management, and the navigation/retrieval components. Standardized XML interfaces define ingestion of data from the sensors into the database. Similarly, XML interfaces are defined between the analytics components and the database. The analytics components, which can be distributed on different servers, query the database for new data via XML, appropriately process the raw data, and ingest the processed results back into the database. Finally, the navigation tools retrieve data and present it to the user as appropriate. This architecture allows the system to be distributed—easily changed to add or remove sensors, add or replace a particular analytic component, or modify the navigation tools. Thus, for example, the same base architecture is able to support synchronized automated capture of audio, video, and location in one domain while allowing manual capture of data in another domain. This plug-in approach is based on and inspired by [26].

For the rest of this chapter, we will focus on an implementation of the part of this system that involves automatic mobile capture and back-end analysis and navigation of the captured data. Figure 12.2 shows a simplified view of

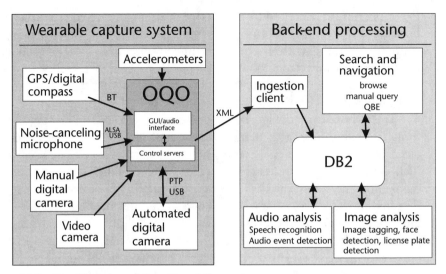

Figure 12.2 Simplified view of the eChronicling system.

the electronic chronicling architecture in Figure 12.1 to focus on this mobile capture system with attendant back-end processing. This system consists of a wearable capture component that can be used for automatic capture in the field. The system was created to support an open-ended assortment of data gathering devices: digital/video cameras, microphones, GPS trackers, skin conductance sensors, and so forth.

Figure 12.3 shows examples of capture devices and two different examples of users sporting a wearable capture system.

There is also flexibility in on-board versus back-end processing. Some of this data may be processed on board a wearable computer; some will be sent real-time to the back-end system for processing; some will have to wait for a more opportune time for ingestion to the back end. The user can annotate this data during capture (if and when possible) and certainly after ingestion.

Once ingested, the data is analyzed by analytics engines. The results of this analysis are elemental events detected at the sensory data level (e.g., detection of a face,

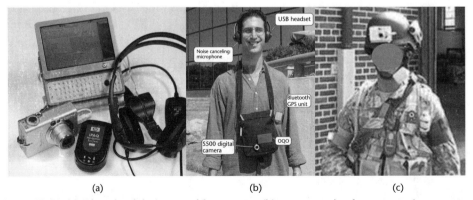

Figure 12.3 (a) Example of devices used for capture; (b) one example of a user wearing a capture system consisting of devices in (a); and (c) another example of a user wearing an extended set of capture devices.

a moving vehicle, somebody speaking a particular word); these results are ingested in the database (which could be different than the one containing the sensor data).

The ingested results of the analysis may be:

1. Fed to another analytics engine that will try, during a second pass, to correlate them (for example, correlate the detected-by-the-image-analysis vehicle with the detected-by-the-audio-analysis motor sound, to enhance the confidence level that a vehicle was indeed present);
2. The results of 1 may be similarly further analyzed to derive domain-specific events. For example, detection of an explosive sound or visible fire is an elemental event in our terminology, while detection of a fuel gas incident of type propane involving toxic release with high severity is a domain-specific event.

Eventually some GUI client will retrieve what has been already analyzed (note that the analysis need not be completed). The client should summarize the information; a user, however, should be able to track down any detected "event" to the media captured. Another set of clients will be notified the moment some domain-specific event gets ingested.

12.4 Event Management

The eChronicling system fundamentally manages "Events"—data entities with an associated time stamp/time interval and often with a location stamp. The following types of events are included:

1. *Elementary events, detected at the sensory level.* There are specified either by the corresponding analytics engine or by a user when browsing the data.
2. *Composite events.* In two flavors:
    a. As in 1, only surmised from more than one medium (e.g., "car in a location" may be inferred from two pictures and the audio of what was recognized to be an engine);
    b. Higher-level, domain-specific events.
3. *Semi-independent.* Typically events are directly related to some media (e.g., "car in certain a location" will be inferred from the images). A soldier can annotate a part of a patrol as "chasing the suspect." During that time interval several images will be displaying the suspect. Others will not. The "event" is associated implicitly (by its duration) with a subset of the pictures/video/audio.

In order to accurately represent events, we must ensure that the devices are properly synchronized so that temporal searches and correspondences are accurate. We take the GPS time to be the master clock and use the OQO clock to compute offsets between GPS readings. We estimated the camera delay by taking a series of automated captures of a graphical display of the OQO's clock and computing the time difference between issuing the capture command and the visible time in the photograph. For manually captured images, we read the capture time embedded in the EXIF data and offset this by a constant determined by photographing the OQO display as before. The second issue is that of labeling each voice annotation and image with a GPS coordinate and correlating images with audio clips. In both cases, we use temporal correspondence to guide the mapping. Currently, we search the GPS log for the nearest location and make the association if the difference is less than 5 seconds.

12.4.1 Storage

Each kind of event is stored in its own table; this provides for cleaner data. Alternatively, we would need to encode extra information in the table (i.e., what kind of event a certain row represents). The downside is that the single-table approach is more extensible; new kinds of events can be specified with less database interference. The current thinking is that this more flexible path should be taken in the future.

User interfaces want to look at the events chronologically, by location, by type, and so forth. Some of this aggregation happens at the database level, and some happens in the client. Since we anticipate an open-ended set of interfaces, what happens where will certainly vary: things that need time-consuming SQL joins (e.g., "retrieve all images of cars grouped by detected plate number, taken the past month, in those two towns") should be computed in the database (or, even, precomputed); simpler data relationships can be managed or cached by the client.

12.4.2 Representation

There are three categories of data associated with events, as seen in Figure 12.4.

1. *Sensor data*. These are immutable.
2. *Annotations*. These are either output of the analytics or comments manually or vocally entered by humans.
3. *Correlations and histories of navigation* of the data by a user using a GUI client.

The constraint here is the ability to do efficient SQL-type joins for data retrieval. In the current incarnation of the system, all entries are stored in the same database (more details in Section 12.5). Sensor data may be stored anywhere— only the URLs are stored in the tables. Also, every datum has a UUID associated with it. This was created to be used as a foreign key to correlate the data in the tables as well as a means to maintain sanity once the sensor data starts getting replicated in order to be cached/distributed/performance-scaled.

12.4.3 Retrieval

During the development of the system, it became clear that the retrieval requirements for the data defied prediction. At a minimum, an SQL query would be needed for each such access as well as a hosting script (which could be in the form of a servlet, a CGI script, a DLL loaded by the client, or anything else).

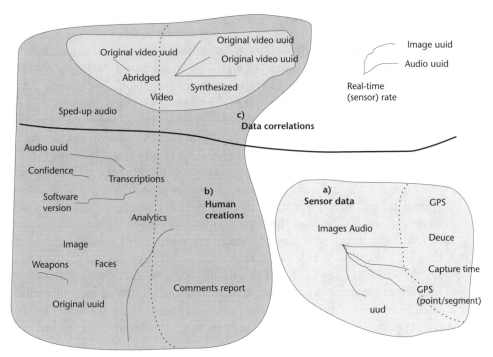

Figure 12.4 (a–c) Overview of types of data associated with events and their representation scheme.

The decision was made to create a very generic http-accessible script and let the client submit the full-blown SQL query (this can be done in a semitransparent way, with the database maintainer making the query available for download by the clients). This way the set of queries is open-ended; the server does not need to be touched as new ones are developed.

12.5 Multimodal Analytics

We focus here on three types of analytics—image classification, face detection, and speech/audio analysis. Image classification helps in searching through the numerous images captured by the user based on concepts associated with the images. Face detection aids in automatically retrieving those images in which there were human faces. Speech/audio analytics help in: (1) transcribing and extracting keywords from the annotations made by the user when on his or her mission; and (2) analyzing environmental sounds such as other people talking, sounds of vehicles, explosions, and so forth. Together, these analytics aim to enhance the user's ability to identify and retrieve interesting events that occurred during missions.

12.5.1 Image Classification

Numerous images or long hours of video data could accumulate very quickly in the context of electronic chronicling applications. Sifting through these visual data for when a building was seen or a vehicle was spotted becomes a challenging task if one only relies on time and space-based navigation tools. Efficient means for accessing these data via the *semantic concepts* they portray becomes indispensable. Our goal is to equip the electronic chronicling system with automatic concept tagging capability for visual data to provide such means of interaction.

Automatic semantic concept tagging in image-based data requires bridging the semantic gap for the domain of application of the chronicling system. As evidenced by the top performing approaches reported in video tagging benchmarking exercises, such as TRECVID, machine learning–based methodologies for tagging are becoming the de facto standard [42]. In our chronicling system, we also train discriminative classifiers using both global and local image features extracted from a collected set of positive and negative example images for each concept of interest.

Figure 12.5 gives an overview of our approach for image clarification. As shown in the figure, for each distinct dataset, where distinction is due to the nature of the images and the image collection process, a *support vector machine* (SVM) [43] classifier is trained using the positive and negative data provided in that set for different types of features extracted from images. Different classifiers for the same dataset are fused to provide one semantic concept classifier for each individual semantic concept tag that the user is interested in, and annotation has been provided for the training data set. Classifiers obtained for any given semantic concept from various data sets are then fused to obtain a single classifier for that concept.

The reason for employing various datasets is to cover the *multiview* mani-festations of the same concept for different applications and context. For example, concept vehicle could have a different visual manifestation in consumer photos than its visual manifestation in military-related applications. In other words, the multiple data sets and therefore models for a given concept are employed to address the issue of visual polysemy for a given concept.

The result of applying the array of semantic concept taggers to a test image is a set of confidence values associated to the image by each of the concept models.

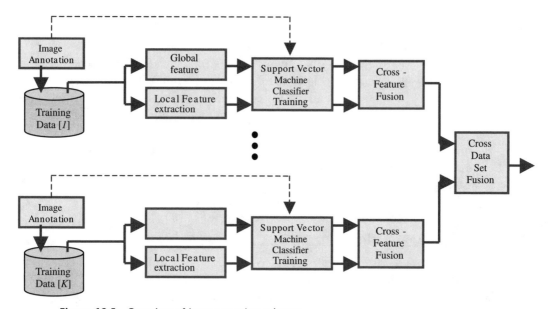

Figure 12.5 Overview of image-tagging scheme.

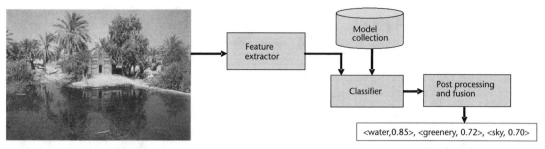

Figure 12.6   Illustration of image tagging.

Only those concept tags are assigned to the image, which have a confidence value greater than a given threshold. This threshold is obtained such that it best matches the satisfaction of a human user. We obtain this threshold from a set of training data beforehand using a utility maximization scheme, referred to as T10U [44]. The confidence values, which pass the threshold, are then converted to posterior probabilities using the sigmoid function according to Platt [45], in order to be used for further fusion with the outputs of other modules of the electronic chronicling system. Figure 12.6 illustrates the process of assigning semantic concept tags to an image.

12.5.2   Face Detection and License Plate Recognition from Images

12.5.2.1   Face Detection

Face detection is a challenging problem because of variations in pose, lighting, expression, and occlusions (e.g., beards, mustaches, glasses, hair, hat) There are a number of techniques that can successfully detect frontal faces, such as Neural Networks [46], statistics of parts [47, 48], and Adaboost learning methods [49–51]. In our system, we implemented multiview face detectors based on Haar and optimized wavelet features by using a cascade of boosted classifiers.

Haar features are very useful for training extremely fast detectors. However, when learning from small datasets, these features are limited to produce robust detectors. In fact, the choice of good features is very important for projecting boosted classifiers when few examples are provided. We observe that discriminative features in general tend to match the structure of the object. For example, the first selected Haar features encode the fact that eyes are darker than nose, and so forth. This suggests us to use an optimization technique—wavelet networks—that creates features (Gabor, Haar, and so forth) whose scale, orientation, and position aligns with the object local structure. So, our approach is to learn optimized features for each specific object instance in the training set and then bring together these features into a bag of features. Then we project a cascade classifier that selects the most general features from the bag of features for discrimination.

First, the input image is scanned across location and scales. At each location, an $M \times M$ (20 × 20 in our system) image subwindow is considered. Following [51], a boosted collection of five types of Harr-like features (Figure 12.7) and a bag of

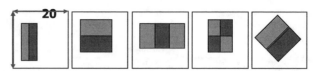

Figure 12.7 Example of Harr-like features for face detection.

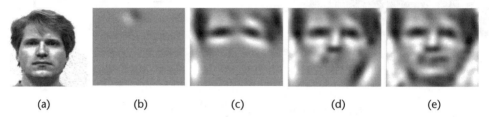

(a)   (b)   (c)   (d)   (e)

Figure 12.8 Example of optimized wavelet features for face detection. (a) Original image and (b)–(e) wavelet presentation.

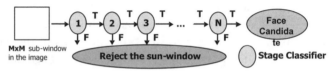

Figure 12.9 Cascade of classifiers for face detection.

Figure 12.10 Examples of face detection results.

optimized wavelet features (Figure 12.8) are used to classify image windows. Then, cascade classifiers are employed to evaluate the input subwindow (Figure 12.9). If the first classifier of the cascade returns false, then computation on the subwindow stops and the subwindow is nonface. If the classifier returns true, then the subwindow is evaluated by the next classifier in the cascade. Face detection is declared only if a subwindow passes all the classifiers with all returning true. Since most subwindows in an image are nonfaces, they are rejected quickly. This process greatly improves computational efficiency as well as reducing the false positive rate [49, 51]. Figure 12.10 shows some example results.

Figure 12.11 Examples of license plate recognition result. (Part of licence plate has been occluded for the purpose of privacy.)

12.5.2.2 License Plate Detection

We also integrated the license plate recognition into our system. The license plate recognition technology is licensed from Hi-Tech Solutions [52]. This technology could be deployed at the entrance to a facility where it catalogs the license plate of each of the arriving and departing vehicles. First, the license plate is located in the image. Then the optical character recognition (OCR) solutions are used to recognize the number of the license plate. Some examples are shown in Figure 12.11.

12.5.3 Audio and Speech Analytics

The audio and speech analysis system is an invaluable resource for labeling relevant and interesting audio events. Such technology is useful for applications where rapid search and indexing capabilities are paramount for the early identification of consistent routines, new trends, or isolated events. For businesses, the rapid location of such audio events can present new opportunities and trends while exposing potential threats that would otherwise remain concealed.

The audio and speech analysis system exploits multiple independent acoustic event detectors. The advantage of using independent detectors is that a change made (or a critical failure) for a detector will not influence other operational audio event detectors. Figure 12.12 presents the basic structure of the acoustic processing system tailored for analyzing audio from military scenarios. In this system the audio is preprocessed with a bulk audio processing component that downsamples

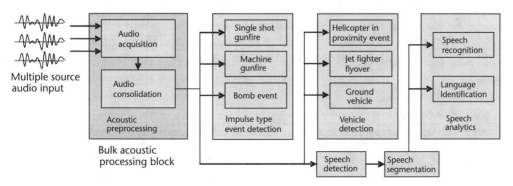

Figure 12.12 Structure of the acoustic event extraction system.

the audio and removes uninteresting audio such as silence. This has the added benefit of reducing the processing load on the following processing components. This processed audio is passed to the impulse, vehicle, and speech analytics elements accordingly. The speech analytics block includes both speech recognition and language identification. Other technologies that could be included are speaker recognition, emotion detection, dialect detection, and background noise profiling. Each of the technologies used in this system will be briefly described.

- *Impulse type event detection:* The task here is to detect acoustic events that occur as a result of a projectile or bomb explosion. These types of acoustic events are impulsive in nature and generally have high signal energy when in close proximity. A two-phase approach was implemented to detect such events. The first phase used hidden Markov models (HMMs) trained on cepstral- and energy-based features in a similar manner to [40] to detect the candidate events. These candidates were then submitted to a second phase test whereby events that were below a specified power level were disregarded.

- *Acoustic-based vehicle presence detection:* Acoustic analysis may be used to determine the presence of different types of utility vehicles, both land-based and airborne. The system established for this evaluation was designed to detect land-based vehicles. A car, for example, generates sound from the combustion engine, moving parts on the vehicle, and the tires. An HMM structure trained on cepstral-based features was established to detect the presence of a vehicle. At the time of the experiment, it is important to note that no noise or channel compensation was applied to these features. Preliminary studies examined the misclassification of vehicles (or when the system was making the classification errors). It was observed that some types of diesel engine vehicles were being confused with other diesel engine vehicle types. Similarly, nondiesel (car fuel) engines were misclassified as other vehicles of the nondiesel type.

- *Speech analytics:* The *speech recognition* block is comprised of two main parts: namely, a speech segmentation block followed by a state-of-the-art speech recognition engine. The speech recognition engine is based on an HMM framework. Some of the significant developments for improved speech recognition performance include subspace precision and mean (SPAM) models [53] and fMPE [54]. For this particular evaluation, the speech recognition system was evaluated within the framework of keyword spotting. (A keyword spotting system would have a considerable advantage.) The *language identification* component utilized speech segmentation boundaries identified by the speech recognition preprocessing engine. The identified segment was phonetically transcribed and later classified according to a binary tree classifier structure [55, 56]. The language identification system was trained on seven different languages.

12.5.4 Multimodal Integration

After the processes of image classification, object detection, and speech analysis, the processes results are integrated to achieve more accurate results. To integrate all the different features and results, several approaches such as multilayer HMMs method and rule-based combination of a hierarchy of classifiers can be used. For example, without multimodal integration, the license plate recognition is running on all the input images. By combining with the image classification results, the license plate recognition is skipped for indoor images. This process reduces the computation cost and decreases the number of false positives.

12.6   Interface: Analysis and Authoring/Reporting

We have implemented several navigation interfaces that enable end users to search, retrieve, and filter events and data of interest to them, either from their own missions and experiences or those of other people. These interfaces are essentially multidimensional and allow navigation based on space, time, or events of interest. Our aim has been not only to retrieve events of interest, but also to allow users to relive relevant experiences, especially those shared by others.

Figure 12.13 shows a screen shot of the multidimensional navigation tool for retrieving, searching, and filtering the captured and analyzed data in the electronic chronicle. The interface consists of a document/data summary view (the left column in Figure 12.13), a set of browsing controls (top right of Figure 12.13), and a map (bottom right of Figure 12.13). The document summary shows a thumbnail view of the document/data, if available, and metadata about the document/data.

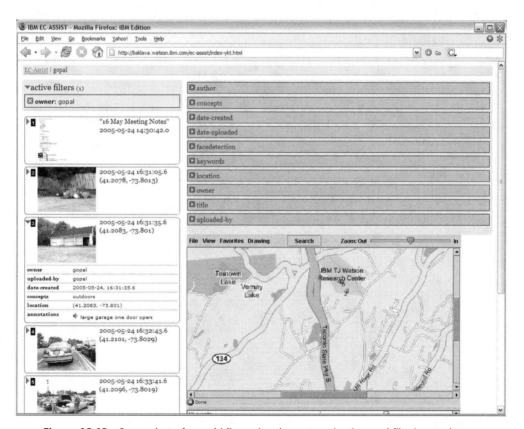

Figure 12.13   Screenshot of a multidimensional-event navigation and filtering tool.

A document (captured from the user's PC) and several images (captured in the field) are seen in Figure 12.14. All these have time stamps associated with them, and images also have GPS locations associated with them. The pull-down on a document gives further information associated with the document. For example, the image numbered three in Figure 12.14 shows further information including the author, date of creation, and so forth. Notice that the concept associated with this image is "outdoors," obtained by the analytics discussed in Section 12.5.1. Also notice that the image has associated "annotations." In this case there is an audio file, which can be played by clicking on the speaker icon. The transcription of the speech in the audio is also seen ("large garage one door open"). This is obtained by analyzing the recorded audio, as discussed in Section 12.5.2.

The browsing controls on the right enable the user to filter the vast quantities of captured data based on their interest. The user can navigate the data by space, time, and a variety of metadata, including image concepts, presence or number of faces, type of document, keywords from speech, audio events, author/creator of the data, and so forth. The tool allows multiple filtering criteria to be combined. When location information is available, the data is also shown on a map. In Figure 12.13, the geographic location of documents is shown with the number associated with the particular document. If the user is interested in browsing data or events that occurred only in a particular geographic area, he or she can simply draw a box

Figure 12.14 Example result: User views all data from a trip without filtering. Notice the ability to view the data by space, time, concepts, and keywords.

to indicate the area of interest on the map, which results in displaying only data or events from that area.

The interface in Figure 12.13 is implemented within a Web browser running on the user's machine. The Web page requests XML data via an HTTP call to the database server. The interface is generated from the XML via XSLT and JavaScript. The XSLT stylesheet processes the XML to produce the HTML for the visual aspects of the interface. It also generates embedded JavaScript calls to pass information gleaned from the XSLT parser to the client-side JavaScript logic. This allows us to avoid having to parse the XML data ourselves. This interface approach has been inspired by [57].

We have started running data collection and navigation experiments with our system, by collecting data within a 20-mile radius around our research labs. For illustration purposes, we will use data collected in one such trip where a user made a round-trip from our lab location in Hawthorne, New York, to our other lab located in Yorktown, New York. The user collected images, GPS, and audio all along the way (the car was driven by a colleague). The user and colleagues surveyed the Yorktown facility, both outdoors and indoors, had lunch there, and returned to Hawthorne. In addition to the data collected during the trip, the user also input some trip-planning documents and presentations into the database. The following examples show some of the different ways of browsing and retrieving data of interest from this trip.

Figure 12.14 shows one view of the data without any filters applied. In this view, the user sees a timeline and events marked along the timeline to indicate when images were taken. The corresponding locations are also shown in the map (marked by numbers overlaid on the map, along the highway in the middle of Figure 12.14). On the left are a variety of image concepts and keywords detected from the user's speech annotations, as well as from his documents and presentations. All the images captured are seen in this view—the user simply has to move his cursor along the timeline.

Figure 12.15 shows the result of a "face" filter applied to this data—in other words, only the user chooses to see only those images that had faces in them. Figure 12.15 shows a subset of the images in sequence. Notice how only the face images from Figure 12.14 appear in Figure 12.15 and how the images of cars and other objects are eliminated. The view in Figure 12.15 shows the original captured image as well as subimages for each face detected in that image. Figure 12.16

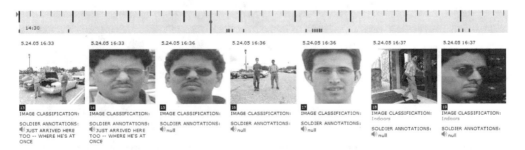

Figure 12.15 Example result: User filters data in Figure 12.12 to view only images with "faces." This view also shows the subimages of detected faces in each original image.

Figure 12.16 Example result: User further filters the images with "faces" to view only those labeled as "indoors."

shows a sequential sample of the result when the user combines two image concepts—"faces" and "indoors." "Indoors" indicates that the image classifier categorized the image as one taken indoors, and "faces" indicates that the image had at least one face detected in it.

Figures 12.17 and 12.18 illustrate the case where a second user combines spatial and speech annotations for navigating the data captured by a first user. The new user first marks out an area in the map corresponding to a portion of the Yorktown facility, which is of interest to this user. Only the data corresponding to this area is highlighted with a rectangle on the map in Figure 12.18. Similarly, only this data appears normally on the timeline, while the rest is grayed out.

From Figure 12.17, the user notices that there is a picture of a garage in the area with a corresponding audio annotation. The user listens to this annotation, recorded at 16:31 on 05–24–2005, which talks about seeing a large garage with

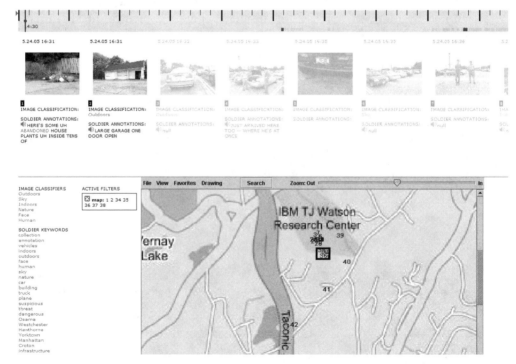

Figure 12.17 Example result: A second user selects the area of the map that is of interest to him by drawing a rectangle. The spatial filtering results in only the data obtained in that area being highlighted along the timeline.

Figure 12.18 Example result: Noticing the presence of the word "garage" in the speech annotation for the image labeled 2 in Figure 12.15; the user searches for other images with the word "garage" in the annotation. The user notices from the images and the annotations that there was a significant change in the garage in the 80 minutes between the two images and decides to investigate further.

one door open. To see if there were any other interesting pictures of this garage, the user further searches by the keyword "garage" in the speech annotation. This leads to the picture shown in Figure 12.18 from the same area of the map taken at 17:54 on the same day with a corresponding audio annotation that talks about how two doors are open and there is a car parked in front of the garage, but with no people in the car or in the garage. The user decides to investigate further.

12.6.1   Experiential Interface

This filtering of discrete events represents but one way to navigate the chronicled events. Figure 12.19 shows a very different eChronicle navigation interface. This is

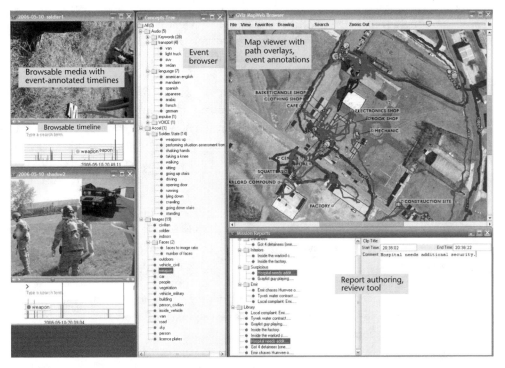

Figure 12.19 A second eChronicle navigation system that allows the viewer to browse, replay, retrieve, and analyze mission experiences.

another multidimensional interface that enables the authorized user to review, analyze, and even relive the mission experiences captured by the user himself or, more importantly, by other people who share their experiences with the user. This multidimensional interface consists of a map viewer, event-annotated timelines, media windows, an event browser, and a report authoring/review tool.

The map viewer provides a map of the area relevant to events of interest and displays path overlays and event annotations. Paths and annotations get updated on the map as a user browses the timeline. The browsable timeline shows selected detected events with associated confidence levels. Browsable media windows show video, audio, and high-resolution images. The media windows get automatically updated as a user browses the timeline—enabling the user to rapidly scan a captured experience. Media can be played at any time, allowing the user to relive the experience. The event browser shows a list of events and concepts that are automatically detected from captured data. Users can select events or event combinations to view on the timeline and map, thus enabling them to rapidly zero in on times and spaces of interest and review the experiences in the spatio-temporal zone of interest to them. Finally, the interface also enables users to author reports by selecting portions of media/events/timeline/map and associating an annotation/ report with these selected portions. The author or other users can then rapidly view these reports through the multidimensional interface, again based on space, time, and events of interest.

12.7   Experiments and System Evaluation

We conducted targeted evaluations of the eChronicling system in a specific domain—soldiers performing patrols of urban areas. While this by no means covers the gamut of application scenarios of the systems, it allowed us to evaluate the system in at least one specific situation. The evaluations were performed by an independent evaluation team and consisted of two types of evaluations: (1) elemental and (2) vignette. The elemental evaluations involved evaluations of the specific elemental event detection capabilities. The vignette evaluations judged the system as a whole by having targeted users perform missions in simulated settings. Both people who performed missions and people who received mission reports (without having been on actual missions) evaluated the value such an eChronicling system provides.

The vignette evaluations were more qualitative in nature and will not be reported here in detail. In summary, the system and the interface illustrated in Figure 12.18 were given very high ratings in the vignette evaluations and seen as considerably enhancing the mission reporting, recall, and intelligence analysis. The multidimensional browsing capability by space, time, and events was received very positively. The ability to easily relive a relevant experience was of special interest. This capability allowed people to get their own perspective on different aspects of a mission, independent of automatic event detection and already added great value. Another feature that was found to be of great value is the ability to select events of interest and display them with associated confidence values. This, according to users, was of immense value compared to filtering based on a preset threshold. This allowed users to use their own judgment in deciding which events to explore at what levels of confidence and not be entirely dependent on machine-made decisions. Also, this approach gave visibility to clusters of events on the timeline or on the map and allowed users to find spatio-temporal zones of interesting events.

While these qualitative results were the most important outcome of the evaluations, we also describe below the quantitative evaluations on specific elemental capabilities. These evaluations were certainly useful in estimating the state of event-detection capabilities, but they also had limitations in the way they were conducted. Firstly, the data set was very limited (for example, only a total of 25 images were used to evaluate the image tagging and object detection capabilities) and was not statistically rich enough to draw a conclusion on the event detection capabilities. Secondly, the evaluation required a binary answer on the presence or absence of an event or concept and hence was driven by hard thresholds used in the analytics. This did not account for the ability to show confidence values for events to humans in a visual event-browsing scenario as described earlier. With this preamble, we go into the specifics of the elemental evaluation results.

12.7.1 Image Tagging Performance and Observations

The automatic semantic concept tagger was trained from three distinct datasets for the following semantic concepts: *outdoors, indoors, vegetation, vehicle_civil (car, truck, bus), vehicle_military, person (soldier, locals), weapon, building*. The three datasets had representative images in the following three contexts: (1) news video data used in the TRECVID benchmark, (2) personal photo collection, and (3) military data collected both on and off field (Web). The depiction of the same concept in these different datasets could be drastically different. The reason why only the military related dataset was not used was due to the few images collected on field and the questionable quality of the content harvested off the Web. The other two data sets were used to enhance the visual examples for the given concepts. Note that not all concepts were modeled using all three different sets. For example, concepts *indoors* and *outdoors* were only modeled using the news video dataset, whereas three different sets of models for the vehicle-related concepts were used, one from each of the three different datasets.

Each training and testing data was represented by the color correlogram and color moments features, and SVM classifier with RBF kernels were trained for each of the concepts using each of the representations. Fusion across feature models was done using simple yet good performing averaging mechanism, and fusion across datasets for same concept was also done using averaging. Note that in order to address the issue of visual polysemy, in addition to fusing models of the same concept across datasets, we also fused the models of subcategories of concepts. For example, for concept *vehicle* the results of concepts *truck, car,* and *bus* were fused together to provide the confidence value for *vehicle*.

For testing, 25 images were provided for automatic tagging. Figure 12.20 shows three such images and the tags automatically associated to them. The image on the left is being tagged without any mistake, neither false positive nor false negative. In the middle image the person is wrongly identified as soldier (although it is likely), and in the image on the right vehicles have been missed and soldier and building are

Figure 12.20 Tags associated with the left-hand side image: Car, Outdoors, Soldier, Vegetation, Vehicle_Civil. Tags associated with the middle image: Indoors, Soldier. Tags associated with the right-hand side image: Building, Outdoors, Soldier.

wrongly tagged (again, absence of building is questionable). Table 12.1 shows the results of the evaluation. According to the table, the most challenging category of concepts appeared to be *weapon*; this is both due to the various forms of the appearance of the weapon and the context in which a weapon occurs. Note that our approach does not locate the manifestation of the concept in the image, and neither does it count the number of occurrences of the concept; but it only predicts the degree of its presence or absence in the image. The results could also be attributed to the quality and quantity of the training data provided in the three different sets. The concept, however, appeared abundantly in the vignette evaluations. So, the threshold for the binary decision seemed to have been a major factor in the elemental test. For concepts such as *outdoors*, which is a highly recurrent concept, there were plenty of positive examples provided in the news data set, and therefore a good model was obtained, whereas for concept *weapon*, there was not enough training data. Concepts such as *vehicle*, which had a fair amount of training data in both the news video set and the personal photo collection one, performed poorly due to the different depiction of such concept in the military context.

In the image elemental test, the face detection and license plate detection were evaluated. The test images were captured at 25 viewpoints by inserting the following factors in images: distance, background clutter, occlusion, and angle of view. The face detection and LPR results are presented in Table 12.2. The failures of the face detection are mainly caused by small face size, occlusions, and lighting changes. The failures of the license plate recognition are mainly caused by small license plate size, the bad viewpoints, and non-U.S. license plates.

Table 12.1 Image-Tagging Performance Table

Concept	Outdoors	Indoors	Building	Vegetation	People	Weapon	Vehicle
% pos. ID over all instances of presence	95.2	100	86	100	29	0	50
% miss over all instances of presence	4.7	0	14	0	71	100	50
% pos. ID over all ID'd	91	80	100	67	40	0	57
% neg. ID over all ID'd	9	20	0	33	60	0	43

Note that results are not reported for all tags.

Table 12.2   Face Detection and License Plate Recognition Elemental Results

	Face	License Plate
Ground Truth	43	19
Correctly Detected	10	2
False Positives	4	2
Detection Rate	23.3%	10.5%

We now present the audio analytics results for the elemental evaluation data set. For this set we evaluated the performance of impulse detection, land vehicle detection, language identification, keyword spotting, and speech detection. To simplify scoring across all audio domains, we proposed a single consistent metric that has a range from 0 to 100% performance. This figure of merit (FOM) is calculated as the number of correct hits divided by the total number of event labels (or the summation of the total hits, misses, false alarms, and substitutions).

$$\text{FOM} = \frac{\#\text{hits}}{\#(\text{hits} + \text{misses} + \text{false alarms} + \text{subst})} \times 100\%$$

In this evaluation each detection system is evaluated in isolation, and the results are given in Table 12.3.

The results show promise, considering the audio environment is relatively harsh. Of special mention is the impulse detection result. For the impulse detection, a two-phase approach was used: a first phase event location pass followed by a second phase that checks if a minimum average energy threshold was reached for the audio event. This reduced the number of false alarms from 53 to 1. (The single false alarm was the starting whistle for the session.) The minimum energy criterion improved the FOM from 12% to 37%. As a consequence of the second pass, a number of low power impulse audio candidates identified from phase 1 were rejected in phase 2.

The vehicle detection system performed poorly. A possible explanation may be mismatch between the audio data used to create the models and the audio actually recorded in the elemental evaluation. The training audio was derived by recording cars on closed asphalt pavements, while the evaluation audio was staged on gravel tracks. To perform better vehicle detection, it would be useful to examine more robust classifier features and to train the system on diverse scenarios.

The language identification system performed well under the recording conditions. The system was trained to differentiate between seven languages. The language identification setup classified the audio segments identified by the acoustic segmentation block produced as a byproduct of the speech recognition engine. It is interesting to note that for the vignette data, the language identification tool correctly located foreign language speech in background music on multiple occasions.

Table 12.3   Audio Analysis Elemental Results

Analytics Type Detected	*Headset Microphone FOM*
Impulse	37
Vehicle	3
Language	45
Keyword	23
Speech	68

Keyword spotting performed reasonably, considering that it is based on a large vocabulary continuous speech recognition (LVCSR) system. Although the hit rate was relatively low for the keyword spotting configuration, there was only one false alarm identified for nearly 320 spoken keywords. The benefit is that if the system identifies a hit, it is highly likely the word event occurred.

The speech detection component performed sufficiently. It is essential that the speech detection and segmentation element performs reliably, because it is the first component in the line-up to parse the audio. Errors introduced here are propagated to the follow-on speech analytics. Various delayed decision techniques may be introduced to minimize this effect.

12.8   Conclusions and Future Work

This chapter presented an overview of our electronic chronicling system and architecture that enables mobile and pervasive data capture, assimilation, and retrieval. The main contribution of this work lies in developing an end-to-end architecture unifying a variety of sensors, logging software on PCs, analytic engines, data management, and navigation tools. By building such a system with components that are state-of-the-art, we are able to explore unprecedented capabilities that are greater than the sum of the parts. We view this still as the early stage for research on such pervasive electronic chronicling systems. The initial results presented here show promise that this approach could indeed impact a variety of applications involving field operations, mobile workforce management, and situation analysis.

Multiple conclusions have been derived from the image tagging unit of the reported electronic chronicling system. First, those concepts that are related to specific objects could not be supported adequately using the reported approach to image tagging. We plan to extend generic approach for semantic concept modeling to a palette of approaches specifically designed for different classes of semantic concepts, for example, a salient region-based approach could be more suitable for modeling object-related concepts.

In addition, one difficulty we faced in designing the concept model was manually annotating the data training data sets prior to concept tagger training. It is a very tedious task to obtain the ground truth data for the concepts of interest. We are planning to devise methodologies to leverage socially obtained tags via the specific applications offered on the Web.

Audio analytics provides substantial information toward an eChronicling system.

Follow-on work will involve finer grained classification of the audio events. Some examples include the identification of the type of gunshot or vehicle.

Our future work will continue to extend the individual analytic capabilities and corresponding search and navigation tools. Future work will focus on the long-term data association based on multimodal integration.

Finally, there is room for significant research on the underlying data and event representation models, and techniques for event and activity mining.

Acknowledgments

This work was partially supported by DARPA under the ASSIST program under contract NBCHC050097. The authors would like to acknowledge the input and influence of the rest of the people in the EC-ASSIST project team at IBM, Georgia Tech, MIT, and UC Irvine. The authors especially thank Milind Naphade for providing the models trained for LSCOMLite ontology.

References

[1] Jain, R., "Media Vision: Multimedia Electronic Chronicles," *IEEE Multimedia,* July 2003.

[2] Pingali, G., and R. Jain, "Electronic Chronicles: Empowering Individuals, Groups, and Organizations," *Proc. of IEEE International Conference on Multimedia and Expo,* Amsterdam, July 2005.

[3] Bush, V., "As We May Think," *Atlantic Monthly,* Vol. 176, No. 1, January 1945, pp. 641–649.

[4] Gemmell, J., et al., "MyLifeBits: Fulfilling the Memex Vision," *ACM Multimedia'02,* 2002, pp. 235–238.

[5] Gemmell, J., ACM SIGMM CARPE (Continuous Archival and Retrieval of Personal Experiences) home page, http://www.sigmm.org/jgemmell/CARPE.

[6] Caskey, S., et al., "Infrastructure and Systems for Adaptive Speech and Text Analytics," *Proc. of 2005 International Conference on Intelligence Analysis,* McLean, VA, May 2–6, 2005.

[7] NSF Workshop on Personal Information Management, January 27–29, 2005, http://pim.ischool.washington.edu/

[8] Starner, T., "Wearable Computing and Context Awareness," Ph.D. thesis, Massachusetts Institute of Technology Media Lab, April 30, 1999.

[9] Rhodes, B., and T. Starner, "Remembrance Agent: A Continuously Running Automated Information Retrieval System," *Proc. of First International Conference on the Practical Application of Intelligent Systems and Multi Agent Technology,* London, April 1996.

[10] Hori, T., and K. Aizawa, "Capturing Life Log and Retrieval Based on Context," *Proc. of IEEE ICME 2004,* June 2004.

[11] Dumais, S., et al., "Stuff I've Seen: A System for Personal Information Retrieval and Re-Use," *Proc. of ACM SIGIR '03,* Toronto, Canada, August 2003, pp. 72–79.

[12] Freeman, E., and S. J. Fertig, "Lifestreams: Organizing Your Electronic Life," *Proc. of AAAI Fall Symposium: AI Applications in Knowledge Navigation and Retrieval,* 1995.

[13] Plaisant, C., et al., "Lifeline: Visualizing Personal Histories," *Proc. of ACM Conference on Computer Human Interaction (CHI'96),* Vancouver, 1996.

[14] Rekimoto, J., "Timescape: A Time Machine for the Desktop Environment," *Proc. of ACM Conference on Computer Human Interaction (CHI'99),* 1999.

[15] Guven, S., M. Podlaseck, and G. Pingali, "PICASSO: Pervasive Chronicling, Access, Search, and Sharing for Organizations," *IEEE International Conference on Pervasive Computing,* Hawaii, March 2005.

[16] Huynh, D., D. Karger, and D. Quan, "Haystack: A Platform for Creating, Organizing, and Visualizing Information Using RDF," *The 12th Int. World Wide Web Conference,* May 2003.

[17] Kim, P., M. Podlaseck, and G. Pingali. "Personal Chronicling Tools for Enhancing Information Archival and Collaboration in Enterprises," *ACM Workshop on Continuous*

Archival and Retrieval of Personal Experiences (CARPE 2004), New York, October 2004.

[18] Moore, C., "Diving into Data," *InfoWorld,* October 25, 2002. http://www.infoworld.com/article/02/10/25/021028feundata_1.html.

[19] Collet, C., G. Vargas, and H. Ribeiro, "Towards a Semantic Event Service for Distributed Active Database Applications," *Proc. of 9th International Conference on Database and Expert Systems Applications (DEXA'98),* Vienna, Austria, August 1998.

[20] Coupaye, T., C. Roncancio, and C. Bruley, "A Visualization Service for Event-Based Systems," *Proc. of 15emes Journees Bases de Donnees Avancees,* Toronto, Canada, August 2003.

[21] Fiege, L., G. Muhl, and F. Gartner, "A Modular Approach to Building Structured Event-Based Systems," *Proc. of the 2002 ACM Symposium on Applied Computing (SAC'02),* Madrid, Spain, 2002, pp. 385–392.

[22] *Proc. of International Workshop on Distributed Event-Based Systems (DEBS'05),* Columbus, OH, June 10, 2005, http://www.cs.queensu.ca/~dingel/debs05/index.html.

[23] Pingali, G., et al., "LucentVision: Converting Real World Events into Multimedia Experiences," *Proc. of IEEE International Conference on Multimedia and Expo,* New York, July 2000.

[24] Collins, R., et al., "A System for Video Surveillance and Monitoring," *VSAM Final Report,* Technical Report, CMURI-TR-00-12, May 2000.

[25] Greiffenhagen, M., et al., "Design, Analysis and Engineering of Video Monitoring Systems: An Approach and Case Study," *Proc. of IEEE,* Vol. 89, No. 10, October 2001, pp. 1498–1517.

[26] Hampapur, A., et al., "Smart Video Surveillance: Exploring the Concept of Multiscale Spatiotemporal Tracking," *IEEE Signal Processing,* Vol. 22, No. 2, March 2005, pp. 38–51.

[27] Lipton, A., et al., "Critical Asset Protection, Perimeter Monitoring, and Threat Detection Using Automated Video Surveillance," white paper, ObjectVideo, 2002.

[28] Haritaoglu, I., D. Harwood and L. Davis, "W4: Real Time Surveillance of People and Their Activities," *IEEE Trans. on Pattern Analysis Machine Intelligence,* Vol. 22, No. 8, August 2000, pp. 809–830.

[29] Horprasert, T., D. Harwood, and L. Davis, "A Statistical Approach for Real-Time Robust Background Subtraction and Shadow Detection," *Proc. IEEE Frame-Rate Workshop,* Kerkyra, Greece, 1999.

[30] Stauffer, G., "Learning patterns of Activity Using Real-Time Tracking," *IEEE Trans. on Pattern Analysis Machine Intelligence,* Vol. 22, No. 8, August 2000, pp. 747–757.

[31] Remagnino, P., et al., *Video Based Surveillance Systems Computer Vision and Distributed Processing,* Norwell, MA: Kluwer, 2002.

[32] VACE: Video Analysis and Content Exploitation [YLOnline]. http://www.ic-arda.org/InfoExploit/vace/.

[33] Blanz, V., and T. Vetter, "Face Recognition Based on Fitting 3D Morphable Model," *IEEE PAMI,* Vol. 25, No. 9, September 2003, pp. 1063–1074.

[34] Phillips, J., et al., "Face Recognition Vendor Test 2002 P," *Proc. of IEEE International Workshop Analysis and Modeling of Faces and Gestures (AMFG'03),* 2003.

[35] *Human ID at a Distance,* U.S. Government, DARPA Project, 2002.

[36] *Combat Zones That See,* U.S. Government DARPA Project, 2003.

[37] Bregman, A., *Auditory Scene Analysis: The Perceptual Organization of Sound,* Cambridge, MA: MIT Press, 1994.

[38] Potamianos, G., et al., "Recent Advances in the Automatic Recognition of Audio-Visual Speech," *Proc. of the IEEE,* Vol. 91, No. 9, 2003, pp. 1–18.

[39]   Rabiner, L., and B. Juang, *Fundamentals of Speech Recognition*, Upper Saddle River, NJ: Prentice-Hall, 1993.

[40]   Kristjansson, T., B. Frey, and T. Huang, "Event Coupled Hidden Markov Models," *Proc. of IEEE International Conference on Multimedia and Expo (ICME)*, 2000.

[41]   Levas, A., et al., "Exploiting Pervasive Enterprise Chronicles Using Unstructured Information Management," *Proc. of IEEE International Conference on Pervasive Services (ICPS 2005)*, July 2005.

[42]   Adams, B., et al., "Semantic Indexing of Multimedia Content Using Visual, Audio and Text Cues," *EURASIP Journal of Applied Signal Processing*, Special Issue, February 2003.

[43]   Vapnik, V., *The Nature of Statistical Learning Theory*, New York: Springer, 1995.

[44]   Shanahan, J., and N. Roma, "Boosting Support Vector Machines for Text Classification Through Parameter-Free Threshold Relaxation," *CIKM*, 2003, pp. 247–254.

[45]   Platt, J., "Probabilities for Support Vector," in *Advances in Large Margin Classifiers*, A. Smola et al., (eds.), Cambridge, MA: MIT Press, 1999, pp. 61–74.

[46]   Rowley, H., S. Baluja, and T. Kanade, "Neural Network-Based Face Detection," *PAMI*, Vol. 20, 1998, pp. 22–38.

[47]   Schneiderman, H., and T. Kanade, "A Statistical Method for 3D Object Detection Applied to Faces and Cars," *ICCV*, 2000.

[48]   Sung, K., and T. Poggio, "Example-Based Learning for View-Based Face Detection," *PAMI*, Vol. 20, 1998, pp. 39–51.

[49]   Jones, M., and P. Viola, "Fast Multi-View Face Detection," *CVPR*, 2003.

[50]   Li, S., et al., "Statistical Learning of Multi-View Face Detection," *ECCV*, 2002.

[51]   Viola, P., and M. Jones, "Rapid Object Detection Using a Boosted Cascade of Simple Features," *CVPR*, 2001.

[52]   SeeCar license plate recognition, Hi-Tech Solutions. http://www.htsol.com.

[53]   Goel, V., et al., "Discriminative Estimation of Subspace Precision and Mean (SPAM) Models," *Proc. of ISCA Eurospeech*, 2003, pp. 2617–2620.

[54]   Povey, D., et al., "FMPE: Discriminatively Trained Features for Speech Recognition," *Proc. of IEEE International Conference on Acoustics, Speech and Signal Processing*, Vol. 1, 2005, pp. 961–964.

[55]   Campbell, W., et al., "Advanced Language Recognition Using Cepstra and Phonotactics: MITLL System Performance on the NIST 2005 Language Recognition Evaluation," *Proc. of IEEE Odyssey Speaker and Language Recognition Workshop*, 2006.

[56]   Navratil, J., "Recent Advances in Phonotactic Language Recognition Using Binary-Decision Trees," *Proc. of ISCA Interspeech*, 2006.

[57]   SIMILE project's Longwell. http://simile.mit.edu/longwell/index.html.

CHAPTER 13

Systems Issues in Distributed Multimodal Surveillance

Li Yu and Terrance E. Boult

13.1 Introduction

To be viable commercial multimodal surveillance systems, the systems need to be reliable, robust, and must be able to work at night (maybe the most critical time). They must handle small and nondistinctive targets that are as far away as possible. Like other commercial applications, end users of the systems must be able to operate them in a proper way. In this chapter, we focus on three significant inherent limitations of current surveillance systems: the effective accuracy at relevant distances, the ability to define and visualize the events on a large scale, and the usability of the system.

Surveillance systems inherently have the human tightly coupled in the system loop. In most cases, the human operator is the decision-maker who will act upon (or not act upon) the feedback of the system. In the aspect of usability, video surveillance systems are different from other software applications in that they are time critical and emphasize the accuracy of users' responses. Decisions often need to be made accurately in real time. Furthermore, what is of interest to a particular surveillance system user can vary greatly, and the security forces using the system are not, in general, advanced computer users. Therefore, the usability issue becomes crucial to a multimodal surveillance system.

In the first part of this chapter, we address the usability issue of a video surveillance system by presenting a novel paradigm: understanding the images in the graphical user interfaces (UI-GUI). This paradigm integrates the GUI design aspect of a surveillance system with underlying vision modules. Improvement of usability can also improve the performance of the human activity detection/recognition algorithm. We show experiments with real data and compare it to HMM-based specification of activities. Not only does the UI-GUI approach provide for faster specification of activities, but it also actually produces statistically more accurate specifications and is significantly preferred by the users.

The second part of the chapter takes a more general systems view, reviewing what was necessary to develop solutions that support commercial intelligent camera networks deployed with hundreds of sensors. The two most fundamental issues for the distributed surveillance systems are tied to the resolution needed for the tasks and the communication. The chapter reviews the key system components in effective distributed video surveillance, and then discusses the major open issues, including hardware-accelerated algorithms needed for increasing resolution while reducing power, and the issues of mobile surveillance.

The chapter reviews issues in tradeoffs between target size, field of view and response time and cost. It then discusses active visual surveillance, which uses data from computer-controlled pan/tilt/zoom (PTZ) units combined with state-of-the-art video detection and tracking to, in a cost-effective manner, provide active assessment of potential targets. This active assessment allows an increase in the number of pixels on target and provides a secondary viewpoint for data fusion, while still allowing coverage of a very large surveillance area. This active approach and multisensor fusion, not a new concept, was developed as part of the DARPA Video Surveillance and Monitoring (VSAM) program in the late 1990s. While we have continued to expand upon it since that time, there has been only limited academic research and, until 2003, no commercial video surveillance that provided these important abilities.

The final system issue discussed is that of data/network communication. While some systems assume analog cable feeds back to a central location, this is quite costly and does not scale extremely well. We discuss the communication issues for a distributed system and the extensions we provided to the original DARPA VSAM protocol to support adaptive bandwidth management to allow hundreds of cameras on wireless links.

13.2   User Interfaces

Visualization and usability are crucial to the success of video surveillance systems. Video surveillance systems are not only measured by statistic results of their vision module (e.g., detection and false alarm rate), but also by the degree of user satisfaction of the whole system. Surveillance systems inherently have the end users tightly coupled in the system loop (Figure 13.1). In most cases, the human is the decision maker who will act upon (or not act upon) the feedback of the system. As shown in Figure 13.1, a typical surveillance system takes frames from video cameras as input, processes frames with vision modules, and generates system feedbacks (in the forms of alerts, e-mail notification, and so forth). Generally,

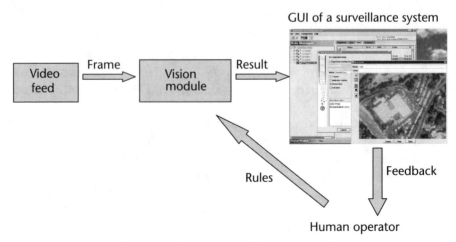

Figure 13.1 A generic architecture of a video surveillance system. Human operators are tightly coupled with the system loop.

human operators are involved in two types of tasks: define rules for vision module and monitor the system's feedbacks. Well-designed graphical user interfaces (GUI) can help users understand the strength and limitation of the system, and thus operate the system in a proper way. On the other hand, poorly designed GUIs can compromise the performance of the whole system if end users do not know how to interpret what is displayed on the monitor and thus make poor decisions.

Research has been done on how to evaluate the usability of a software system. The usability of a general graphical user interface (GUI) is categorized into five areas: overall reaction, learning, screen, terminology, and system capability [1]. To evaluate a system, users are asked to perform tasks with the system and make quantitative evaluation in the five categories.

Video surveillance systems are different from general systems in that they are more time-critical and put more emphasis on accuracy of users' responses. Decisions need to be made accurately in real time. Therefore, the usability issue becomes crucial to a video surveillance system. There are several measurements we consider specifically important to a video surveillance system: Learning to operate the system, user tasks performed in a straight-forward manner, correcting mistakes, considerations for both experienced and inexperienced users' needs, and the detection/false alert rate of the vision module. We will discuss these metrics in detail in the following paragraphs.

What is of interest to a particular surveillance system user can vary greatly, and the security forces using the system are not, in general, advanced computer users. Therefore, a system needs to provide straightforward and adaptable commands for end users. Some current systems use complex, hand-coded mathematical models to represent suspicious events [2, 3]. Although they are powerful in event detection perspective, those models require substantial understanding and experience in building and adjusting the models. Besides, these current systems are usually set up for certain scenarios, and when a scenario changes, manual adjustment and substantial changes are often required. Our user studies show how difficult these approaches are for end users.

System feedback contributes to several of the metrics we mentioned. System feedbacks tell users how the system responded to the user's inputs, and thus either confirm or disagree with the user's operation. However, sufficient information does not necessarily guarantee a good GUI design. Too many visual feedbacks can be distracting, making end users confused and thus reducing their confidence in the system. Real-time markup is an effective visualization approach to provide end users with how the vision engine is working. On a live video display window, bounding boxes are drawn around the detected targets. Figure 13.2 shows a snapshot of real-time markup. Although informative, real-time markup can be confusing when there are noises in the scene and bounding boxes are flying around everywhere. This is an example that too-detailed feedback can actually harm the usability of the system by exposing the vulnerability of the system and thus reducing users' confidence.

The false alarm rate is always a big consideration for video surveillance. When an alert happens, the authority will be notified and security staff will be called upon to investigate the situation. As a result, if the system cries wolf enough times, the operator will begin to lose trust and may disable the system altogether.

Figure 13.2 Detected targets are marked up with bounding boxes. The box with thick lines denotes the correct target, while the ones with thin lines are noises.

The false alarm rate can be reduced not only by improving vision algorithms, but also by smartly restricting users from making "bad" rules as well. Figure 13.3 shows a scenario where operators are interested in any objects appearing at the intersection. So an area of interest (AOI) rule is defined as in Figure 13.3(a) to report when there is anything appearing inside the AOI marked up by a small rectangle. However, there are trees that cause many false alerts. A proper visualization of the alerts could include both objects inside AOI and AOI itself as shown in Figure 13.3(a). This way, human operators can easily compare alerts with the rules they have defined, and thus find out that leaves on the trees are causing trouble. Having seen that, operators can readily change the rule by shrinking the size of AOI [Figure 13.3(b)].

                 (a)   (b)

Figure 13.3 AOI rules. (a) Describes an AOI rule defined at the intersection. A false alert is caused by waving leaves on the tree. (b) The AOI rule is modified to exclude the tree.

13.2.1   UI-GUI: Understanding Images of Graphical User Interfaces

Our goal is to address the usability issue of a video surveillance system by presenting a paradigm to integrate the GUI design aspect of a surveillance system with underlying vision engines. A well-designed GUI cannot only improve the usability but also the performance of the vision modules. The paradigm is called understanding images of graphical user interfaces (UI-GUI, pronounced as ōō ʼē g ōō ʼ ē). The basic idea of UI-GUI is that a video surveillance system visualizes the states of the vision engine onto the GUI. Based on what they see on the screen, users define rules. The visual rules, as part of the GUI, are extracted and used to generate rule logics. At run time, detection and/or tracking results of the vision module will be matched up against the rule logics. Alerts are triggered if rules are violated.

A simple example of UI-GUI is Area of Interest (AOI) as shown in Figure 13.3. Defining an AOI rule is easy for end users because users can directly draw the rule on the GUI and understand how the rule works by watching how an alert is triggered when someone walks into the AOI from the GUI. Thus what is displayed on the GUI matches up with their expectations. Figure 13.4 shows how UI-GUI works for the AOI rule. After the rule is defined on the GUI, its metadata (coordinates of four corners of the AOI in this example) will be used to generate a logic in the vision engine: if there is any target detected in the AOI, an alert will be triggered. At run time, targets' coordinates generated from the tracking system will be matched up to the rule logic.

Simple rules, such as AOI and tripwire, only generate simple logics, and thus cannot handle temporally or spatially complex scenarios, for example, a drop-off event (shown in Figure 13.6). In the next section, we will talk about how UI-GUI is applied to specifying and recognizing complex activities in a multimodal surveillance system.

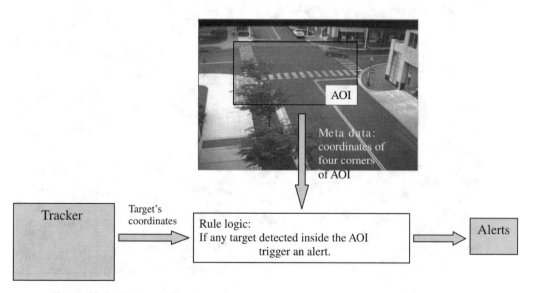

Figure 13.4   How an AOI rule works.

13.2.2 Visualization and Representation

A satellite image or map can be used as an integration console for all kinds of sensors (PTZ, omnidirectional, fixed camera, and so forth). Targets from different sensors are geo-registered onto a satellite map and visualized as icons. Different types of targets are represented as different icons. Different colors are assigned to indicate the degrees of priority. Rather than just showing current location, a short history is graphically represented as trails. The size of an icon decreases as time passes. The larger the icon, the more recent the target was at that position. Observed targets use icons with detailed texture, while predicted locations are marked with uncertainty cones. Different intensity can be applied to the icons to represent certainty of the predicted geo-location. High intensity means strong confidence, while low intensity means weak confidence. Implicit in the spatio-temporal nature of the trajectory, more subtle semantics can be inferred. Because a target's trail represents the path it goes through in a certain amount of time, the longer the trail, the higher speed at which a target is moving. Figure 13.5 shows an example of how targets are visualized on a map interface.

Visualization can be changed based on different applications. In Figure 13.6, a target's trajectory is visualized as a sequence of icons shrinking as time goes by. This representation provides more details about where exactly the target was in the past.

Currently systems use humans to interpret their system output and recognize activities of interest. A human operator is quickly fatigued, and not as efficient as machines when performing voluminous routine tasks. Thus, we seek to use computer vision techniques to recognize what is happening on the map. While

Figure 13.5 A map-based surveillance console where targets' present, history, and predicted locations are visualized on the map.

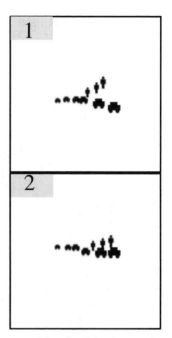

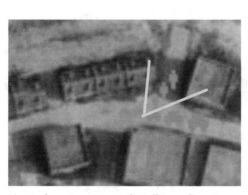

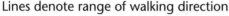
Lines denote range of walking direction

Figure 13.6   A user-drawn drop-off event (on the left) and two examples of training images (on the right). 1. The human's trajectory is rotated 10° from the original drawing. 2. The human's trajectory is rotated 40°.

previous work has sought representations that analyze the internal state produced by the vision subsystems [2–4], our goal is to understand the state of running systems by analyzing the semantics contained in their GUIs with what we call a UI-GUI-based system.

As the visual representation is clear and straightforward, security operators can simply draw an activity of interest on a map-based GUI. We present an approach to recognizing activities by understanding the images of GUIs using an appearance-based recognition algorithm. The recognition system is trained from user-specified representations. Training on a real scene is not necessary, but because it just takes GUI images, training on real data from the running systems' GUIs can be used as training sets as well.

13.2.3   Advanced UI-GUI Recognition Algorithm

To match GUI images with training data, we consider a parametric appearance representation [5], with some customized training approaches. For any given vision task, images that coarsely sampled from all possible appearance variations are used as the training dataset. The image set is compressed as a low-dimensional eigenspace. Activities of interest are represented as manifolds in the eigenspace. At recognition time, an unknown image will be projected to the subspace:

$$f_c = [e_1, e_2, \ldots\ldots, e_k]^T (\tilde{i}_c - c) \tag{13.1}$$

where $\{e_k \mid k = 1, 2, \ldots, K\}$ are eigenvectors, and c is the image average.

Then the minimum distance between f_c and the manifold f_q is computed as:

$$d_r = \overset{min}{q} \, \| f_c - f_q \| \tag{13.2}$$

If the distance d_r is sufficiently small, we consider the activity recognized, and the closest point on the manifold encodes the parameters of that activity, such as trajectory rotation angle. If the distance between the unknown image and one manifold is sufficiently small, we consider the activity represented by manifold recognized, and the closest point on the manifold encodes the parameters of that activity, such as trajectory rotation angle.

Training images will be generated for combinations of different variations of trajectory parameters. Since the parametric appearance algorithm [5] used can recover an object by coarsely sampling the space of the object's appearances, it is not necessary to generate every possible pattern of activities. Training data is directly generated from users' drawings by perturbing icon locations (e.g., by Gaussian distribution). Rotation is applied to a trajectory if it is orientation-independent. However, by perturbing a trajectory, some unintended activity images might be included. The resulting expanded training set can be viewed and edited by the user. An example of training image generation is shown in Figure 13.6.

13.2.4 Sensor Fusion with UI-GUI: GUI Is API

The previous sections described the UI-GUI concept with respect to recognizing trajectories in the primary surveillance system. The approach has other uses within our surveillance system. One of the sensors of interest for an application included a radar system to detect boats approaching the facility at night. It would take considerable effort to obtain a specification of the interface to the radar and integrate it into the system. Furthermore, most low-cost radar, intended for use on small boats, does not produce geo-located targets, but are optimized for human viewing of the radar image.

By considering the GUI display of commercial marine radar as an image, we were able to very rapidly integrate the radar into our visual surveillance system. Since the radar was operating using its manufacturer's supplied interface, no effort was needed to develop an interface to set system parameters. We considered the radar display as an image and applied standard background subtraction to the highly processed radar data. The result was that within two weeks we were able to obtain the radar targets and display them as trajectories on the primary overhead map. Because of the limited screen resolution, the radar was processed on a second machine and the target locations and images were sent over a network to the main control station.

As almost any surveillance sensor will have a GUI designed to allow the security force to quickly assess the setting, our approach can quickly integrate new sensors, even those with propriety interfaces. While we reduced the radar to target locations for fusion, an alternative would have been to treat the radar GUI image directly with appearance-based activity recognition.

13.2.5 Formal User Study on UI-GUI

Despite the importance of usability, we are not aware of any user studies done to evaluate the usability of video surveillance systems. In addressing this problem, we conducted formal user studies to compare UI-GUI and traditional HMM-based approach, and obtained usability evaluation and recognition results with real users and real data.

13.2.5.1 Design

The user studies are designed to see whether users can quickly learn the approach, and specify activities of interest in a satisfactory way. In the user studies, we also compared UI-GUI with a widely used state-based approach (HMMs) in terms of recognition results, and time needed for learning and defining a rule. Our HMM recognizer is based on the hidden Markov model toolkit (HTK) [6].

Participants were asked to use one approach (UI-GUI or HMMs) first, and come back a week later to use the other one. The order of the approaches is assigned so that each approach had an equal number of users. With each approach, participants were trained with a brief tutorial of the approach, followed by a short demonstration. The training session finished when the participants felt they understood the training material. After training, subjects were asked to specify three activities: "follow," "meet," and "concurrent approach" (Figure 13.7). Generated rules were tested against a small set of real data, and subjects were asked to change their specifications, if their rules were not fired.

We predicted that with our UI-GUI, subjects were quicker and more accurate in specifying and adapting activities than with HMMs. We also predicted that the recognition results using UI-GUI were better than those with HMMs.

13.2.5.2 Procedure and Materials

At the start of each experiment, the subjects were assigned a unique identifier, which was recorded with their experiment results. No records of names or other personal data that could be used for identification were maintained. The subjects

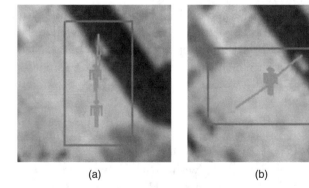
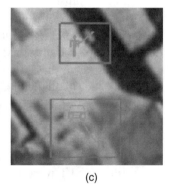

| (a) | (b) | (c) |

Figure 13.7 Examples of three types of activities. Part (a) shows the "follow" event. A person follows the other one toward the parking lot. Part (b) shows the "meet" event. Two people meet in front of a building. Part (c) shows the "concurrent approach." A car is entering a parking lot while a person is walking out of a building.

were asked to complete a background questionnaire and consent forms. Background questionnaires were used to gather information on subjects' educational background, knowledge and use of computers, and experience with the state-based approach and drawing tools. Then the subjects were assigned one approach. With each approach, there were four sessions:

1. *Training session:* Participants were trained with a brief tutorial of the approach. The UI-GUI tutorial is in Appendix D. The HMM tutorial is the first four pages of [7]. Participants read the tutorial until they felt they had understood the material. Then a short demonstration with verbal explanation was given. Participants were provided with a set amount of hands-on practice time.

2. *Prequestionnaire sessions:* Subjects were asked to evaluate the system after training. The questionnaire was a slightly revised version of the Generic User Interface Questionnaire (QUIS), version 5.0 [1]. The questionnaires included overall reaction, learning, features of the screen, terminology, and system capabilities.

3. *Task session:* Subjects were asked to specify three types of activities: "meet," "follow," and "Concurrent Approach," as presented in Figure 13.7.

4. *Postquestionnaire session:* After finishing using each interface, subjects were asked to evaluate it with post questionnaires.

In the activity specification experiment, each subject was asked to use HMMs and UI-GUI to specify several activities of interest. First, subjects were assigned to one of the systems. They were shown a video clip containing the activity of interest. Then they were asked to specify the activities in the video. The activities we asked the subjects to specify are "follow," "meet," and "concurrent approach." The correctness of users' specification and time needed to complete a specification were recorded to evaluate accuracy and efficiency. After specification, one example of the activity of interest was run against the subject's specification. The subject was allowed to revise his or her specification according to the recognition result. The adaptation process could be repeated until the subject felt comfortable. After subjects finished using one interface, they were asked to complete the same questionnaires. Then they were asked to come back to evaluate the other system with the same procedure and same tasks. By letting the same subject evaluate both interfaces, we obtained and compared the same subject's opinion on both systems. To minimize the fact that the order in which subjects used the systems may have influenced their evaluation, we needed to make each system have approximately the same number of first-time users.

We predicted that the average time a subject needed to learn UI-GUI was shorter than HMMs. We predicted that a subject needed less time in defining the rules with UI-GUI than with HMMs. We predicted that the recognition results with UI-GUI were generally better than those with HMMs. We predicted that subjects felt more comfortable with UI-GUI than with HMMs (measured by pre- and postquestionnaires).

13.2.5.3   Participants

Twenty graduate students with engineering majors participated in the user study. All subjects used computers more than 20 hours per week and were familiar with Windows operating system. Among the subjects, 19 attended the UI-GUI experiment, and 10 of them used UI-GUI as the first system. Nineteen attended the HMM experiment, and ten of them used HMM as the first system.

13.2.5.4   Types of Data Collected

We collected several types of data in the user studies, including: subjects' background questionnaire, prequestionnaires and postquestionnaires, subjects' specification of three activities, and time needed for training and for each specification task. The background questionnaire was used to gather the information of subjects' educational background, knowledge of computers, and previous experience with HMM and drawing tools. The prequestionnaires and postquestionnaires included overall reaction, learning, features of the screen, terminology, and system capabilities. Open-ended questions were also included to pull out the information about what subjects liked or disliked about the system, what they thought was easy or difficult about the system, and their other comments about the two systems.

13.2.5.5   Recognition Results and Discussion

Nineteen out of 20 participants used UI-GUI, and 10 of them used UI-GUI as their first approach. All of the nineteen subjects completed all three modeling tasks. Their average training time was 3 minutes, 30 seconds, and the average time to specify an activity was 11 seconds; 19 subjects used HMMs, and 10 used HMMs as the first approach; 12 subjects finished modeling "follow," and 7 finished modeling "meet" and "concurrent approach." The average training time for all HMM users was 13 minutes, and the average time for modeling an activity was 25 minutes.

Test data include 10 "follow," "meet," and "concurrent approach"; 853 noise events for UI-GUI; and 1,088 noise events for HMMs. After the user studies, subjects' rules were run against the test data. The median and mean ROC curves are shown in Figure 13.8. The average UI-GUI recognition results of those subjects who successfully modeled the activities with HMM are similar to the average UI-GUI recognition results of all the participants. From Figure 13.8(a), we can see that the recognition results with UI-GUI are generally better than those with HMMs. The curve of "concurrent approach" using HMMs is an exception. However, Figure 13.8(d) indicates that the average ROC curve of "concurrent approach" with UI-GUI is higher than that of HMMs, and the standard deviations of HMMs are generally larger than that of UI-GUI.

13.2.6   Questionnaire Analysis

The questionnaires we used in the user studies are based on GUIS version 5.0 [1]. There are five categories in the questionnaires: overall reaction, learning, screen,

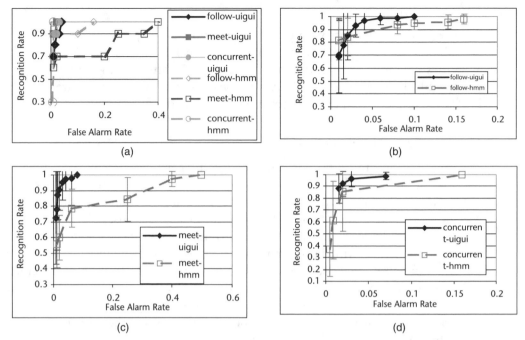

Figure 13.8 ROC curves of event recognition with different approaches. (a) Median ROC curves of "follow," "meet," and "concurrent approach" specified by subjects. (b) Mean ROC curves with standard deviation of "follow" using UI-GUI and HMMs. (c) Mean ROC curves with standard deviation of "follow" using UI-GUI and HMMs. (d) Mean ROC curves with standard deviation of "concurrent approach" using UI-GUI and HMMs.

terminology, and system capability. Nineteen subjects participated the user evaluation for both UI-GUI and HMM-based approaches. Before evaluations, we asked the subjects to consider not only the software they learned or used for the user study, but also the approach underlying the software. Especially to our interest, in the category of overall reaction, there are questions to measure the degree of easiness, satisfaction, and adequate power of the approach they were using. In the category of system capability, one question is to measure the easiness of correcting users' mistakes in modeling. Another question is to tell to what degree both experienced and inexperienced users' needs are taken into consideration.

The ratings of UI-GUI are statistically significantly higher than those of HMM. The average rating of the whole questionnaire for UI-GUI is 64.96% higher than that for HMM. Table 13.1 shows the comparison of items of specific interest to us. We can infer from the five items that the major advantages of UI-GUI are easy to learn, easy to adapt, easy to use, and convenient for both experienced and inexperienced users.

Paired t-tests show that post-questionnaire of UI-GUI has significant higher ratings than prequestionnaire in the category of overall reactions of the approach ($p = 0.03236$). Subjects also had significant different ratings on the degrees of easiness ($p = 0.015$) and satisfaction ($p = 0.014$), which means that generally subjects felt that the UI-GUI is actually even easier and more satisfactory than their expectations. On the contrary, for HMM, average ratings of postquestionnaire are significantly lower than prequestionnaire in the categories of overall reaction,

Table 13.1 Usability Comparison of UI-GUI and HMM

	UI-GUI Mean	UI-GUI Variance	HMM Mean	HMM Variance	P-Value	UI-GUI-HMM % Difference
Easiness	8.63	0.36	3.68	5.23	1.73117E-07	134.29%
Learning to Operate the System	8.79	0.18	3.68	8.67	4.30324E-06	138.571%
Tasks Can Be Performed in a Straightforward Manner	8.32	1.67	3.72	8.8	0.000699082	123.409%
Correcting Your Mistakes	8.68	0.23	3.33	7.53	2.94967E-06	160.526%
Experienced and Inexperienced Users' Needs Are Taken into Consideration	8.05	1.5	2.78	7.12	5.33779E-07	189.895%

learning, and system capabilities. Paired t-test reveals that the difference in the question measuring degree of adaptation is also significant ($p = 0.0296$).

The ratings of postquestionnaire of HMM, however, are generally lower than those of prequestionnaire. Especially in the category of correcting mistakes, postquestionnaire ratings are 15.493% lower than those of prequestionnaire. The p-value is 0.029646. In the category of "experienced and inexperienced users' needs are taken into consideration," postratings are −15.254% lower, and the p-value is 0.057917. In the category of "tasks can be performed in a straightforward manner," postratings are −26.374% lower, and the p-value is 0.066939. We may infer that participants were even more disappointed in HMM after they actually used the approach.

13.2.7 User Interface System Issues Summary

In the first part of this chapter, we focus on the user interface aspects of distributed video surveillance. We present a novel paradigm to integrate GUI design with underlying vision modules: Understanding Images of Graphical User Interfaces (UI-GUI). Formal user studies had to be done to evaluate the usability of our approach and compare it with HMM-based one. Using UI-GUI, 19 users specified the events of interest more than 10,000% faster than with HMMs, and the ROC curves for the resulting recognition were generally statistically significantly better.

The UI-GUI approach is not limited to rule definition and inference in surveillance system. Some commercial applications do not provide API access to their raw data, but are optimized in their GUIs. With UI-GUI, any GUI can be used as API. By grabbing images of GUIs, we can access the inner states of an application by interpreting its GUI images. For example, we integrated a radar unit into the surveillance system by reinterpreting the radar screen output rather than processing the raw data or developing a complex communication protocol. We believe the UI-GUI has a broader impact.

13.3   System Issues in Large-Scale Video Surveillance

Intelligent video surveillance is a systems-level problem with five major components:

1. Low-level detection/processing algorithms;
2. Higher-level algorithms for combining data and selecting particular events of interest;
3. User-interface;
4. Sensor/hardware/computation architecture;
5. Software/communication architecture.

The second half of this chapter is a high-level review of issues that, within the video community, are often overlooked. The majority of vision work in the surveillance area has focused on components 1 and 2, where image processing and vision dominate the work. The previous part of the chapter focused on the third component, user interfaces, showing how it also could play a significant role in event detection and specification. But to address large-scale deployments, with hundreds of cameras, we cannot ignore the last two components, the sensor/hardware/computation architecture and the software architecture. Not only do these impact the systems level decisions, they also impact or limit the choices of early components. This part of the chapter draws on nearly a decade of net-worked intelligent video surveillance systems development by Dr. Boult[1] and students across multiple university labs and many companies, including designing and deploying large-scale distributed intelligent video surveillance systems, (see [8–12]). Many of the key systems architecture issues were actually established and validated in the DARPA VSAM program from 1996 to 1998 (see [13–18]). The basic paradigms of background subtraction, simple motion models, and stabilized planar motion for subtraction were all explored. There have been lots of good papers since then on the detection/tracking and motion work, with whole workshops dedicated to it. This section of the chapter looks at the other systems aspects, which are so often ignored. There are no fancy equations, just a discussion of the issues and recommendations of what are the critical things to be resolved.

13.3.1   Sensor Selection Issues

The first issue that must be addressed in an intelligent sensor system is the selection of the sensor(s)—if you cannot sense the target, no amount of postprocessing is going to find it. The issues here include selection of the imaging technology and the lens system. For the basic sensor technology, the choices are visible sensors (CCD/CMOS), an intensified low-light sensor, or a thermal or LWIR sensor. For each technology there is also the potential choice of the resolution of the sensor, with

1.   This work started while Dr. Boult was at Lehigh University and has been funded over the years by multiple contracts from DARPA, ONR, ARMY, and SBIRS with Remote Reality and Securics, Inc. It also includes his experience while the founding CTO of Guardian Solutions and founder/CEO of Securics, Inc. The views expressed are the author's, not necessarily those of the funding agencies, or current or past employers.

analog sensors supporting CIF (320×240), NTSC/QCIF (640×480), and with digital sensors supporting these plus 1-, 3-, or 5-megapixel resolutions at 8, 10, or 12 bits per pixel. A major issue in these decisions is the cost of the system, which is often not as trivial as it sounds. You cannot just look at sensors (e.g., a LWIR sensor costs $30,000 to $80,000) but must also consider the lighting and cabling/installation costs. Since surveillance is often most critical at night, a true cost comparison must include the costs of lighting along with the sensors' resolutions, which can quickly shift the economic balance. For example, installation of a single wide-area lighting system may cost $30,000 to $70,000 per pole, depending on the location and requirements. Running cables back to a central site has equally surprising costs, especially if it must be dedicated coax or fiber for analog video transmission. IP cameras are an option, but the compression involved often limits the target sizes for detection and tracking and introduces enough noise that it limits their usefulness for intelligent video surveillance. While rugged distributed computers are not inexpensive, a system using them can be considerably cheaper if they can use existing network infrastructure or wireless.

Lens choices and minimum target size impact both the field of view and range to target. To be viable commercial video surveillance systems, the systems need to reliably and robustly handle small and nondistinctive targets from great distances. The need for detection of small targets at a distance is a conflict of security concerns versus cost. Distance translates to response time—the goal of security is not only to record events but also to respond to them while they are occurring. Therefore, it is necessary to detect events far enough in advance to respond. While one could increase standoff distance by increasing the focal length of the imaging system, this results in a narrowing FOV and reduction of the overall imaged area, which means that protecting a reasonable area requires numerous cameras, proving to be generally cost prohibitive. Figure 13.9 shows the impact of the minimum size detection target on the number of sensors needed to cover the staging area of an airfield. While minimum target size is clearly a function of the algorithms, it is also clear that sensor resolution impacts this as well; for a fixed minimum target size, a megapixel sensor covers considerably more area than an NTSC or CIF sensor.

In a real system, end users would investigate each alarm, and in many of the current government deployment projects, the requested goal is to produce fewer

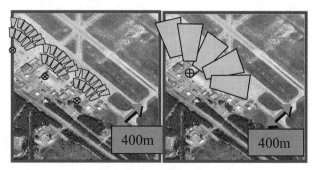

Figure 13.9   Impact of minimum target size on number of sensors required to secure an area. Left assumes minimum target size of 48 pixels (6×8-pixel human target) and requires 33 sensors to cover the staging area. Right diagram is assuming a minimum size of 12 pixels (3×4-pixel human target) and requires only 5 sensors. If false alarms accumulate independently, the impact of the number of sensors on overall system false alarms is obvious.

than three false alarms (FA) per day. For these military applications, undetected targets could be, literally, deadly, so the misdetection (MD) rates also need to be low, with stated goals in multiple programs for a less than 5% misdetection on very distant targets (1 km to 2 km). Again, the overall minimum target size has a significant impact on system false alarm rates, though for almost any algorithm, the FA/MD rates are strongly impacted by the choice of minimum target size. Having a formal model that allows one to make such trade-offs is therefore an important requirement for a video surveillance algorithm.

With each NTSC video containing approximately 1,020 potential target regions per camera per hour, achieving acceptable false alarm/misdetection rates places very strong demands on the low-level processing of the system. In [9, 10] we investigated formal methods for analyzing FA and MD rates for this type of problem. These papers are, to our knowledge, still the only work to formally model pixel group aspects that allow meaningful FA/MD for a video surveillance system. These papers analyzed the grouping that allowed us to address the signal-level FA and impacts of random noise. However, they did not address nuisance alarm (NA) rates, where lighting, water, or brush produce real changes that are significant, but not interesting motion. To address this class of nuisance alarms, we added saliency models, similar in spirit to those used by [19]. However, even with salience models, birds, bugs, and other animals are a significant nuisance issue. If one is using only 6 to 12 pixels on target, then distinguishing a crawling human from a deer or other large animal is quite difficult.

While sensors and low-level processing are probably the most advanced components of intelligent video systems, even at this low level we see two important open issues: no illumination sensors, and increasing sensor resolution. The no-illumination case is the extreme, requiring long wave IR (LWIR), but the costs of those sensors continue to drop. In some settings, for example, harbor security, they may be the only effective choice as lighting is not an option. But for near-shore and terrestrial settings, small amounts of light may be reasonably expected just from local lights at a great distance. The new EMCCDs, with .001 lux color imaging (color with one-fourth moonlight with no added illumination) offer some new potentials. Both LWIR and EMCCD, however, need different front-end processing, as they change both the signal and noise models.

13.3.2 Computation and Communication Issues

The second question, increasing sensor resolution, comes back to the FOV/range tradeoff, but increasing resolution impacts two things—computation and communication. Real-time video processing is a very computationally intensive task and, if the system is not designed well, can swamp most computation and communication systems.

The early VSAM systems at CMU mostly used tower-type 500 MHz PCs or full SGI/Unix workstations, computing CIF data at around 10 to 15 fps, while Sarnoff used custom hardware boards (which eventually led to their Arcadia chips) to achieve 30 fps on NTSC video. In comparison, Dr. Boult's VSAM algorithms were computing 30 fps at 640×480 using an embedded 233-MHz system.

As algorithms improve from those early versions, they have required considerably more computing; ObjectVideo uses dual 3-GHz processors for 4 QCIF channels, and Guardian Solutions uses 2.4 GHz for 4 QCIF channels. More significantly, as higher-resolution sensors are starting to be used, they demand considerably more powerful approaches.

High-end PC-class processors are still the dominant forms of computing intelligent video surveillance. ObjectVideo recently introduced ObjectVideo on board versions using a TI DaVinci DSP to provide their core computation at 15 fps on CIF video. Their goal is reduced cost and reduced system size, not increased performance nor the support for larger sensors.

Moving to higher-resolution sensors is, however, critical to improving overall system performance. This is probably the most significant open issue—how to accelerate the detection/tracking techniques using hardware accelerations such as FPGA and/or local DSPs. To be cost-effective, it will involve tighter integration of PC-based megapixel cameras with cameralink interfaces, which significantly limit distance between cameras and the processing. We need to move the processing closer to the pixels. In the lab at UCCS, as we move to programs using 3-, 5-, and, soon, 16-megapixel sensors, we are focusing on FPGAs to address this. In our FIINDER (FPGA-enhanced intensified image network detectors with embedded recognition) system, we have been using the Elphel 333 network cameras with a Spartan 3, an Etrax processor, 64M of memory and a 3-megapixel sensor. We have designed a version of our Cascaded Haar-wavelet, which, in simulations, requires an average of 6 ms (worst case is, however, 300 ms), and will be porting it to a new 5-megapixel version of the camera. We are also porting our QCC-based detection and tracking into the FPGA/Etrax system. This approach is not simply taking previous video surveillance techniques and adapting them to an FPGA; it is looking at what we can do well in hardware and developing new approaches to exploit that ability.

This then brings us back to the sensor architecture. An effective low-level detection can detect/track small targets, but assessment and identification need more resolution, which suggests a multicamera system where the detected targets are then handed off to a pan-tilt-zoom (PTZ) sensor. As is well-demonstrated in the original DARPA VSAM effort, an effective way to address this issue of identification would require the cameras to be calibrated and use geo-spatial coordinates, passed through the network, to pass control from one sensor to another. With a PTZ in a lower security setting, the system can also use a temporal stop-and-stare approach to trade probability, or time to detection, against cost. Instead of five sensors, the left side of Figure 13.9 could be five temporal stops on an active PTZ tour. But the more significant issue for PTZs is simply to get enough pixels on target to support moving the problem from detection to either recognition or assessment. This concept was well demonstrated in the original VSAM effort where wider field of view static cameras provided control information. Commercial systems, including from GuardianSolutions, Phillips/Bosch, and, more recently, Sarnoff, provide active tracking with feedback from the actual PTZ sensor to control tracking.

For a single camera with small numbers of targets, these solutions are sufficient. But in a complex environment, especially if the PTZ is being used for automated recognition/assessment, the PTZ will likely have multiple targets to which it must attend.

Another sensor issue, which is truly multimodal, but generally not a focus in vision systems, is that pragmatic distributed surveillance will almost assuredly require integration with other sensors. Integration can be for two primary reasons: saving resources or improved overall system detection abilities. In our earlier work for the Army Smart Sensor Web (SSW), we extended the video-based system to integrate with wireless sensors that provide passive IR tripwires and magnetic, seismic, and acoustic triggering. The first goal was to reduce the processing power demands of the video system, allowing it to be in a lower-power state until one of the other sensors triggered. Then the multisensor integration provided for a reduced overall false-alarm rate as it could help distinguish target types using acoustic, magnetic, or seismic signatures in addition to video features. The multisensor integration for that system was done by McQ Associates using a simple templated fusion; for example, using estimated target size from video plus a class estimate from the McQ acoustic sensors, the system could report individual or multiple dismounted soldiers, even when the video through the trees was barely able to determine if there were one or more soldiers.

The queuing and added nonvisible features were also incorporated when Dr. Boult developed the commercial version of the system. We chose a particularly simple set of interfaces, one using serial port receivers with simple ASCII protocols, which was the basis for the SSW effort, and the second a contact sensor interface, as such sensors are extremely common in standard security installations for door alarms, window alarms, and PIR motion sensors. In the commercial case, power savings was not an issue, but it did play an important role in both cueing PTZs where to look and in filtering other events. Examples included having PTZs respond to any door openings, to ensure good close-up video of potential subjects of interest, and also programming the zones so that motion at glass doors or windows only registered as an alarm when the door or window has recently been opened. The latter was important in port settings, where many of the buildings had almost entirely glass walls with counter-obstructed views, so the only visible data would be people moving in front of glass walls, which was difficult visually to distinguish from people moving just behind the glass walls.

System design to incorporate these nonvisual sensors is important and can often reduce both system costs and false alarms. To a vision researcher this might seem like cheating and would not be an acceptable research step. It would have been more interesting to solve the glass wall problem with a sophisticated algorithm. But when you need systems to work in the real world, at reasonable prices, simple multimodal sensors integration can be far more effective.

13.3.3 Software/Communication Architecture

At the core of any distributed system is its software and communication architecture. The details of the software architecture and how the data follows within the processes are critical. There are many models—cooperating independent processes, multithreaded daemons, databased models, process-driven models. Our own work has not significantly advanced the art there since the original VSAM days. Thus we focus our discussion on the network/communication.

Due to the potentially massive amounts of video data and the need for this to be real-time, design must address some means of communicating target information and cannot simply use standard streaming video protocols. Dr. Boult was part of the team to define the original VSAM communication protocol [20].

The VSAM protocol represents key target properties as well as image data. It was sufficient for the dozen or so sensors used in the VSAM but had limitations that prohibited its use in larger systems, including using a single central coordination node and a fixed packet structure. Dr. Boult enhanced that protocol as he developed architectures for wireless video [11] to support hundreds of nodes. The most critical extension for scalability was adaptive bandwidth control. The overall scalable network architecture is shown in Figure 13.10. The sensor-processing modules (SPM) do the actual video processing and detect/track targets, and could be just about any subsystem, but make an important assumption. We presume their target detection produces descriptors other than just an image, for example, GPS-location of the target, size and shape descriptors, as well as a small image chip containing the target. The GPS and size/shape are critical for maintaining tracking information, for user-interface display, and for sensor-to-sensor handoff and fusion. Each target also has priority information based on the sensor's threat assessment and time since last transmission. The SPMs then send these packets into the distributed architecture with each description of the target including its geo-location and a localized image chip of an area around each target. The archive gateway module (AGM) provides traffic routing, reliable multicast support, archiving to support replay on lightweight nodes, and traffic bandwidth adaptation. The operator control units (OCU) are a display and control user interface. On a regular schedule, the SPM also sends out reference images that are a copy of the whole scene. If there is plenty of bandwidth and few targets, then this model allows the local image chips to behave as sprites moving over the

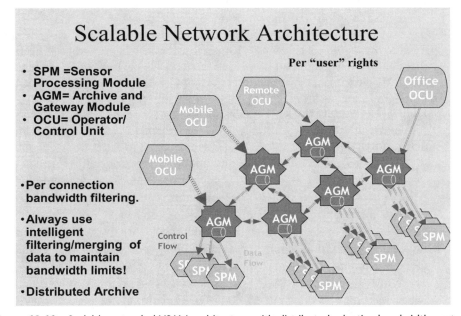

Figure 13.10   Scalable extended VSAM architecture with distributed adaptive bandwidth control.

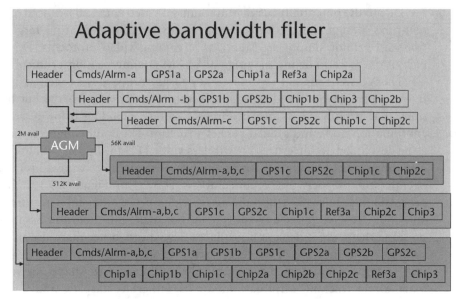

Figure 13.11   An AGM module for adaptive data fusion and filtering.

background in an mpeg-4–like manner. But unlike mpeg-4, the data is packaged for analysis, causing possible dropping and reassembly. For alarms and events of significant interest, the system still provides standard mpeg-4 recorded traffic, but they are not intended for real-time transmission.

An example of bandwidth adaptation is shown in Figure 13.11, where three packets are coming into an AGM node. Having the bandwidth done at the application layer means it cannot address the actual issues of network QoS, as other clients and/or traffic are not under the application layer control. But it has the advantage that it can run on unmodified networks and can make better decisions about the priority of the VSAM traffic. The system can approximate the available bandwidth because it has an initial maximal allocation and monitors the actual channel usage. The upstream links in different directions have different available effective bandwidth, for example, one has 56 Kb/s available, one has 512 Kbps, and the last has 2 Mbps. The three packets waiting for transmission are combined into a single packet, reducing the overhead. For the 2-Mb link in this example, there is sufficient room for all the data. But the 512-Kbps link cannot fit in all the data based on priority and size, so the adaptive bandwidth filter in the AGM removes enough data to keep the expected transmission rate down. For the link with only 56K, only the alarms, target descriptors and a few chips can be sent.

Using this extended protocol and adaptive bandwidth filtering has significantly extended the size of an effective system, allowing large deployments on limited bandwidth. We have had a dozen sensors effectively monitored over a single dial-up line. The commercial version of this developed by Dr. Boult for GuardianSolutions supports a sensor network with over 100 sensors, including 88 cameras covering 2 km^2 of a major U.S. port—all using a single 802.11B channel running mostly at 2 or 5.5 Mbps.

While the extended VSAM approach was critical for bandwidth adaptation, it is designed for a closed system. For security reasons, every node is verified and has to have special software. This is fine for dedicated solutions, but it is not always

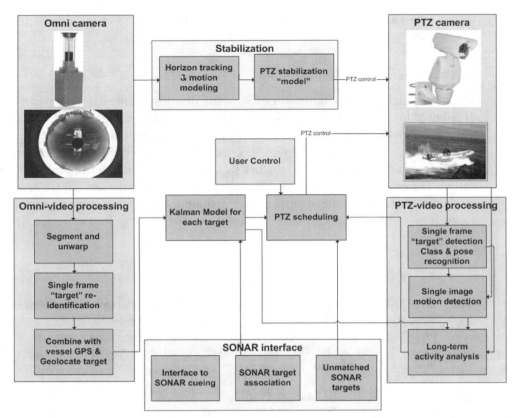

Figure 13.12   SEE-PORT architecture. Each of the stabilization modules, the omni-video processing module, and the PTZ video processing module is implemented in an independent embedded computer. Both visible and thermal omni-cameras will be used, and eventually 4 omni-cameras and 4 to 6 PTZs will be used on larger vessels. Current Development is using 1.2 megapixel visible and 640×480 LWIR but designing algorithms for 16-megapixel visible sensor. Communication will be established via SOAP interfaces that can then provide a direct port, and protocol specification, for the real-time streaming data.

what is needed. While the extended VSAM architecture provides an effective transport and scalability, it does not address sensor discovery or larger multisensor management issues. For a collection of reasons, DOD has been moving their overall network architectures toward service-oriented architectures, requiring multiple programs to follow those guidelines. For two of our ongoing ONR projects, we have moved to a SOAP-based protocol. For the SEE-PORT project (see Figure 13.12), this provides for a scalable and flexible way for other sensors to be integrated and for multiple users to view the data and potentially control the PTZs. Note that the SOAP-based XML is *not* necessarily being used to transport the data; rather, it defines the discover and access protocols, which in turn can provide a direct IP/PORT/Protocol mapping (often a choice of them), which provides the actual transport. Using this, one might discover a sensor that can provide streaming multimimed jpeg (mmjpeg) video (good for a browser), stream mpeg-4, an extended VSAM protocol or a simple target protocol in XML (e.g., just x, y, ID). The point of the SOAP-based interfaces is to simplify larger scale system interactions and integration. But an open issue is how to also fold in the other network issues such as bandwidth, security, and reliability, when dealing with an abstraction layer as removed as SOAP.

13.3.4 System Issues Summary

The second half of this chapter reviewed some of the systems issues necessary for distributed intelligent video surveillance. These issues include sensor choices, low/no light systems, FOV/Resolution trade-off, scaling computation for megapixel sensors, and communication issues. All are critical for an effective system. Rather than being an after-thought, these issues should be considered up front in a system design. The field has developed a good toolbox of vision techniques, but we need to work harder at making meaningful system assumptions and then deriving the right algorithms and systems components given those constraints.

References

[1]   Chin, J., V. Diehl, and K. Norman, "Development of an Instrument Measuring User Satisfaction of the Human-Computer Interface," *Proc. of SIGCHI Conference on Human Factors in Computing Systems*, 1998, pp. 213–218.

[2]   Ivanov, Y., and A. Bobick, "Recognition of Visual Activities and Interactions by Stochastic Parsing," *IEEE Trans. on Pattern Analysis and Machine Intelligence*, Vol. 22, No. 8, August 2000.

[3]   Oliver, N., B. Rosario, and A. Pentland, "A Bayesian Computer Vision System for Modeling Human Interactions," *IEEE Trans. on Pattern Analysis and Machine Intelligence*, Vol. 22, No. 8, 2000, pp. 831–843.

[4]   Davis, J., and A. Bobick, "The Representation and Recognition of Action Using Temporal Templates," *Proc. of Computer Vision and Pattern Recognition (CVPR'97)*, 1997, pp. 928–934.

[5]   Nayar, S., H. Murase, and S. Nene, "Parametric Appearance Representation," in *Early Visual Learning*, S. K. Nayar and T. Poggro, (eds.), London, U.K.: Oxford University Press, 1996.

[6]   htk3, Cambridge University Engineering Dept. (CUED), http://htk.eng.cam.ac.uk.

[7]   Fosler-Lussier, E., "Markov Models and Hidden Markov Models: A Brief Tutorial," http://www.cse.ohio-state.edu/fosler/papers/tr-98–041.pdf, December 1998.

[8]   Boult, T., et al., "Frame-Rate Multi-Body Tracking for Surveillance," *Proc. of the DARPA IUW*, 1998.

[9]   Gao, X., et al., "Error Analysis of Background Adaptation," *Proc. of IEEE Conf. on Computer Vision and Pattern Recognition*, June 2000.

[10]   Boult, T., et al., "Into the Woods: Visual Surveillance of Non-Cooperative and Camou-flaged Targets in Complex Outdoor Settings," *Proc. of IEEE*, Vol. 89, October 2001, pp. 1382–1402.

[11]   Boult, T., "Geo-Spatial Active Visual Surveillance on Wireless Networks," *Proc. of IEEE Applied Imagery Pattern Recognition (AIPR) Workshop*, October 2003.

[12]   Boult, T., et al., "Omni-Directional Visual Surveillance," Special issue of *Image and Vision Computing on Surveillance*, 2004.

[13]   Kanade, T., et al., "Advances in Cooperative Multi-Sensor Video Surveillance," *Proc. of the DARPA IUW*, 1998, pp. 3–24.

[14]   Grimson, W., et al., "Using Adaptive Tracking to Classify and Monitor Activities in a Site," *Proc. of IEEE Conf. on Computer Vision and Pattern Recognition*, 1998.

[15]   Stauffer, C., and W. Grimson, "Adaptive Background Mixture Models for Real-Time Tracking," *Proc. of IEEE Conf. on Computer Vision and Pattern Recognition*, 1999.

[16]   Ben-Ezra, M., S. Peleg, and M. Werman, "Robust, Real-Time Motion Analysis," *Proc. of DARPA Image Understanding Workshop*, November 1998.

[17]   Cohen, I., and G. Medioni, "Detecting and Tracking Moving Objects in Video from an Airborne Observer," *Proc. of DARPA Image Understanding Workshop*, 1998, pp. 217–222.

[18]   Haritaoglu, I., D. Harwood, and L. Davis, "Who, When, Where, What: A Real Time System for Detecting and Tracking People," *Proc. of the Third Face and Gesture Recognition Conference*, 1998, pp. 222–227.

[19]   Wixson, L., "Detecting Salient Motion by Accumulating Directionally-Consistent Flow," *Proc. of IEEE Tran. PAMI*, August 2000, 774–781.

[20]   Lipton, A., T. Boult, and Y. Lee, *Video Surveillance and Monitoring Communication Specification Document 98-2.2*, Technical Report, CMU, September 1998, http://www.vast.uccs.edu/~tboult/vsam_protocol_98_22.ps.gz.

CHAPTER 14

Multimodal Workbench for Automatic Surveillance Applications

Dragos Datcu, Zhenke Yang, and L. J. M. Rothkrantz

Noticeable developments have lately been achieved on designing automated multimodal smart processes to increase security in people's everyday lives. As these developments continue, proper infrastructures and methodologies for the aggregation of various demands that will inevitably arise, such as the huge amount of data and computation, become more important. In this chapter, we introduce a multimodal framework with support for an automatic surveillance application. The novelty of the attempt resides in the modalities to underpin data manipulation as a natural process but still keeping the overall performance at high levels. At the application level, the typical complexity behind the emerging distributed multimodal systems is reduced in a transparent manner through multimodal frameworks that handle data on different abstraction levels and efficiently accommodate constituent technologies. The proposed specifications include the use of shared memory spaces (XML data spaces) and smart document-centered content-based data querying mechanisms (XQuery formal language). We will also report on the use of this framework in an application on aggression detection in train compartments.

14.1 Introduction

The challenge to build reliable, robust, and scalable automated surveillance systems has interested security people ever since the first human-operated surveillance facilities came into operation. Moreover, since the bombings in London and Madrid in 2005, research in methods to detect potentially unsafe situations in public places has taken flight.

Given the size and complexity of the sensing environment surveillance systems have to cope with, including the unpredictable behavior of people interacting in this environment, current automated surveillance systems typically employ diverse algorithms (each focusing on specific features of the sensor data), many sensors (with overlapping sensing area and able to communicate with each other through a network), and different types of sensors (to take advantage of information only available in other modalities).

It is our belief that in modern surveillance applications, satisfactory performance will not be achieved by a single algorithm, but rather by a combination of interconnected algorithms. In this chapter, we present a framework for automated surveillance designed to facilitate the communication between different algorithms. This framework is centered on the shared memory paradigm, the use of which allows for loosely coupled asynchronous communication between multiple processing components. This decoupling is realized both in time and space. The shared memory in the current design of the framework takes the form of XML data spaces. This suggests a more human-modeled alternative to store, retrieve, and process data. The framework enhances the data handling by using a document-centered approach to tuple spaces. All the data is stored in XML documents, and these are subsequently received by the data consumers following specific XML queries. In addition, the framework also consists of a set of software tools to monitor the state of registered processing components, to log different type of events, and to debug the flow of data given any running application context.

The remainder of this chapter is structured as follows. Section 14.2 starts with an overview of the related work, examining the existing approaches and motivating the key ideas of each approach. Then we give an overview of our framework and discuss its main building blocks. Section 14.4 shows the framework in action with an example of a surveillance application built on top of the framework. The application uses several cameras and microphones to detect unusual behavior in train compartments. Finally, Section 14.5 summarizes our contribution.

14.2   Related Work

14.2.1   Video-Based Approaches in Automated Surveillance Research

The bulk of existing work on automated surveillance has largely concentrated on using only video. Video-based surveillance related algorithms have been extensively investigated [1]. The underlying algorithms consist of methods ranging from simple background extraction algorithms to more complex methods, such as dynamic layer representations [2] and optical flow methods [3].

Depending on the characteristics of the sensing environment, more specific methods have been employed. The performance of face detection algorithms [4], for example, depends on the size of the surveillance area, whereas the different methods for people or (more generally) object detection and tracking depend on the amount of occlusion and the motion characteristics of the objects involved. Even in visual event detection, different methods (e.g., video event graphs [5] and HMMs [6]) have been developed.

Some researchers approach the surveillance problem with special kinds of visual sensors and their specific algorithms. For example, in [7] the authors integrate normal camera, infrared (IR), and Doppler vibrometers (LDVs) in their multimodal surveillance system for human signature detection.

14.2.2   Audio-Based Approaches in Automated Surveillance Research

As sound/speech is not always present, it is impractical to do continuous sound-based tracking. Therefore, sound-based surveillance algorithms are usually sound event detectors or recognizers. These algorithms usually consist of a training phase in which various features are extracted from training sound signals to obtain characteristic acoustic signatures of the different types of events. During operation, a classification system matches sound signals against the trained acoustic signatures to detect events [8–10]. The work of [11] adapts a multilevel approach in which audio frames are first classified into vocal and nonvocal events. Then a further classification into normal and excited events is performed. The events are modeled using a Gaussian mixture model with optimized parameters for four different audio features.

14.2.3   Multimodal Audio-Video–Based Approaches

Sound-based algorithms are most effective when used in combination with sensors from different (in particular the visual) modalities. Audio can be used complementarily to help resolve situations in which video-based trackers lose track of people due to occlusion by other objects or other people. The combined approach yields better and more robust results, as demonstrated by different researchers using decentralized Kalman filters [12], particle filters [13], and importance particle filters [14]. In general, multimodal approaches of surveillance applications consist of unimodal, modality-specific, low-level feature extraction algorithms and higher-level multimodal fusion algorithms. For example, in [15] the authors use Kalman filtering to combine stand-alone audio and video trackers. Other techniques used in higher-level fusion are Bayesian networks [16–18], rule-based systems [19], and agent-based systems. A different approach is adopted in [20], where a system for tracking a moving object was developed, using a probabilistic model describing the joint statistical characteristics of the audio-video data. Typical of this approach is that the fusion of the data is accomplished at a low level, exploiting the correlations between the two modalities. The model uses unobserved audio and video variables to describe the observed data in terms of the process that generates them.

14.2.4   High-Level Interpretation

At a higher level, automated surveillance research is focused on the semantic interpretation of the events in their spatial and temporal relationships. In [21] an automatic surveillance system is discussed that performs labeling of events and interactions in an outdoor environment. It consists of three components: an adaptive tracker; an event generator, which maps object tracks onto a set of predetermined discrete events; and a stochastic parser. The system performs segmentation and labeling of surveillance video of a parking lot and identifies person-vehicle interactions. In [22] human behavior models are used to obtain interpretation.

Generally, high-level approaches are heavily dependent on the context in which the system is applied. There are at least two projects for which extensive research has been conducted in applying video- and audio-processing algorithms for improving passenger safety and security in public transportation systems. In the PRISMATICA project [18] a surveillance system for railway stations is designed to integrate different intelligent detection devices. In the PRISMATICA system, detection of events is done in geographically distributed visual, audio, and other types of devices and using different algorithms. Bayesian networks are used for presenting and fusing the data by these devices. In the ADVISOR project [19], a surveillance system was developed for detecting specific behavior (such as fighting or vandalism) in subway stations. Different methods have been defined to compute specific types of behaviors under different configurations. All these methods have been integrated in a coherent framework.

14.2.5 Frameworks

In most related research, the focus has been on the algorithms to achieve the automated surveillance. In most cases, the overall framework to make the algorithms work with each other emerges as a result of the demands of the algorithms. From a system engineering perspective, this approach is far from optimal, as very important functions such as communication, resolution needs, resource management (e.g., handling distributed data sources or data sharing) and exception handling (if a sensor might break down) rely on a good overall framework. Chapter 13 describes these issues in more detail.

In recent years, application-specific frameworks based on emerging technologies, such as peer-to-peer environments, message-based middleware, and service-oriented approaches, have been developed. Most use XML or XML-based technologies to support data manipulation and to facilitate component interaction. XMIDDLE [23], for example, is a mobile computing middleware using XML and document type definitions (DTD) or schemas to enhance data with typed structures and document object models (DOM) [24] to support data manipulation. It further uses XPath [25] syntax to address the data stored in a hierarchical tree-structured representation. Other examples are Jini and Java Media Framework (by Sun Microsystems), Vinci [26], and HP Web Services Platform and e-Speak (by Hewlett Packard). Finally, there have been frameworks developed for multimodal fusion applications, for example, [27, 28]. Table 14.1 gives a comparison of existing frameworks. We have tried to make an overview of the frameworks used by the different researchers and briefly indicate their characteristics. We omit detailed descriptions due to lack of space.

14.3   General Model Description for the Multimodal Framework

The multimodal framework being described in this paper is centered on the shared memory paradigm. It introduces a novel technique in the way data is handled by different purpose data consumers. Compared with the traditional way of implying

Table 14.1 Comparison of Existing Frameworks

Framework	Ref.	1	2	3	4	5	6	7	8	9
XMIDDLE		0	−	+	0	+	+	+	+	+
JMF/Jini	[29]	0	−	+	0	+	+	+	+	+
Vinci	[26]	0	−	+	+	+	+	+	+	−
e-Speak		0	−	+	0	+	+	+	+	+
iROS	[28]	+	?	?	−	+	+	+	+	−
ADVISOR	[16, 19]	+	+	+	+	+	0	−	−	−
PRISMATICA	[18]	+	+	+	+	+	+	−	−	−
KNIGHT	[17]	−	+	?	+	?	0	−	−	−

The comparison is based on: modality dependency (1), applied-for surveillance applications (2), scalability (3), performance (4), transparency (5), modularity (6), repository services (7), coordination services (8), and security services (9). The qualifications +, 0, −, and ? indicate good, neutral, bad, and unknown, respectively.

direct connections between the system components, each connection having its own data format, the new approach suggests a more human-modeled alternative to storing, retrieving, and processing the data. The data is conferred on an underlying structure that complies with eXtended Markup Language (XML). The shared memory in the current design of the multimodal framework takes the form of XML data spaces.

The use of shared memory allows for loosely coupled asynchronous communication between multiple senders and receivers. The communication decoupling is realized both in time and space. The specification fully complies with the requirements of data manipulation in a multidata producer/consumer context where the availability of data is time-dependent and some connections might be temporarily interrupted.

The information is structured in documents using XML standard. An XML schema [30] is used to validate each existing XML document prior to extracting meaningful information. Furthermore, binary data can be easily interchanged via XML documents after converting it using XML MIME protocol. The multimodal framework also consists of a set of software tools to monitor the state of registered processing components, to log different types of events, and to debug the flow of data, given any running application context.

Depending on the type of application to be built on top of the multimodal framework, a routing algorithm has been designed to manage the data transfer among existing shared memories on different physical networks. This capability is highly required, commonly for multimodal applications that involve distinct wireless devices. Considering the case study of an automatic surveillance application, this capability allows wireless devices, such as PDAs or mobile phones equipped with video cameras, to communicate with the system core and to send useful video data.

The framework specifications solely emphasize the presence and role of all its components through existing technologies and standards and not on the implementation details. Yet several proposed technologies present a certain degree of freedom in some functional aspects for the implementation phase. Although the multimodal framework has been designed by taking into consideration the further development of an automatic surveillance–oriented application, it can be adopted as a basis for any kind of complex multimodal system involving many components and heavy data exchange. By making use of the framework specifications, a common description of possible processing components, along with their interconnections for a surveillance application, is provided as grounds for eventual specific examples that may be given throughout the chapter. Because the workbench implementation itself relies on the philosophy of shared XML data spaces, special attention is given to examples of how to integrate the two underlying technologies for modules and XML data management.

The examples aim at studying the usability and extensibility of concepts in formulating proper data and command maneuvers to ensure a logical and natural data flow through the network of data processing nodes.

The algorithms in the illustrated processing components related to the example of automatic surveillance application are not described in full detail due to space and topic considerations. Indeed, they cover a broad range of research areas from different computing fields, and barely any standards exist to favor one method

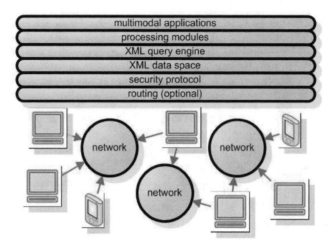

Figure 14.1   The multimodal framework diagram.

above the other. In restricted cases, some algorithms are employed for exemplification purposes, though.

One distinct remark concerning the multimodal framework is that it should support platform-independent interconnection of the processing modules. This requirement is essential for heterogeneous working environments, while it also leaves open space for the programming languages, tools, and technologies to be used at the implementation phase. An overview of the current multimodal framework is shown in the diagram in Figure 14.1.

If the processing modules are envisaged as services in the multimodal workbench, then a multimodal application specifies, along with the service set, a working plan on how the services are used to get the data through different abstraction levels to obtain the desired semantic information. The multimodal framework defines a way to represent the work plan through a monitor application and uses that to get a detailed state analysis over its fulfillment. Each service registered in the framework publishes its input and output XML-formatted specifications using Web Services Description Language (WSDL) [31].

The service-oriented overview is given in Figure 14.2. Related technologies to define distributed services in distributed object environments (DOEs) include Jini [29], defined by Sun Microsystems in the late 1990s; service-oriented architecture (SOA); common object request broker architecture (CORBA) from OMG; and distributed component object model (DCOM) from Microsoft.

14.3.1   XML Data Spaces

In the proposed multimodal workbench, the processing of data takes place in a distributed manner. Several technologies, such as RPCs, remote method invocation (RMI), or common object request broker architecture (CORBA), which have been researched in the past to overcome the problems imposed by the distributed computing environment, still present some inaccuracies. Due to its capabilities to solve the typical problems regarding synchronization, persistence, and data communications, the extensively researched tuple spaces algorithm has been used

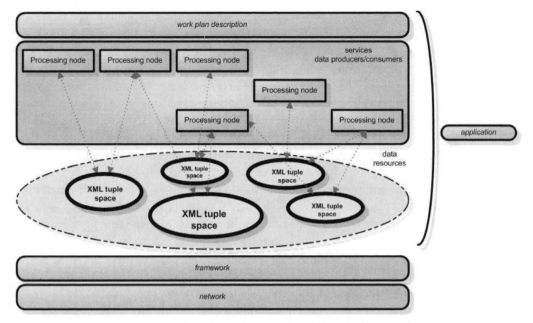

Figure 14.2   The service-oriented overview on the multimodal framework.

ever since its introduction by Yale University in the mid-1980s. The goal of the original Linda system was to achieve coordination within various parallel applications. The greatest advantage of the tuple spaces is the decoupling of sender and receiver components in both time—by removing the existence simultaneity restriction—and space—by removing the address information requirement. One component, being part of the shared memory space, needs to know neither the location of the component it plans to send data to nor anything about its availability. Later Linda versions include Melinda, a multiple-data-space Linda-based algorithm; Limbo, for adaptive mobile applications; Lime and eLinda, which specify fully distributed tuple spaces; and an additional form of output operation. The other eLinda extensions regard the programmable matching engine (PME) as a more flexible criterion for tuple addressing and support for multimedia through the Java Media Framework (JMF). JavaSpaces from Sun and TSpaces [32] from IBM are recent Linda variants that store and exchange data in the form of objects.

JavaSpaces uses transaction terms to denote a set of tuple-space operations that can be rolled back in faulty cases and by lease term, to enable the automatic removal of tuples after the time expiration.

TSpaces allows for programming language–independent services described either using WSDL or proprietary description languages (IDLs) and for a general-purpose mechanism to associate tuple-space events with actions via specific rules. sTuples [33] proposes an architecture for semantic tuple spaces that extends a secure communication framework to support a description-logic reasoning engine and a Web ontology language.

A recent improvement for the tuple-spaces algorithms allows for XML-document handling. A few implementations already exist, such as xSpace [34], Rogue Wave, Softwares Ruple [35] or XMLSpaces.NET [36]. xSpace advances xSpaces Query Language, an XPATH syntaxlike querying language using both content and structure information with support for both XML Schema and DTDs. XML Spaces.NET implements the tuple-space algorithms on the .NET platform.

Our implementation of the multimodal workbench is an adapted version of Ruple XML spaces. The Ruple Alpha V2 release includes XML Spaces, http and Simple Object Access Protocol (SOAP) for data accessibility, MIME attachments, and a subset of XML Query Language (XQL) for document retrieval.

The design of our XML space follows the simplicity criterion so as to provide the multimodal workbench with fast data manipulation and, at the same time, to present high enough complexity for data storage and retrieval. All the documents written in spaces supported by the multimodal framework are in XML format. The four basic operations (Figure 14.3) on XML data spaces supported by the multimodal framework are: *write*—for sending an XML document that has an expiration time to an XML space; *read*—for retrieving an XML document from an XML space based on an XML query; *readMultiple*—for retrieving a set of XML documents based on an XQuery XML query; and *take*—for reading and deleting an XML document that matches an XML query.

The case of querying XML documents from several XML tuple spaces is illustrated in Figure 14.4. Such an operation demands distinct connections to each of the XML tuple spaces involved in the query. After the retrieval of the documents from the data space using the basic tuple operations, the proper document is selected by using a query expression. This query specifies how the XML document to be taken should look in terms of specific attribute values. The last step is the extraction of audio/video data from the queried XML document. Subsequently, the framework uses base-64 decoding for reconstructing the original audio/video data.

For an application context in which several processing services are working simultaneously as data consumers and producers, there can be assumed the existence of a set of XML spaces accommodating the integration of the different nodes in the

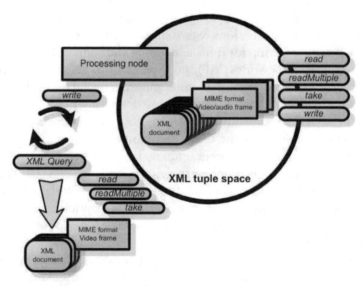

Figure 14.3 The basic operations used to handle XML documents in XML tuple spaces.

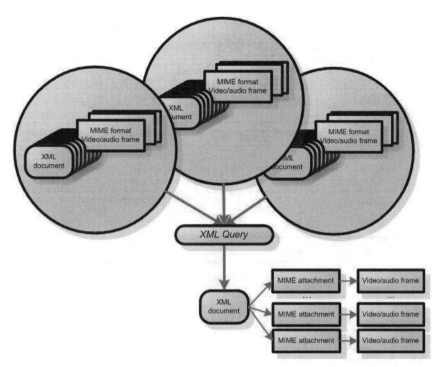

Figure 14.4   Audio/video data retrieval based on XML query from multiple XML tuple spaces.

multimodal framework. Our framework does not have any restrictions for the number of XML data spaces. Moreover, given a distributed running context, the current XML data flows are optimized through different data spaces, enabling high-speed data propagation in the services network. A general integration of XML spaces is presented in Figure 14.5. Every two data spaces are asynchronously connected through at least one common node in the network.

The connecting node can be part of different physical networks, thus acting as a bridge between XML tuple spaces lying in distinct networks. This is the first case of Figure 14.5 (physical network layer I), in which the three XML tuple spaces reside in three different physical networks (*Net1*, *Net2*, and *Net3*). As can be seen, there are processing nodes (*PN1–PN2, PN8–PN9,* and *PN11–PN12*) that coexist on the same hardware device and are connected to the same XML tuple space. The nodes with no direct visibility can still exchange data through a sequence of interconnected XML spaces.

The second case (physical network layer II) illustrates the example of a unique physical network (*net*) that supports all the XML spaces in the multimodal framework. Our implementation of XML tuple spaces respects the general specifications also followed by other papers. Along with the implementation of the four basic XML space operations, our framework exports certain functions to the processing modules and the monitor application in the multimodal workbench.

More exactly, the *creation of a new space* operation allows for initiating a new empty XML tuple space by both the services and the monitor application. In the same way, it is possible to remove an XML space from the workbench with

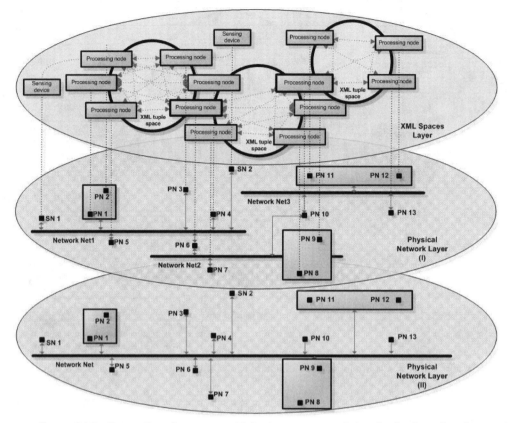

Figure 14.5 Connections between multiple data spaces and the distribution of nodes on the physical network.

the *deletion of a space* operation. In this case, all the XML documents that existed up to the time of deletion are also deleted. Any processing module can subscribe to an XML data space by calling the *registration to a space* operation.

The reverse operation is also supported for unregistering from an XML space through the *unregistering from a XML space* operation. The set of XML tuple spaces can be searched for the existence of a certain XML space by using the *searching for a space* operation. The function provides several characteristics regarding the XML space, such as the time of XML space creation, the entity that created it, the maximum capacity, and the number of XML documents in the space, and performance-related measures, such as the frequency and the workload of operations applied on the space. The list of all existing spaces can be retrieved using the *retrieve the XML space list* operation.

14.3.1.1 XML Document Lease

The framework automatically checks for the availability of the documents as it is specified by the life-span parameter in the XML document lease. Before writing to an XML data space, each XML document is assigned a lease specifying an expiration time. The parameter determines for how long the document is available in the XML data space. Every time the XML document list is checked and an XML documents, lease time is found to be expired, it is removed from the XML document set of the XML data space. The XML space ensures an efficient flexible document exchange between the sender node and the destination node. Occasionally, connected devices can provide data to the multimodal framework and also receive processing results.

14.3.1.2   Multimedia Data Support

An XML document may include audio/video data as an attachment. The framework includes functions for manipulating the multimedia data (Figure 14.6). The base-64 [37] binary-to-text encoding scheme is used to convert an audio or video frame byte sequence to a sequence of printable ASCII characters. This operation allows for the integration of binary data into text-based XML files. The reverse of the encoding is requested at the retrieval of an XML document from an XML tuple space, prior to the actual audio/video data processing.

Due to the multimodal nature of the target application, the current framework involves operations that fuse data of different types. In order to make room for a proper classification of data fusion operations accommodated by the multimodal framework, the types of data on different abstraction scales have to be introduced. On the first level of abstraction is the raw data (feature level). This represents unchanged data, as it is received from input-sensing devices, such as video frames from video cameras or audio signals from microphones. In our implementation of the workbench, the low-level data is stored in *raw data XML spaces* (Figure 14.7).

The second type assumes an intermediate level that includes all the data derived from input, raw data. It represents all kinds of variables that are to be used as inputs to other processing modules, yet do not have a self-contained meaning to the application-context from the user's point of view. For example, facial features, as the location of facial characteristic points, other computed parameters, as distances or angles between them, or the phoneme set, in the case of speech analysis, are considered to belong to the class of intermediate data. The intermediate abstraction-level data is stored in *intermediate data XML spaces* (Figure 14.7).

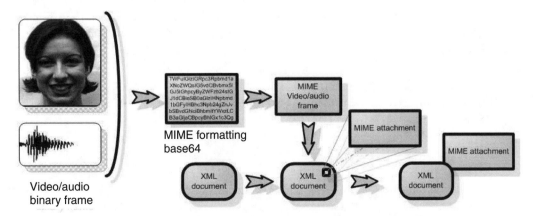

Figure 14.6   The transformation of binary data to base-64 formatted attachments.

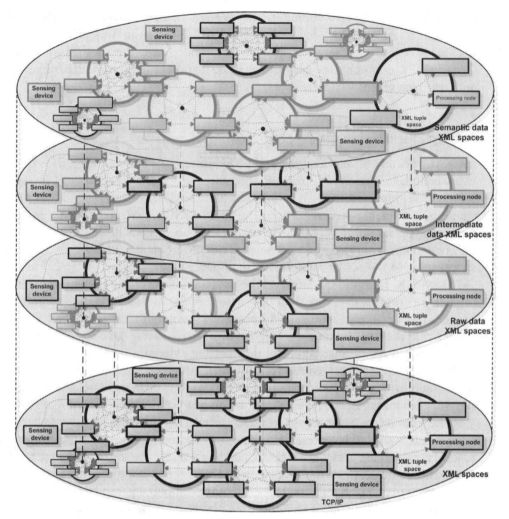

Figure 14.7   The transformation of data to higher levels of abstraction through different XML spaces.

The third type is high-level, semantic data. This class contains highly processed data extracted by following close analysis on raw, intermediate, or other high-level data. This data has a high level of abstraction and is meant to explain the state and behavior of various entities in the application context. The high-level data is stored in *semantic data XML spaces* (Figure 14.7). To enumerate some examples, these include the number of faces or persons from the view, along with meaningful information, such as the emotions shown by everyone, the talk, and the gaze for each person, as well as information related to other existing objects or external events, such as background voices.

According to the levels of abstraction of data on which fusion is accomplished, there are three types of data-fusion transformations through which the processing modules in the multimodal framework are structured: low-level, intermediate, and high-level data fusion. Low-level data fusion implies operations on raw data directly received from the sensing devices, such as video cameras or microphones, and the output represents intermediate or high-level data on the abstraction scale.

Intermediate data fusion operates on either intermediate-level or both low-level and intermediate-level data. High-level data fusion takes into account all the operations, considering high-level data or combinations of high-level with intermediate and low-level data. The processing modules implying fusion of data at a certain abstraction level are depicted in Figure 14.7 as *processing nodes* that are active in a certain XML space layer denoting a data abstraction layer and are connected to other XML spaces from other layers.

14.3.2   Querying Data from XML Data Spaces

The video/audio data as well as other derived information within the multimodal framework is XML document–centered and the retrieval is based on XML queries. Query languages present an efficient way to support content-based information extraction and updates. Prior to running the processing on data, the registered modules perform such a query on a specific XML shared data. The query is generated according to the specifications of XQuery format [38]. XQuery is a typed, functional language for querying XML documents. It is currently developed by the XML Query Working Group of the World-Wide Web Consortium (W3C). The XQuery specification set provides a high-level description of the XML syntax for a query, as well as the underlying semantics. XQuery expressions consist of a set basic building blocks defined as unicode strings containing keywords, symbols, and operands.

There are several XML-related standards that XQuery shares common interoperability with. Along with other requirements, the XPath describes how to manipulate the XML document content generated in correspondence with the associated XML schema. The XML schema contains all the necessary information on how the XML document is formatted. Similar ways of specifying the structure of an XML document are document type definitions (DTD) included in the original XML specification or document structure descriptions (DSD).

The requirements of data manipulation within our multimodal framework specify the presence of XML schema along with the XML document description for each shared data space. This is a requirement for the multimodal framework and not necessarily for XQuery, as the W3C specifications state that the existence of XML schema is not mandatory.

A major advantage of using XQuery is the FLWOR expression, providing iteration and binding of variables to intermediate query representations. The XML document is parsed by an XML parser that generates an XML information set. In our implementation of the multimodal workbench, we use Xerces [39] for parsing XML files, using the DOM/SAX parser. After parsing, the document is further validated against an XML schema, and the result is an abstract structure called post-schema validation infoset (PSVI). In our workbench implementation, an interface makes the connection between variables and functions in the framework and in XQuery expressions through the XQuery expression context (Figure 14.8).

The XML schema is retrieved from the same tuple space as the one containing the XML documents. If an XML document does not comply with its XML schema, an error is raised, as it may contain flawed formatted data, and further processing is no longer performed.

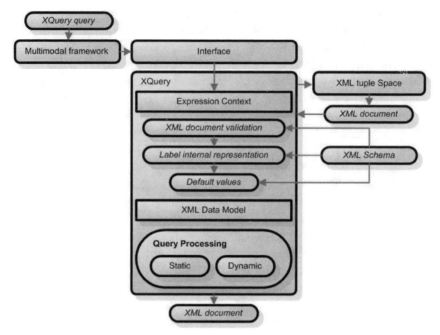

Figure 14.8   Interface between the multimodal framework and the XQuery.

Once the document is checked to determine whether it has the correct format, an internal representation is generated and labeled as indicated by the same schema. The third step that makes use of document schema focuses on possible missing attributes to assign default values.

The XML data model (XDM) resulting from the information set or PSVI undergoes a sequence of XQuery query processing steps, including static and dynamic evaluation phases, to determine the values of the inner expressions.

14.3.3   Comparison of the Multimodal Framework with iROS Framework

The iROS event heap (CSLI, Stanford University) [28] is an adaptive middleware infrastructure that provides coordination of data and interactions in ubiquitous computing environments. Among other features, it supports transparent communication, content-based addressing, limited data persistence, and logical/physical centralization. Table 14.2 presents the list of functional characteristics for both the iROS and our framework.

Ad hoc networks, as well as fixed networks, are well supported by our multimodal framework. There are no requirements for network configurations or address information of any kind to be known by the user in order to access the framework. In the case of iROS, there is a need to know the address of the centralized component (iROS DataHeap) for the connection. Our multimodal framework works also in multihop networks, if routing support is also employed. Our multimodal framework can seamlessly work over different physical

Table 14.2 Comparison Between Our Multimodal Framework and iROS

Characteristic/Framework	iROS	Our Multimodal Framework
Approach	Centralized	Decentralized
Network type	Fixed	(multihop) ad hoc
Network protocols	TCP	UDP
Routing	Standard	Advanced
Binary audio/video data support	—	OK
XML support	—	OK
Semantic query	—	OK
Data structure validation	—	OK
Physical networks connectivity	—	OK
Fault tolerance	—	OK
Service support	—	OK
Monitoring tools	—	OK
Security	—	OK
Decoupling (time, space)	—	OK
Event handling	OK	OK
Limited data persistence	OK	OK
Log support/activity tracking	—	OK
Update operation support	—	OK
Replication	—	OK

networks. In that case, the nodes of the framework act as bridges on distinct networks. Each such bridge extends the accessibility range of the framework in existing hardware environments. iROS can work in limited network segments, the accessibility range being restricted to the network where the centralized component runs.

Our multimodal framework is up as long as there is at least one node to support it. iROS functionality is conditioned by the working state of the centralized component (iROS DataHeap). If a problem occurs at that layer, then the functionality of the entire framework is broken.

In our framework, the transfer of data and control packages is based on the UDP protocol. This feature naturally fits with the working conditions in wireless networks, where the sudden breakdown of connections can occur. The framework keeps track of each network disconnection and makes sure the data packages are temporarily stored until the destination can be reached again and the data is resent. iROS makes use of TCP protocol to connect clients to the iROS DataHeap. Additional work must be done on setting functional components to ensure the integrity of the data in such conditions.

One important characteristic of our multimodal framework is the support for XML documents as well as the support for querying data based on the XML content. Together with the support for binary data as video and audio frames, these options provide proper handling of multimodal contents for multimodal applications.

Another property of our multimodal framework is the validation of the data against the structure and content of the documents. The validation process is realized by taking into account the XML schema for the XML documents.

14.3.4 General Description Model of the Multimodal Workbench for the Automatic Surveillance Application

The context awareness of our automatic surveillance application is achieved by applying a set of data processing operations on input raw data received from sensing devices, such as cameras and microphones. Figure 14.9 depicts a typical diagram of the necessary processing modules for supporting the automatic surveillance application. The processing modules range from face, gaze, and gesture to sound, sound location, speech recognition, and emotion analysis and aim at extracting useful information from audio and video data at different levels of abstraction.

The *behavior recognition* module has the role of detecting any type of aggression that takes place in the scene being analyzed. The analysis focuses on each person and on the groups by considering semantic information related to the gestures, text from the conversations, emotions shown by each person, and sound patterns, such as breaking glass or shouting. The aggression-related high-level data is stored in the XML data space *scene.person.semantic.aggression*. The information stored in this XML space is of great importance for the feedback of the entire surveillance session. The actual feedback is supported by two types of actions. The first type represents the set of actions to be followed when a certain kind of aggression is detected by the system. It can take the form of video recording

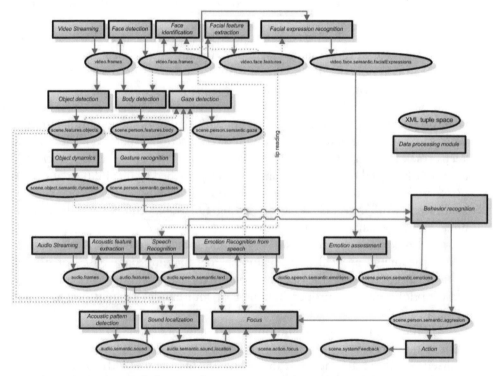

Figure 14.9 A typical diagram of processing modules and shared data spaces for an automatic surveillance application.

the whole session or generating SMS, e-mail, or other notifications. The relevant information is stored in XML space *scene.system Feedback*.

The second type represents the actions concerning the commands to the pan/tilt/zoom (PTZ) cameras. Along with sound, sound location, gaze, and face, the aggression information is further taken into account for determining the region of interest for surveillance. The processing takes place in the *focus* module and consists of a prediction of the system over the next possible places in the scene where aggression might take place.

A typical example of an aggression scene assumes the detection of a sound event that is captured and, with the relevant information, stored in the *audio. semantic.sound* space. Based on relevant information from the XML space *scene.action.focus*, the system feedback is provided as motor commands to the video cameras to focus on that part of the scene. At the next processing moment, the aggression actions are captured and interpreted by the system, and eventually the specific feedback is generated according to the steps to be followed in such contexts. Listing 14.1 illustrates typical examples of querying data from the tuple-space *video.frames*, which contains an XML document storing video frames received from multiple video cameras. The XML document has the XML schema as indicated in Listing 14.2.

The query Q1 retrieves an XML-formatted result comprising all the video frames as they are stored in the original document. The second query (Q2) introduces a video frame selection based on two constraints pointing to data frames of type *jpeg* received from camera *camera1*. The resulting partial XML-formatted document contains only the XML element *data* out of each compound element *frame*. The query Q3 represents an alternative way of writing the query Q2, their results being the same. The last example, Q4, iterates through all the video frames of type *jpeg* that are received in the last 60 seconds from video camera *camera1*. For each such video frame, it calls a face-detection function declared locally.

The face-detection routine is managed by our multimodal workbench as a service that has the WSDL specification, as exemplified in Listing 14.3. The connection between the published processing routine, represented using a *multimodal framework* component in the diagram (Figure 14.8), and the *XQuery* XML query processor is realized through the component *interface*. The interface assumes an optimal matching and translation of the input and output parameters of the routines existing in the workbench and those in the XQuery expressions.

Depending on the specific requirements for the routines of each data processing module, the type of system of the XQuery/XPath data model (XDM) may be augmented in an actual implementation to enhance the use of data variables of distinct types (i.e., face variables). Our implementation of the face detection routine [40] published by the *face detection* module is based on the Adaboost algorithm with the relevance vector machine (RVM) as a weak classifier, using Viola & Jones features.

The output of the query for retrieving the data for face detection may be an XML document satisfying XML schema in Listing 14.4 that contains data on all the detected faces in the ascending order of the recording time.

Listing 14.1   Examples of Using XQuery Language to Retrieve Data from Tuple <video.frames>

```
LQ1: doc("video.frames.xml")/videodata/frame

Q2: doc("video.frames.xml")/videodata/frame[cameraID="camera1" and dataType="jpg"]/data

Q3:
for $video_frame in doc("video.frames.xml")/videodata/frame
where $video_frame/cameraID="camera1" and dataType="jpg"
return $video_frame/data

Q4:
<faces>
{

        ...
        let $current_time:= current-time()
        for $vf in doc("video.frames.xml")/videodata/frame
        where xs:time($vf/time)>$current_time-60
        order by xs:time($vf/@time)
        return
                <face faceID="{local:getFaceID($vf)}">
                  <personID>unidentified</personID>
                  $vf/cameraID
                  $vf/time
                  local:detectFaces($vf/width, $vf/height, $vf/dataType, $vf/data)
                </face>

}
</faces>
```

Listing 14.2   XML Schema for Tuple <video.frames>

```
<xs:schema xmlns=http://www.w3.org/2001/XMLSchema">
<xs:element name="videodata" type="videodata-type">
<xs:complexType name="videodata-type">
 <xs:sequence>
  <xs:element name="frame" type="frame-type" minOccurs="1" maxOccurs="unbounded"/>
 </xs:sequence>
</xs:complexType>
<xs:complexType name="frame-type">
 <xs:sequence>
  <xs:element name="cameraID" type="xs:string" minOccurs="1" maxOccurs="1"/>
  <xs:element name="time" type="xs:time" minOccurs="1" maxOccurs="1"/>
  <xs:element name="width" type="xs:integer" minOccurs="1" maxOccurs="1"/>
  <xs:element name="height" type="xs: integer" minOccurs="1" maxOccurs="1"/>
  <xs:element name="dataType" type="xs:string" minOccurs="1" maxOccurs="1"/>
  <xs:element name="data" type="xs:string" minOccurs="1" maxOccurs="1"/>
 </xs:sequence>
<xs:attribute name="frameID" type="xs:string" use="required"/>
</xs:complexType>
</xs:schema>
```

The example in Listing 14.5 shows the emotional information selection from both video and audio data for refining the assessment of emotions for the persons in the scene. Each audio frame is synchronized with the video frames (Figure 14.10), with respect to the individual recording timestamps. The result of the script contains the time-ordered emotion information from audio and video structured using a tag to identify the human subject of the emotional analysis. In terms of XQuery specifications, the operation is a join on partial XML data from the two XML documents. The XML document storing emotional information about faces is stored in the XML space *video.face.semantic.facialExpressions* and has the XML schema as illustrated in Listing 14.6.

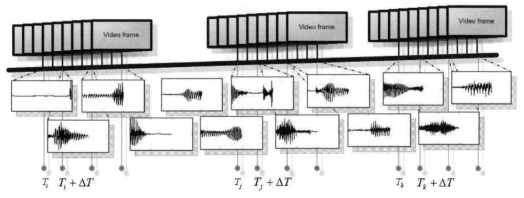

Figure 14.10   The time-span correspondence between video and audio frames.

Listing 14.3   Example of WSDL Statement for Face Detection Processing Module

```
< wsdl:definitions >
        <xsd:element name = "faceDetection" >
         <xsd:complexType >
          <xsd:sequence >
                <xsd:element name = "width" type = "string"/>
                <xsd:element name = "height" type = "string"/>
                <xsd:element name = "dataType" type = "string"/>
                <xsd:element name = "data" type = "string"/>
          </xsd:sequence >
         </xsd:complexType >
        </xsd:element >

        < wsdl:message name = "getWidthRequest" >
                <wsdl:part name = "width " type = "xs:unsignedShort"/>
        </wsdl:message >

        < wsdl:message name = "getHeightRequest" >
                <wsdl:part name = "height" type = "xs:unsignedShort"/>
        </wsdl:message >

        < wsdl:message name = "getDataTypeRequest" >
                <wsdl:part name = "dataType" type = "xs:string"/>
        </wsdl:message >

        < wsdl:message name = "getDataRequest" >
                <wsdl:part name = "data" type = "xs:string"/>
        </wsdl:message >

        <wsdl:message name = "getFaceResponse" >
                <wsdl:part name = "face" type = "faceDetection"/>
        </wsdl:message >

        <wsdl:portType name = "FaceDetectionLib" >
                <wsdl:operation name = "detectFaces" >
                        <wsdl:input message = "getWidthRequest"/>
                        <wsdl:input message = "getHeightRequest"/>
                        <wsdl:input message = "getDataTypeRequest"/>
                        <wsdl:input message = "getDataRequest"/>
                        <wsdl:output message = "getFaceResponse"/>
                </wsdl:operation >
        </wsdl:portType >
        ⋮
</ wsdl:definitions >
```

Listing 14.4 XML Schema for Tuple <video.face.frames>

```
<xs:schema xmlns="http://www.w3.org/2001/XMLSchema">
<xs:element name="faces" type="faces-type">
<xs:complexType name="faces-type">
 <xs:sequence>
  <xs:element name="face" type="face-type" minOccurs="1" maxOccurs="unbounded"/>
 </xs:sequence>
</xs:complexType>
<xs:complexType name="face-type">
 <xs:sequence>
  <xs:element name="personID" type="xs:string" minOccurs="1" maxOccurs="1"/>
  <xs:element name="cameraID" type="xs:string" minOccurs="1" maxOccurs="1"/>
  <xs:element name="time" type="xs:time" minOccurs="1" maxOccurs="1"/>
  <xs:element name="width" type="xs:integer" minOccurs="1" maxOccurs="1"/>
  <xs:element name="height" type="xs:integer" minOccurs="1" maxOccurs="1"/>
  <xs:element name="dataType" type="xs:string" minOccurs="1" maxOccurs="1"/>
  <xs:element name="data" type="xs:string" minOccurs="1" maxOccurs="1"/>
 </xs:sequence>
 <xs:attribute name="faceID" type="xs:string" use="required"/>
</xs:complexType>
</xs:schema>
```

Our implementation for the facial expression recognition routines published by the *facial expression recognition* module includes RVM and support vector machines (SVM) [41] and Bayesian belief networks (BBNs) [42] using a face model consisting of distance and angle features between typical facial characteristic points (FCPs) on the human face. By using the monitor application at runtime, the multimodal workbench exposes all the available services, and it offers the possibility to choose which face detection module/service will be used for the current analysis session.

The XML document with emotional information on speech is stored in XML space *audio.speech.semantic.emotion* and has the XML schema as in Listing 14.7. In our multimodal workbench, the implementation of the emotion recognition from speech routine published by the *emotion recognition from speech* module is based on the GentleBoost classifier [43], using an optimal utterance segmentation.

A distinct type of emotion recognition is the assessment of stress level. The implementation of the stress recognition system [44] is realized as an embedded component of the *emotion recognition from speech* module.

The generated XML document represents an intermediate result and constitutes the input of the *emotion assessment* processing module that stores the final emotion-related information for each person in the scene into the XML data space *scene.person.semantic.emotions*.

The actual routine published by the *emotion assessment* processing module either would remove the ambiguity, in the case of contradictory data, or would increase confidence in the result, in the case of semantic agreement. Though the semantic fusion case can be naturally sketched as a distinct module located in the *semantic emotion recognition* module, various designs present it as part of either the *facial expression recognition* module or the *emotion recognition from speech* module, depending on the algorithms used and the aim of the analysis.

Our implementation of the routine for fusing audio and video semantic emotion data makes use of dynamic Bayesian belief networks (BBNs) to model the human emotions and to catch their inner temporal dependencies in a probabilistic framework.

Listing 14.5 Data Preparation for Semantic-Level Data Fusion for Emotion Recognition Based on Audio-Video Data

```
emotion>
{
        for $es in doc(''audio.speech.semantic.emotions.xml")//audio/frame
        let $ev in doc(''video.face.semantic.facialExpressions.xml")//video/frame [
        $es/personID = personID and xs:time($ev/time) > = xs:time($es/time/start) and
        xs:time ($ev/time) < = xs:time($es/time/stop) and
          xs:time($es/time/start) > = xs:time($time_start) and
        xs:time($es/time/stop) < = xs:time($time_stop)]
        order by $ev/time
        return
                <person personID = "{$ev/personID}">
                        <video id = {$ev/@frameID}>
                                $ev/time
                                $ev/emotion
                        </video>
                        <audio id = {$es/@frameID}>
                                $es/time
                                $es/emotion
                        </audio >
                </person >
}
<emotion>
```

Listing 14.6 XML Schema for *video.face.semantic.facialExpressions.xml*

```
<xs:schema xmlns = "http://www.w3.org/2001/XMLSchema"> 6
<xs:element name = "video" type = "video-type">

<xs:complexType name = "video-type">
 <xs:sequence>
  <xs:element name = "frame" type = "frame-type" minOccurs = "1" maxOccurs = "unbounded"/>
 </xs:sequence>
</xs:complexType>

<xs:complexType name = "frame-type">
 <xs:sequence>
  <xs:element name = "personID" type = "xs:string" minOccurs = "1" maxOccurs = "1"/>
  <xs:element name = "time" type = "xs:time" minOccurs = "1" maxOccurs = "1"/>
  <xs:element name = "emotion" type = "emotion-type" minOccurs = "1" maxOccurs = "1"/>
 </xs:sequence>
 <xs:attribute name = " frameID" type = "xs:string" use = "required"/>
</xs:complexType>

<xs:complexType name = "emotion-type">
 <xs:sequence>
  <xs:element name = "happiness" type = "xs:integer" minOccurs = "1" maxOccurs = "1"/>
  <xs:element name = "sadness" type = "xs:string" minOccurs = "1" maxOccurs = "1"/>
  <xs:element name = "fear" type = "xs:string" minOccurs = "1" maxOccurs = "1"/>
  <xs:element name = "disgust" type = "xs:string" minOccurs = "1" maxOccurs = "1"/>
  <xs:element name = "surprise" type = "xs:string" minOccurs = "1" maxOccurs = "1"/>
  <xs:element name = "anger" type = "xs:string" minOccurs = "1" maxOccurs = "1"/>
 </xs:sequence>
</xs:complexType>

</xs:schema>
```

Listing 14.7   XML Schema for *audio.speech.semantic.emotions.xml*

```
<xs:schema xmlns = "http://www.w3.org/2001/XMLSchema">
<xs:element name = "audio" type = "audio-type">

<xs:complexType name = "audio-type">
 <xs:sequence>
  <xs:element name = "frame" type = "frame-type" minOccurs = "1" maxOccurs = "unbounded"/>
 </xs:sequence>
</xs:complexType>

<xs:complexType name = "frame-type">
 <xs:sequence>
  <xs:element name = "personID" type = "xs:string" minOccurs = "1" maxOccurs = "1"/>
  <xs:element name = "time" type = "interval-type" minOccurs = "1" maxOccurs = "1"/>
  <xs:element name = "emotion" type = " emotion-type" minOccurs = "1" maxOccurs = "1"/>
 </xs:sequence>
 <xs:attribute name = "frameID" type = "xs:string" use = "required"/>
</xs:complexType>

<xs:complexType name = " interval-type">
 <xs:sequence>
  <xs:element name = "start" type = "xs:time" minOccurs = "1" maxOccurs = "1"/>
  <xs:element name = "stop" type = "xs:time" minOccurs = "1" maxOccurs = "1"/>
 </xs:sequence>
</xs:complexType>

<xs:complexType name = "emotion-type">
 <xs:sequence>
  <xs:element name = "happiness" type = "xs:integer" minOccurs = "1" maxOccurs = "1"/>
  <xs:element name = "sadness" type = "xs:string" minOccurs = "1" maxOccurs = "1"/>
  <xs:element name = "fear" type = "xs:string" minOccurs = "1" maxOccurs = "1"/>
  <xs:element name = "disgust" type = "xs:string" minOccurs = "1" maxOccurs = "1"/>
  <xs:element name = "surprise" type = "xs:string" minOccurs = "1" maxOccurs = "1"/>
  <xs:element name = "anger" type = "xs:string" minOccurs = "1" maxOccurs = "1"/>
 </xs:sequence>
</xs:complexType>

</xs:schema>
```

14.4   The Automatic Surveillance Application

In this section, we present a multimodal surveillance experiment based on the framework to illustrate the feasibility of our approach and to show how the framework can be used. The setting of the experiments is inside a Dutch international train (Benelux train) compartment. The scenarios for the recordings involved a number of hired actors and train conductors playing specific (normal as well as unusual) scenarios. The data was used as input to a surveillance application built on top of the framework.

14.4.1   Goal

Benelux train compartments are equipped with several microphones and cameras. Currently, the audio and video data captured by these sensors is transmitted to a central location, where operators have to monitor the data manually and take appropriate action when unusual events occur.

Figure 14.11 shows the interface an operator is confronted with. Our goal is to use the framework presented in this chapter to build a system to automate the manual inspection process currently performed by the human operators. More specifically,

Figure 14.11   User interface for train compartment surveillance operator.

the system should detect unusual behavior in the train compartment and notify the operators. It is still the task of the operator to take the appropriate actions.

14.4.2   Experiment Setup

In order to capture realistic data, professional actors and a train conductor were asked to play specific scenarios in the train compartment. We used four cameras and four microphones in the compartment to capture these scenarios.

As Benelux train compartments are already equipped with cameras, we used these preinstalled cameras to capture video. The microphones, however, were not located at the positions we preferred. Therefore, we installed four microphones to capture audio data. As can be seen from Figure 14.12, the microphones do not cover the entire compartment. So, most of the unusual scenarios were played near the microphones.

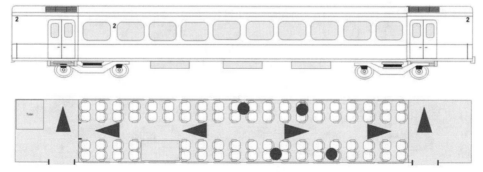

Figure 14.12   Side view of the train (top) and top view of the interior of a Benelux train compartment and the location of the sensors (bottom). The circles indicate the position of the omnidirectional microphones, and the triangles indicate the cameras and their directions.

Table 14.3 The Mapping Between the Behavior to Recognize and the Framework Modules Involved

Behavior to Recognize	Services Involved
Fighting	Gesture recognition
	Emotion recognition from speech
Graffiti and vandalism	Sound event recognition
	Gesture recognition
Begging	Gesture recognition
	Motion tracking
	Speech recognition
Sickness	Gesture recognition
	Emotion recognition from speech
	Speech recognition

The unusual scenarios we asked the actors to play fell in the category of the behaviors we want our system to detect, namely: fighting (including aggression toward the conductor and disturbance of the peace), graffiti and vandalism, begging, and sickness. The modules used from the framework to detect these behaviors include a face-recognition component, gesture-recognition component, face-detection component, facial-expression-recognition component, and emotion-recognition from speech component. The mapping between the behavior to recognize and the modules involved is given in Table 14.3.

The resulting surveillance application that was built on top of the framework consists of several interconnected detection modules made available by the framework. Each module is specialized in handling a specific task. For example, the sound-event-detection module detects whether there is someone shouting. The final module, called the aggression-detection module, detects unusual behavior by fusing the results of the different detection modules (Figure 14.13).

In the application, we have made a distinction between primary and secondary modules. Primary modules require real-time data from the sensors and are continuously active. Primary modules typically feature extraction or object-

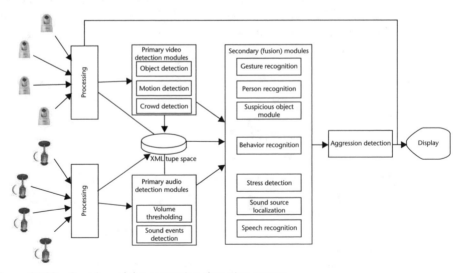

Figure 14.13 Overview of the aggression detection system.

detection algorithms, processing data in the raw data XML and intermediate data XML spaces. The secondary modules operate in the semantic data XML space, do not require real-time attention, and are typically triggered by the results of primary modules. For example, the object-detection primary module fuses data from the cameras and detects a stationary object. This triggers the suspicious-object secondary module that queries the object's history from the XML tuple space to determine the risk of the object.

14.5 Conclusion

The main contribution of this paper is twofold: it proposes a framework for automated multimodal surveillance designed to facilitate the communication between different processing components, and, second, it presents the preliminary results of an ongoing surveillance project built on top of this framework. This framework is centered on the shared memory paradigm, the use of which allows for loosely coupled asynchronous communication between multiple processing components. This decoupling is realized both in time and space. The shared memory in the current design of the framework takes the form of XML data spaces. This suggests a more human-modeled alternative to store, retrieve, and process the data. The data is conferred an underlying structure that complies with eXtended Markup Language (XML). Because the framework implementation itself relies on the philosophy of shared XML data spaces, special attention is given on how to integrate the two underlying technologies of the processing components and XML data management.

In addition, the framework also consists of a set of software tools to monitor the state of registered processing components, to log different types of events, and to debug the flow of data given any running application context. Depending on the type of application to be built on top of the multimodal framework, a routing algorithm has been designed to manage the data transfer among existing shared memories on different physical networks.

Although the framework has been designed by taking into consideration the further development of an automatic-surveillance-oriented application, it can be adopted as a basis for any kind of complex multimodal system involving many components and heavy data exchange. The specification fully complies with the requirements of data manipulation in a multidata producer/consumer context, where the availability of data is time-dependent and some connections might be temporarily interrupted.

So far we have a prototype implementation for the presented multimodal workbench, and our goal is to increase the number of processing components by implementing new algorithms or by modifying existing ones to fit into the framework.

References

[1] Foresti, G., et al., "Active Video-Based Surveillance System: The Low-Level Image and Video Processing Techniques Needed for Implementation," *IEEE Signal Processing Magazine*, Vol. 22, No. 2, 2005, pp. 25–37.

[2]   Tao, H., H. Sawhney, and R. Kumar, "Object Tracking with Bayesian Estimation of Dynamic Layer Representations," *IEEE Trans. on Pattern Analysis and Machine Intelligence,* Vol. 24, No. 1, 2002, pp. 75–89.

[3]   Shin, J., et al., "Optical Flow-Based Real-Time Object Tracking Using Non-Prior Training Active Feature Mode," *ELSEVIER Real-Time Imaging,* Vol. 11, 2005, pp. 204–218.

[4]   Kim, T., et al., "Integrated Approach of Multiple Face Detection for Video Surveillance," *Proc. of the International Conference of Pattern Recognition (ICPR 2002),* 2002, pp. 394–397.

[5]   Hakeem, A., and M. Shah, "Multiple Agent Event Detection and Representation in Videos," *Proc. of the 20th National Conference on Artificial Intelligence,* 2005, pp. 89–94.

[6]   Zhang, D., et al., "Semi-Supervised Adapted HMMs for Unusual Event Detection," *Proc. of the 2005 IEEE Computer Society Conference on Computer Vision and Pattern Recognition (CVPR'05),* Vol. 1, 2005, pp. 611–618.

[7]   Zhu, Z., W. Li, and G. Wolberg, "Integrating LDV Audio and IR Video for Remote Multimodal Surveillance," *Proc. of the 2005 IEEE Computer Society Conference on Computer Vision and Pattern Recognition (CVPR'05),* Vol. 3, 2005, 10, pp. 20–26.

[8]   Clavel, C., T. Ehrette, and G. Richard, "Events Detection for an Audio-Based Surveillance System," *Proc. of IEEE International Conference on Multimedia and Expo (ICME 2005),* 2005, pp. 1306–1309.

[9]   Goldhor, R., "Recognition of Environmental Sounds," *Proc. of IEEE Int. Conf. on Acoustics, Speech and Signal Processing (ICASSP'93),* Vol. 1, 1993, pp. 149–152.

[10]   Härmä, A., M. McKinney, and J. Skowronek, "Automatic Surveillance of the Acoustic Activity in Our Living Environment," *Proc. of the IEEE Int. Conf. on Multimedia and Expo (ICME 2005),* 2005.

[11]   Pradeep, K., et al., "Audio Based Event Detection for Multimedia Surveillance," *Proc. of the IEEE International Conference on Acoustics, Speech, and Signal Processing (ICSSP06),* 2006, pp. 813–816.

[12]   Spors, S., R. Rabenstein, and N. Strobel, "Joint Audio-Video Object Tracking," *IEEE International Conference on Image Processing (ICIP),* Thessaloniki, Greece, 2001.

[13]   Zotkin, D., R. Duraiswami, and L. Davis, "Joint Audio-Visual Tracking Using Particle Filters," *EURASIP Journal on Applied Signal Processing,* Vol. 2002, No. 11, 2002, pp. 1154–1164.

[14]   Gatica-Perez, D., et al., "Audio-Visual Speaker Tracking with Importance Particle Filters," *Proc. of the International Conference on Image Processing (ICIP 2003),* Vol. 3, 2003, III, pp. 25–28.

[15]   Talantzis, F., P. Aristodemos, and P. Lazaros, "Real Time Audio-Visual Person Tracking," *IEEE International Workshop on Multimedia Signal Processing,* 2006.

[16]   Hongeng, S., F. Bremond, and R. Nevatia, "Bayesian Framework for Video Surveillance Application," *15th International Conference on Pattern Recognition (ICPR'00),* Vol. 1, 2000, p. 1164.

[17]   Javed, O., et al., "A Real-Time Surveillance System for Multiple Overlapping and Nonoverlapping Cameras," *Proc. of the International Conference on Multimedia and Expo (ICME 2003),* 2003.

[18]   Velastin, S., et al., "A Distributed Surveillance System for Improving Security in Public Transport Networks," *Special Issue on Remote Surveillance Measurement and Control,* Vol. 35, No. 8, 2002, pp. 209–213.

[19]   Cupillard, F., et al., "Video Understanding for Metro Surveillance," *Proc. of the IEEE International Conference on Networking, Sensing, and Control,* Taipei, Taiwan, 2004.

[20]   Beal, M., H. Attias, and N. Jojic, "A Graphical Model for Audiovisual Object Tracking," *IEEE Trans. on Pattern Analysis and Machine Intelligence* (Special section on graphical models in computer vision), 2003, pp. 828–836.

[21] Ivanov, Y., C. Stauffer, and A. Bobick, "Video Surveillance of Interactions," *IEEE Workshop on Visual Surveillance (ICCV 2001), 2001.*

[22] Buxton, H., and S. Gong, "Visual Surveillance in a Dynamic and Uncertain World," *Artificial Intelligence,* Vol. 78, Nos. 1–2, 1995, pp. 431–459.

[23] Zachariadis, S., et al., "XMIDDLE: Information Sharing Middleware for a Mobile Environment," *Proc. 24th Int. Conf. on Software Engineering ,* 2002, p. 712.

[24] Apparao, V., et al., "Document Object Model (DOM) Level 1 Specification," W3C Recommendation, http://www.w3.org/TR/1998/REC-DOM-Level-1-19981001, World Wide Web Consortium, 1998.

[25] Clark, J., and S. DeRose, "XML Path Language (XPath)," Technical Report, http://www.w3.org/TR/xpath, World Wide Web Consortium, 1999.

[26] Agrawal, R., et al., "Vinci: A Service-Oriented Architecture for Rapid Development of Web Applications," *Proc. of the 10th International Conference on World Wide Web,* 2001, pp. 355–365.

[27] Datcu, D., and L. Rothkrantz, "A Multimodal Workbench for Automatic Surveillance," *Proc. of EUROMEDIA 2004,* April 2004, pp. 108–112.

[28] Ponnekanti, S., et al., "Portability, Extensibility and Robustness in iROS," *Proc. IEEE International Conference on Pervasive Computing and Communications (Percom 2003),* March 2003.

[29] Jini, http://www.jini.org/.

[30] Fallside, D. C., "XML Schema," Technical Report, http://www.w3.org/TR/xmlschema-0/, World Wide Web Consortium, 2000.

[31] Booth, D., and C. Liu, "Web Services Description Language (WSDL)," Candidate Recommendation, http://www.w3.org/TR/wsdl20-primer/, World Wide Web Consortium, 2006.

[32] Fontoura, M., et al., "TSpaces Services Suite: Automating the Development and Management of Web Services," *12th International World Wide Web Conf.,* 2003.

[33] Khushraj, D., O. Lassila, and T. Finin, "sTuple: Semantic Tuple Spaces," *Proc. of the Conference on Mobile and Ubiquitous Systems: Networking and Services (Mobi-Quitous'04),* 0–7695–2208–4/04, 2004.

[34] Bellur, U., and S. Bodre, "xSpace—A Tuple Space for XML and Its Application in Orchestration of Web Services," *SAC'06,* 2006, pp. 766–772.

[35] Thompson, P., "Ruple: An XML Space Implementation," http://www.idealliance.org/papers/xmle02/dx_xmle02/papers/04–05–03/04–05–03.html, 2002.

[36] Tolksdorf, R., F. Liebsch, and D. Nguyen, "XMLSpaces.NET: An Extensible Tuplespace as XML Middleware," *Proc. of the .NET Technologies Workshop,* 2004.

[37] Network Working Group RFC 3548: The Base16, Base32, and Base64 Data Encodings, 2003, http://tools.ietf.org.html/rFc3548.

[38] Boag, S., et al., "XQuery 1.0: An XML Query Language," Candidate Recommendation, http://www.w3.org/TR/xquery/, World Wide Web Consortium, 2006.

[39] Xerces Parser, xml.apache.org/xerces-c/pdf.html.

[40] Wong, W., et al., "Using a Sparse Learning Relevance Vector Machine in Facial Expression Recognition," *Euromedia2006,* University of Ghent, April 2006, pp. 33–37.

[41] Datcu, D., and L. Rothkrantz, "Facial Expression Recognition with Relevance Vector Machines," *IEEE International Conference on Multimedia and Expo (ICME '05),* July 2005.

[42] Datcu, D., and L. Rothkrantz, "Automatic Recognition of Facial Expressions Using Bayesian Belief Networks," *Proc. of IEEE SMC 2004,* October 2004, pp. 2209–2214.

[43] Datcu, D., and L. Rothkrantz, "The Recognition of Emotions from Speech Using GentleBoost Classifier," *CompSysTech'06,* June 2006.

[44] Rothkrantz, L., R. van Vark, and D. Datcu, "Multi-Medial Stress Assessment," *Proc. of IEEE SMC 2004,* October 2004, pp. 3781–3786.

CHAPTER 15

Automatic 3-D Modeling of Cities with Multimodal Air and Ground Sensors

Avideh Zakhor and Christian Frueh

15.1 Introduction

Three-dimensional models of urban environments are useful in a variety of applications, such as urban planning, training, and simulation for disaster scenarios; surveillance applications; and virtual heritage conservation. A standard technique for creating large-scale city models in an automated or semiautomated way is to apply stereovision techniques on aerial or satellite imagery [1]. In recent years, advances in resolution and accuracy have also rendered airborne laser scanners suitable for generating digital surface models (DSM) and 3-D models [2]. Although edge detection can be done more accurately in aerial photos, airborne laser scans are advantageous in that they require no error-prone camera parameter estimation, line or feature detection, or matching. Previous work has attempted to reconstruct polygonal models by using a library of predefined building shapes, or combining the DSM with digital ground plans or aerial images [2–10]. While submeter resolution can be achieved using this technique, it is only capable of capturing rooftops, rather than building facades.

There have been several attempts to create models from a ground-based view at a high level of detail, in order to enable virtual exploration of city environments. While most approaches result in visually pleasing models, they involve an enormous amount of manual work, such as importing the geometry obtained from construction plans selecting primitive shapes and correspondence points for image-based modeling, or complex data acquisition. There have also been attempts to acquire close-range data in an automated fashion by using either images [11] or 3-D laser scanners [12, 13]. These approaches, however, do not scale to more than a few buildings, since data has to be acquired in a slow stop-and-go fashion.

In this chapter, we review a fast, automated method capable of producing textured 3-D models of urban environments for photorealistic walk-throughs, drive-throughs, and fly-throughs [14–16]. The resulting models provide both high details at ground level and complete coverage for a bird's eye view. The data flow diagram of our approach is shown in Figure 15.1. The airborne modeling process on the left provides a half-meter resolution model with a bird's-eye view over the entire area, containing terrain profile and building tops. The ground-based modeling process on the right results in a highly detailed model of the building facades [14–16]. The basic idea is to use two different acquisition processes to generate two separate models, namely ground and aerial, to be fused at a later stage. We acquire a close-range facade model at the ground level by driving a vehicle equipped with laser scanners and a digital camera under normal traffic

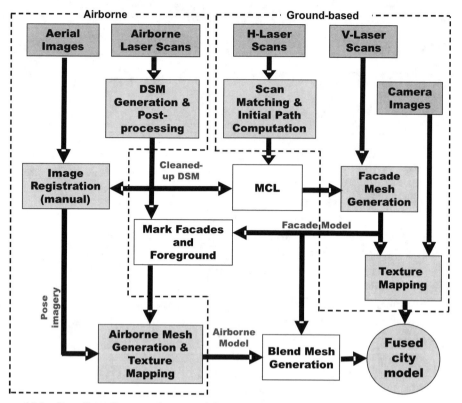

Figure 15.1 Data-flow diagram of our modeling approach. Acquired data is shown in dark gray shaded boxes, airborne modeling steps in light gray boxes on the left, ground-based modeling steps in light gray boxes on the right, and model fusion steps in white boxes.

conditions on public roads. To our knowledge, our ground-based system is the first acquisition system, in which data is acquired continuously, while the acquisition vehicle is in motion, rather than in a stop-and-go fashion. This, combined with automated processing, results in extremely fast model construction schemes for building facades. As for airborne modeling, we create a far-range DSM containing complementary roof and terrain shapes from airborne laser scans, and after post-processing, texture-map it with oblique aerial imagery. The challenge in merging the airborne and ground-based models is registration between the two, for which we use the Monte Carlo localization (MCL) technique. We fuse the two models by first removing redundant parts between the two and then filling the remaining gaps between them. The same MCL process is also used to localize the ground-based acquisition vehicle in constructing the 3-D facade models.

In the remainder of this chapter, we will describe various steps of the above process in more detail and show results on 3-D models of downtown Berkeley. The outline of this chapter is as follows. In Section 15.2, we describe the basic approach to creating textured surface mesh from airborne laser scans and imagery. In Section 15.3, we go over the ground-based data acquisition system and the modeling process; and in Section 15.4, we describe the model merging process. Section 15.5 includes 3-D modeling results for downtown Berkeley, and Section 15.6 discusses possible applications of 3-D modeling to surveillance problems.

15.2   Creating a Textured 3-D Airborne Model

In this section, we describe the generation of a DSM from airborne laser scans, its processing and transformation into a surface mesh, and texture-mapping with color aerial imagery. Our basic approach is to use airborne laser scans to model the geometry of rooftops in order to arrive at rectilinear 3-D models. We then use oblique airborne imagery acquired by a helicopter to texture map the resulting 3-D models. As shown later in Section 15.3.3, we use the DSM resulting from airborne modeling both to localize the ground-based data acquisition vehicle and to register and merge the ground-based model with the airborne one. In contrast to previous approaches, we do not explicitly extract geometric primitives from the DSM [17, 18]. While we use aerial laser scans to create the DSM, it is equally feasible to use a DSM obtained from other sources, such as stereovision or SAR.

15.2.1   Scan Point Resampling and DSM Generation

During the acquisition of airborne laser scans with a 2-D scanner mounted on board a plane, the unpredictable roll and tilt motion of the plane generally destroys the inherent row-column order of the scans. Thus, the scans may be interpreted as an unstructured set of 3-D vertices in space, with the x- and y-coordinates specifying the geographical location, and the z-coordinate the altitude. In order to further process the scans efficiently, it is advantageous to resample the scan points to a row-column structure, even though this step could potentially reduce the spatial resolution, depending on the grid size. To transfer the scans into a DSM, or a regular array of altitude values, we define a row-column grid in the ground plane and sort scan points into the grid cells. The density of scan points is not uniform, and hence, there are grid cells with no scan points and others with multiple ones. Since the percentage of cells without any scan points and the resolution of the DSM depend on the size of a grid cell, a compromise must be made, leaving few cells without a sample, while maintaining the resolution at an acceptable level.

In our case, the scans have an accuracy of 30 cm in the horizontal and vertical directions and a raw spot spacing of 0.5m or less. Both the first and the last pulses of the returning laser light are measured. We have chosen to select a square cell size of 0.5m×0.5m, resulting in about half the cells being occupied. We create the DSM by assigning to each cell the highest z value among its member points, so that overhanging rooftops of buildings are preserved, while points on walls are suppressed. The empty cells are filled using nearest-neighbor interpolation in order to preserve sharp edges. Each grid cell can be interpreted as a vertex, where the x,y location is the cell center, and the z-coordinate is the altitude value, or as a pixel at (x,y) with a gray level proportional to z. An example of a resampled DSM is shown in Figure 15.2(a).

15.2.2   Processing the DSM

The DSM generated via the previous steps contains not only the plain rooftops and terrain shape, but also many other objects such as cars, trees, and so forth. Roofs,

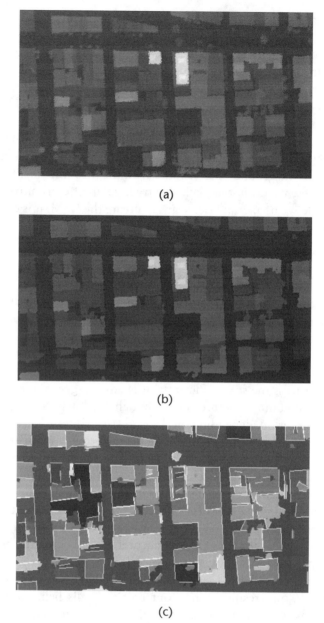

(a)

(b)

(c)

Figure 15.2 Processing steps for DSM: (a) DSM obtained from scan point resampling; (b) DSM after flattening roofs; and (c) segments with RANSAC lines in white.

in particular, look bumpy, due to a large number of smaller objects, such as ventilation ducts, antennas, and railings, which are impossible to reconstruct properly at the DSM's resolution. Furthermore, scan points below overhanging roofs cause ambiguous altitude values, resulting in jittery edges. In order to obtain a more visually pleasing reconstruction of the roofs, we apply several processing steps to the DSM obtained in Section 15.2.1.

The first step is aimed at flattening bumpy rooftops. To do this, we first apply a region growing segmentation algorithm to all nonground pixels based on depth discontinuity between adjacent pixels. Small, isolated regions are replaced with ground level altitude, in order to remove objects such as cars or trees in the DSM. Larger regions are further subdivided into planar subregions by means of planar segmentation. Then, small regions and subregions are united with larger neighbors by setting their z values to the larger region's corresponding plane. This procedure is able to remove undesired small objects from the roofs and prevents rooftops from being separated into too many cluttered regions. The resulting processed DSM for Figure 15.2(a), using the above steps, is shown in Figure 15.2(b).

The second processing step is intended to straighten jittery edges. We resegment the DSM into regions, detect the boundary points of each region, and use RANSAC [19] to find line segments that approximate the regions. For the consensus computation, we also consider boundary points of surrounding regions in order to detect short linear sides of regions and to align them consistently with surrounding buildings; furthermore, we reward additional bonus consensus if a detected line is parallel or perpendicular to the most dominant line of a region. For each region, we obtain a set of boundary line segments representing the most important edges, which are then smoothed out. For all other boundary parts, where a proper line approximation has not been found, the original DSM is left unchanged. Figure 15.2(c) shows the regions resulting from the above processing steps as applied to Figure 15.2(b), superimposed on top of the corresponding RANSAC lines drawn in white. Compared with Figure 15.2(b), most edges are straightened out.

15.2.3 Textured Mesh Generation

Airborne models are commonly generated from LIDAR scans by detecting features such as planar surfaces in the DSM or matching a predefined set of possible rooftop and building shapes [2, 17, 18]. In other words, they decompose the buildings found in the DSM into polygonal 3-D primitives. While the advantage of these model-based approaches is their robust reconstruction of geometry in spite of erroneous scan points and low sample density, they are highly dependent on the shape assumptions that are made. In particular, the results are poor if many nonconventional buildings are present or if buildings are surrounded by trees. Although the resulting models may appear "clean" and precise, the geometry and location of the reconstructed buildings are not necessarily correct if the underlying shape assumptions are invalid.

As we will describe in Section 15.3, in our application, an accurate model of the building facades is readily available from the ground-based acquisition, and as such, we are primarily interested in adding the complementary roof and terrain geometry. Hence, we apply a different strategy to create a model from the airborne view, namely, transforming the cleaned-up DSM directly into a triangular mesh and reducing the number of triangles by simplification. The advantage of this method is that the mesh-generation process can be controlled on a per-pixel level; we exploit this property in the model fusion procedure described in Section 15.4. Additionally, this method has a low processing complexity and is robust. Since no a priori assumptions about the environment are made or predefined models are required, it can be applied to buildings with unknown shapes, even in the presence of trees. Admittedly, this comes at the expense of a larger number of polygons.

Since the DSM has a regular topology, it can be directly transformed into a structured mesh by connecting each vertex with its neighboring ones. The DSM for a city is large, and the resulting mesh has two triangles per cell, yielding 8-million triangles per square kilometer for the 0.5m × 0.5m grid size we have chosen. Since many vertices are coplanar or have low curvature, the number of triangles can be drastically reduced without significant loss of quality. We use the Qslim mesh simplification algorithm [20] to reduce the number of triangles. Empirically, we have found that it is possible to reduce the initial surface mesh to about 100,000 triangles km^2 at the highest level of detail without a noticeable loss in quality.

Using aerial images taken with an uncalibrated camera from an unknown pose, we texture-map the reduced mesh in a semiautomatic way. A few correspondence points are manually selected in both the aerial photo and the DSM, taking a few minutes per image. Then, both internal and external camera parameters are automatically computed and the mesh is texture-mapped. Specifically, a location in the DSM corresponds to a 3-D vertex in space and can be projected into an aerial image if the camera parameters are known. We utilize an adaptation of Lowe's algorithm to minimize the difference between selected correspondence points and computed projections [21]. After the camera parameters are determined, for each geometry triangle, we identify the corresponding texture triangle in an image by projecting the corner vertices. Then, for each mesh triangle, the best image for texture-mapping is selected by taking into account resolution, normal vector orientation, and occlusions.

In addition to the above semiautomatic approach, we have developed an automatic method for registering oblique aerial imagery to our airborne 3-D models [22]. This is done by matching 2-D lines in aerial imagery with projections of 3-D lines from the city model, for every camera-pose location, and computing the best match with the highest score. We have empirically found techniques such as steepest descent to result in local minima, and hence we have opted to use the above exhaustive search approach. To make an exhaustive search computationally feasible, we assume a rough estimate of our initial orientation, that is, roll, pitch, and yaw to be known within a 10° range, and our initial position, or x, y, and z, to be known within a 20-m range. These values correspond to the accuracy of a mid tier GPS/INS system. To achieve reliable pose estimation within the exhaustive search framework, we need to sample the search space at 0.5° in roll, 0.25° in pitch and yaw, and 10 min x, y, and z. This level of sampling of the parameter space results in about 20 hours of computation on a 2-GHz Pentium 4 machine per image [22]. This is in sharp contrast with a few minutes per image, as achieved by the semiautomatic approach described earlier. We have found the accuracy performance of the two approaches to be close.

Examples of resulting textured, airborne 3-D models using the approach described in this section are included in Section 15.5.

15.3   Ground-Based Acquisition and Modeling

In this section, we describe our approach to ground-based data acquisition and modeling.

15.3.1   Ground-Based Data Acquisition Via Drive-By Scanning

Our mobile ground-based data acquisition system consists of two Sick LMS 2-D laser scanners and a digital color camera with a wide-angle lens [14]. The data acquisition is performed in a fast drive-by rather than a stop-and-go fashion, enabling short acquisition times limited only by traffic conditions. As shown in Figure 15.3, our acquisition system is mounted on a rack on top of a truck, enabling us to obtain measurements that are not obstructed by objects such as pedestrians and cars.

Both 2-D scanners face the same side of the street; one is mounted horizontally and the other vertically, as shown in Figure 15.4. The camera is mounted toward the scanners, with its line of sight parallel to the intersection between the orthogonal scanning planes. The laser scanners and the camera are synchronized by hardware signals. In our measurement setup, the vertical scanner is used to scan the geometry of the building facades as the vehicle moves, and hence it is crucial to determine the location of the vehicle accurately for each vertical scan. In [14], we have developed algorithms to estimate relative position changes of the vehicle, based on matching the horizontal scans, and to estimate the driven path as a concatenation of relative position changes. Since errors in the estimates accumulate, a global correction must be applied. Rather than using a GPS sensor, which is not sufficiently reliable in urban canyons, in [15] we introduce the use of an aerial photo as a 2-D global reference in conjunction with MCL. In Section 15.3.3, we extend the application of MCL to a global edge map and DTM (both derived from the DSM) in order to determine the vehicle's six-degree-of-freedom pose in nonplanar terrain and to register the ground-based facade models with respect to the DSM.

Figure 15.3   Acquisition vehicle.

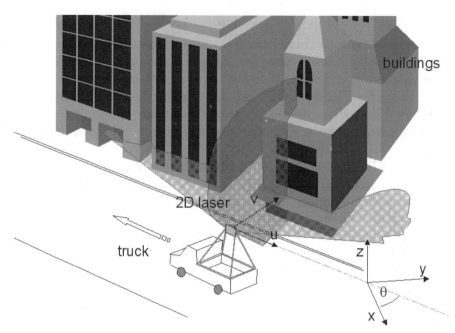

Figure 15.4   Ground-based acquisition setup.

15.3.2   Creating an Edge Map and DTM

In this section, we describe the methodology for creating the two maps that are needed by MCL to fully localize the ground-based acquisition system; the first one is an edge map, which contains the location of height discontinuities in the DSM, and the second one is a digital terrain model (DTM) also obtained from DSM, which contains terrain altitude. In previous work [15], we have applied a Sobel edge detector to a grayscale aerial image in order to find edges that are then used by MCL in conjunction with horizontal scan matches in order to localize the ground-based data acquisition vehicle. However, since in the process of constructing a complete model, we have access to airborne DSM, we opt to use an edge map derived from DSM, rather than one derived from aerial imagery, in order to achieve higher accuracy and faster processing speed [23]. Rather than applying the Sobel edge detector to the DSM, though, we construct an edge map by defining a discontinuity detection filter, which marks a pixel if at least one of its eight adjacent pixels is more than a threshold Δz_{edge} below it. This is possible because we are dealing with 3-D height maps rather than 2-D images. Hence, only the outermost pixels of the taller objects, such as building tops, are marked, rather than adjacent ground pixels, creating a sharper edge map than a Sobel filter. In fact, the resulting map is a global occupancy grid for building walls. While for aerial photos, shadows of buildings or trees and perspective shifts of building tops could potentially cause numerous false edges in the resulting edge map, neither problem exists for edge maps from airborne laser scans.

The DSM contains not only the location of building facades as height discontinuities, but also the altitude of the streets on which the vehicle is driven, and as such, this altitude can be assigned to the z-coordinate of the vehicle in order

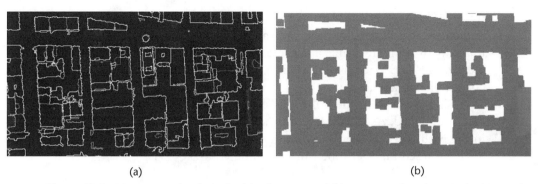

(a)   (b)

Figure 15.5   Map generation for MCL: (a) edge map and (b) DTM. For the white pixels, there is no ground-level estimate available.

to create DTM. Nonetheless, it is not possible to directly use the z value of a DSM location, since the LIDAR captures cars and overhanging trees during airborne data acquisition, resulting in z values up to several meters above the actual street level for some locations. To overcome this, for a particular DSM location, we estimate the altitude of the street level by averaging the z-coordinates of the available ground pixels within a surrounding window, weighing them with an exponential function decreasing with distance. The result is a smooth, dense DTM as an estimate of the ground level near roads.

Figures 15.5(a) and 15.5(b) show an edge map and DTM, respectively, computed from the DSM shown in Figure 15.2(b). The MCL technique to be described in the next section utilizes both maps to localize the ground-based acquisition vehicle.

15.3.3   Model Registration with MCL

MCL is a particle-filtering-based implementation of the probabilistic Markov localization, introduced by Thrun et al. [24] for tracking the position of a vehicle in mobile robotics. Our basic approach is to match successive horizontal laser scans in order to arrive at the initial estimate of the relative motion of the truck on the ground plane [23]. Given a series of relative motion estimates and corresponding horizontal laser scans, we then apply MCL to determine the accurate position of the acquisition vehicle within a global edge map as described in Section 15.3.2 [15, 23]. The principle of the correction is to adjust initial vehicle motion estimates so that horizontal scan points from the ground-based data acquisition match the edges in the global edge map. The scan-to-scan matching can only estimate a 3-DOF relative motion, in other words, a 2-D translation and rotation in the scanner's coordinate system. If the vehicle is on a slope, the motion estimates are given in a plane at an angle with respect to the global (x, y) plane, and the displacement should in fact be corrected with the cosine of the slope angle. However, since this effect is small, for example, 0.5% for a 10%-degree slope, we can safely neglect it, and use the relative scan-to-scan matching estimates, as if the truck's coordinate system were parallel to the global coordinate system. Using MCL with the relative estimates from horizontal scan matching and the edge map from the DSM, we arrive at a series of global pose probability density functions and correction vectors for x, y, and yaw. These corrections are then applied to the initial path estimate using horizontal scan matches only, to obtain an accurate localization of the acquisition vehicle. In Section 15.5, we show examples of such corrections on actual data.

Using the DTM as described in Section 15.3.2, an estimate of two more DOF can be obtained: As for the first, the final $z^{(i)}$ coordinate of an intermediate pose P_i in the path is set to DTM level at $(x^{(i)}, y^{(i)})$ location. As for the second, the pitch angle representing the slope can be computed as

$$pitch^{(i)} = \arctan\left(\frac{z^{(i)} - z^{(i-1)}}{\sqrt{(x^{(i)} - x^{(i-1)})^2 + (y^{(i)} - y^{(i-1)})^2}}\right)$$

that is, by using the height difference and the traveled distance between successive positions. Since the resolution of the DSM is only 1m and the ground level is obtained via a smoothing process, the estimated pitch contains only the low-frequency components, and not highly dynamic pitch changes, for example, those caused by pavement holes and bumps. Nevertheless, the obtained pitch is an acceptable estimate, because the size of the truck makes it relatively stable along its long axis.

The last missing DOF, the roll angle, is not estimated using airborne data; rather, we assume buildings are generally built vertically, and we apply a histogram analysis on the angles between successive vertical scan points. If the average distribution peak is not centered at 90°, we set the roll-angle estimate to the difference between the histogram peak and 90°.

15.3.4 Processing Ground-Based Data

At the end of the above steps, we obtain 6-DOF estimates for the global pose of the acquisition vehicle. This pose can be used to properly stack the vertical laser scans in order to construct a 3-D point cloud of building facades, consisting of all the points in the vertical laser scans. As such, the points corresponding to the horizontal laser scans are not used in the actual model and are only used for localization purposes. Since the resulting 3-D point cloud is made of successive vertical scans, it has a special structure, which can be exploited in the triangulation step.

As described in [16], the path is segmented into easy-to-handle segments to be processed individually as follows: First, we remove foreground objects, such as cars, trees, and pedestrians, by first constructing a 2.5-D depth map from the resulting facade and applying a histogram analysis to separate foreground objects from background that consists of building facades [25]. This method works well in situations where there is sufficient separation between foreground and background. Next, we need to complete the 3-D geometry by filling the holes resulting either from the removal of the foreground objects or from objects, such as glass, that are transparent to the infrared laser scanner [25].

As photorealism cannot be achieved by using geometry alone, we need to enhance our model with texture data. To achieve this, we have equipped our data acquisition system with a digital color camera with a wide-angle lens. The camera is synchronized with the two laser scanners and is calibrated against the laser scanners' coordinate system; hence the camera position can be computed for all images. After calibrating the camera and removing lens distortion in the images, each 3-D vertex can be mapped to its corresponding pixel in an intensity image by a simple projective transformation. As the 3-D mesh triangles are small compared to their distance to the camera, perspective distortions within a triangle can be neglected, and each mesh triangle can be mapped to a triangle in the picture by applying the projective transformation to its vertices.

Since foreground objects have been removed from 3-D geometry, we need to remove the pixels corresponding to foreground objects from camera images in order to ensure that they are not used to texture map background objects such as building facades. A simple way of segmenting out the foreground objects is to project the foreground mesh onto the camera images and mark out the projected triangles and vertices. While this process works adequately in most cases, it could potentially miss some parts of the foreground objects [25]. This can be due to two reasons. First, the foreground scan points are not dense enough for segmenting the image with pixel accuracy, especially at the boundary of foreground objects; second, the camera captures side views of foreground objects, whereas the laser scanner captures a direct view. Hence some foreground geometry does not appear in the laser scans and, as such, cannot be marked as foreground.

To overcome this problem, we have developed a more sophisticated method for pixel accurate foreground segmentation based on the use of correspondence error and optical flow [25]. After segmentation, multiple images are combined into a single texture atlas. We have also developed a number of in-painting techniques to fill in the texture holes in the atlas resulting from foreground occlusion, including copy-and-paste and interpolation [25].

The result of a texture-mapped facade model using the steps described in Section 15.3 is shown in Figure 15.6. Note that the upper parts of tall buildings are not texture-mapped, if they are outside the camera's field of view during the data acquisition process. As will be seen in the next section, the upper parts of buildings are textured by airborne imagery. Alternatively, using multiple cameras could ensure that all parts of building facades are captured for texture mapping purposes.

The texture for a path segment is typically several tens of megabytes, thus exceeding the rendering capabilities of today's graphics cards. Therefore, the facade models are optimized for rendering by generating multiple levels of detail (LOD), so that only a small portion of the entire model is rendered at the highest LOD at any given time. We subdivide the facade meshes along vertical planes and generate lower LODs for each submesh using the Qslim simplification algorithm [20] for geometry and bicubic interpolation for texture reduction. All submeshes are combined in a scene graph, which controls the switching of the LODs depending on the viewer's position. This enables us to render the large amounts of geometry and texture with standard tools such as VRML browsers.

Figure 15.6   Ground-based facade models.

15.4   Model Merging

We now describe an approach to combine the ground-based facade models with the aerial surface mesh from the DSM. Both meshes are generated automatically, and given the complexity of a city environment, it is inevitable that some parts are partially captured, or completely erroneous, thus resulting in substantial discrepancies between the two meshes. Since our goal is a photorealistic virtual exploration of the city, creating models with visually pleasing appearances is more important than CAD properties such as water tightness. Common approaches for fusing meshes, such as sweeping and intersecting contained volume [13], or mesh zippering [26], require a substantial overlap between the two meshes. This is not the case in our application, since the two views are complementary. Additionally, the two meshes have entirely different resolutions: the resolution of the facade models, at about 10 cm to 15 cm, is substantially higher than that of the airborne surface mesh. Furthermore, to enable interactive rendering, it is required for the two models to fit together, even when their parts are at different LOD.

Due to its higher resolution, it is reasonable to give preference to the ground-based facades wherever available and use the airborne mesh only for roofs and terrain shape. Rather than replacing triangles in the airborne mesh for which ground-based geometry is available, we consider the redundancy before the mesh generation step in the DSM: for all vertices of the ground-based facade models, we mark the corresponding cells in the DSM. This is possible since ground-based models and DSM have been registered through the MCL localization techniques described in Section 15.3.3. We further identify and mark those areas in DSM, where our automated facade processing has classified as foreground such objects as trees and cars [16, 25]. These marks control the subsequent airborne mesh generation from DSM; specifically, during the generation of the airborne mesh from DSM, (1) the z value for the foreground areas in DSM is replaced by the ground-level estimate from the DTM and (2) triangles at ground-based facade positions are not created. Note that the first step is necessary to enforce consistency and remove those foreground objects in the airborne mesh, which have already been deleted in the facade models. Figure 15.7(a) shows the DSM with facade areas and foreground, and Figure 15.7(b) shows the resulting airborne surface mesh with the corresponding facade triangles removed, and the foreground areas leveled to DTM altitude.

The facade models to be put in place do not match the airborne mesh perfectly, due to their different resolutions and capture viewpoints. Generally, the above procedure results in the removed geometry to be slightly larger than the actual ground-based facade to be placed in the corresponding location. To solve this discrepancy and to make mesh transitions less noticeable, we fill the gap with additional triangles to join the two meshes, and we refer to this step as *blending*.

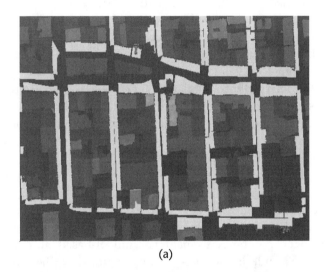

(a)

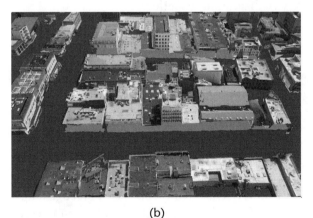

(b)

Figure 15.7 Removing facades from the airborne model: (a) Marked areas in the DSM and (b) resulting mesh with corresponding facades and foreground objects removed.

The outline of this procedure is shown in Figure 15.8. Our approach to creating such a blend mesh is to extrude the buildings along an axis perpendicular to the facades, as shown in Figure 15.8(b), and then shift the location of the loose end vertices to connect to the closest airborne mesh surface, as shown in Figure 15.8(c). This is similar to the way plumb is used to close gaps between windows and roof tiles. These blend triangles are finally texture-mapped with the texture from the aerial photo, and, as such, they attach at one end to the ground-based model, and at the other end to the airborne model, thus reducing visible seams at model transitions.

15.5   Results

We have applied the proposed algorithms on a dataset for downtown Berkeley. Airborne laser scans have been acquired in conjunction with Airborne 1 Inc., at Los Angeles, California; the entire dataset consists of 60 million scan points. We have resampled these scan points to a $0.5\text{m} \times 0.5\text{m}$ grid and have applied the processing steps, as described in Section 15.3, to obtain a DSM, an edge map, and a DTM for the entire area. We select feature points in five megapixel digital images taken from a helicopter and their correspondence in the DSM. This process takes about an hour for 12 images we use for the downtown Berkeley area. Then the DSM is automatically triangulated, simplified, and finally texture-mapped. Figure 15.9(a) shows the surface mesh obtained from directly triangulating the DSM, Figure 15.9(b) shows the triangulated DSM after the geometry processing steps outlined in Section 15.2.2, and Figure 15.9(c) shows the texture-mapped model, applying the texture processing steps outlined in Section 15.2.3. It is difficult to evaluate the accuracy of this airborne model, as no ground truth with sufficient accuracy is readily available, even at the city's planning department. We have admittedly sacrificed accuracy for the sake of visual appearance of the texture-mapped model, for example, by removing small features on building tops. Thus, our approach combines elements of model-based and image-based rendering. While this is undesirable in some applications, we believe it is appropriate for interactive visualization applications.

The ground-based data has been acquired during two measurement drives in Berkeley: The first drive took 37 minutes and was 10.2 km, starting from a location near the hills, going down Telegraph Avenue, and then in loops around the central downtown blocks; the second drive was 41 minutes long and 14.1 km, starting from Cory Hall at the University of California, Berkeley, and looping around the remaining downtown blocks. A total of 332,575 vertical and horizontal scans, consisting of 85 million scan points, along with 19,200 images, were captured during those two drives.

We first apply horizontal scan matching to provide an initial estimate of the driven path. We then correct that using the MCL-based technique of Section 15.3.3. In previous MCL experiments based on edge maps from aerial images with a 30-cm resolution, we had found the localization uncertainty to be large at some locations, due to false edges and perspective shifts. To overcome this, we have had to use a large number of particles, for example, 120,000 with MCL, in order to

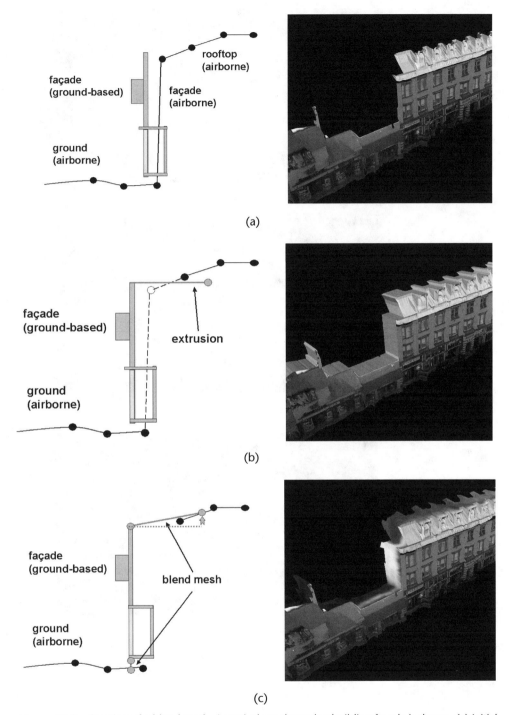

Figure 15.8   Creation of a blend mesh. A vertical cut through a building facade is shown. (a) Initial airborne and ground-based model registered; (b) facade of airborne model replaced and ground-based model extruded; and (c) blending the two meshes by adjusting loose ends of extrusions to airborne mesh surface and mapping texture.

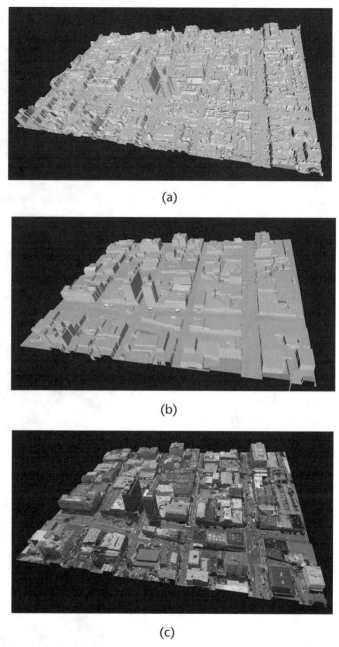

Figure 15.9 Airborne model: (a) DSM directly triangulated; (b) triangulated after postprocessing; and (c) model texture-mapped.

approximate the spread-out probability distribution appropriately and track the vehicle reliably. Even though the edge map derived from airborne laser scans described in Section 15.3.2 has a lower resolution than an edge map resulting from aerial imagery, we can track the vehicle using DSM with as few as 5,000 particles. This is because the edge map derived from DSM is more accurate and reliable than that of aerial imagery.

Having computed an initial path estimate using horizontal scan matching, we apply global MCL-based correction based on DSM edge maps to the yaw angles in path 1 as shown in Figure 15.10(a). We then recompute the path and apply the correction to the x- and y-coordinates, as shown in Figure 15.10(b). As expected, the global correction substantially modifies the initial pose estimates, thus reducing errors in subsequent processing steps. Figure 15.10(c) plots the assigned z-coordinate, clearly showing the slope from our starting position at higher

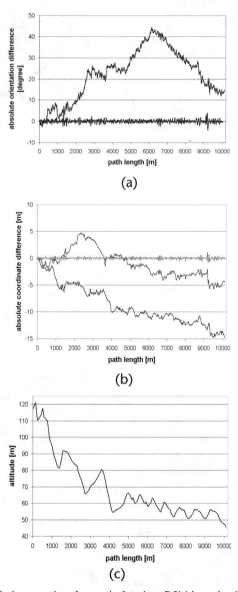

(a)

(b)

(c)

Figure 15.10 MCL global correction for path 1 using DSM-based edge maps: (a) yaw angle difference between initial path and global estimates before and after correction; (b) differences of x- and y-coordinates before and after correction; and (c) assigned z-coordinates. In plots (a) and (b), the differences after corrections are the curves close to the horizontal axis.

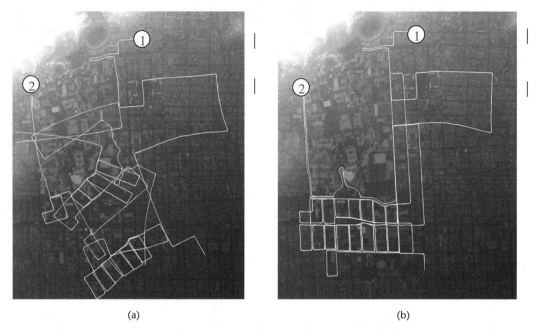

(a)   (b)

Figure 15.11 Driven paths superimposed on top of the DSM (a) before correction and (b) after correction. The circles denote the starting position for paths 1 and 2, respectively.

altitude near the Berkeley Hills down toward the San Francisco Bay, as well as the ups and downs on this slope while looping around the downtown blocks.

Figure 15.11(a) shows uncorrected paths 1 and 2 superimposed on the airborne DSM using horizontal scan matching. Figure 15.11(b) shows the paths after MCL global correction based on DSM. As seen, the corrected path coincides nicely with the Manhattan structure of the roads in the DSM, whereas the uncorrected one is clearly erroneous as it is seen to cross over parts of DSM corresponding to buildings. Figure 15.12 shows the ground-based horizontal scan points for the corrected paths

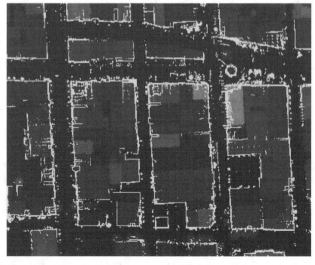

Figure 15.12 Horizontal scan points for corrected paths.

Figure 15.13 Facade model for downtown Berkeley area.

superimposed on top of DSM. As expected, horizontal scan points appear to closely match the DSM edges for the corrected paths.

After MCL correction, all scans and images are geo-referenced. This is used to properly stack vertical laser scans in order to generate a 3-D point cloud, which is then triangulated, processed, and texture-mapped, using the steps described in Section 15.3.4. Figure 15.13 shows the resulting facades for the 12 street blocks of downtown Berkeley. Note that the acquisition time for the 12 downtown Berkeley blocks has been only 25 minutes; this is the time it took to drive a total of 8 km around these 12 blocks under city traffic conditions.

Due to the usage of the DSM as the global reference for MCL, the DSM and facade models are registered with each other, and we can apply the model merging steps as described in Section 15.4. Figure 15.14(a) shows the resulting combined model for the downtown Berkeley blocks, as viewed in a walk-through or drive-through, and Figure 15.14(b) shows a view from the rooftop of a downtown

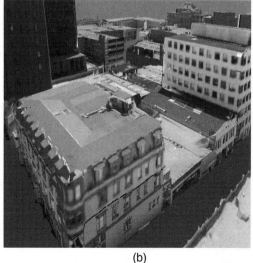

(a)   (b)

Figure 15.14 Walk-through view of the model: (a) as seen from the ground level; and (b) as seen from the rooftop of a building.

Figure 15.15   Bird's-eye view of the model.

building. Due to the limited field of view of the ground-based camera, the upper parts of the building facades are texture-mapped with aerial imagery. The noticeable difference in resolution between the upper and lower parts of the texture on the building in Figure 15.14(b) emphasizes the necessity of ground-based facade models for walk-through applications. Future systems should use multiple ground-based cameras in order to ensure capturing all parts of all buildings. This would, of course, be at the expense of a significant increase in data collection volume. Figure 15.15 shows the same model as Figure 15.14(b) in a view from the top, as it appears in a fly-through. The model can be downloaded for interactive visualization from [27].

Our proposed approach to city modeling is not only automated, but it is also fast from a computational viewpoint. As shown in Table 15.1, the total time for the automated processing and model generation for the 12 downtown blocks is around 5 hours on a 2-GHz Pentium-4 PC. Since the complexity of all developed algorithms is linear in area and path length, our method is scalable to large environments.

Table 15.1   Processing Times for the Downtown Berkeley Blocks

Procedures	Times
Vehicle localization and registration with DSM	164 minutes
Facade model generation and optimization for rendering	121 minutes
DSM computation and projecting facade locations	8 minutes
Generating textured airborne mesh and blending	26 minutes
Total processing time	319 minutes

15.6   Applications of 3-D Modeling to Surveillance

A priori knowledge of the 3-D scene in which a target is located, or on-the-fly 3-D reconstruction of the scene, can be tremendously helpful in identifying and tracking it in surveillance applications. Consider Figure 15.16, where a helicopter is looking for and tracking a person of interest on the ground in downtown Berkeley. If the 3-D model of the target of interest is available at the helicopter, the location, orientation, and velocity of the target can be easily determined by matching the output of the image sensors, or other 3-D sensors focused on the target, with the 3-D model of the scene surrounding it. Similar arguments can be made if ground-based surveillance cameras were randomly scattered on the streets of a city, where each camera station has access to a complete 3-D map of its own surrounding environment. To begin with, the a priori, textured 3-D model of the surrounding environment, together with simple background/foreground separation techniques, can readily be used to separate moving objects within the scene from the stationary background already captured in the city modeling process. Once target identification is carried out, a priori knowledge of the textured 3-D structures of the city areas in which the target is moving can be used to facilitate target localization and tracking process, regardless of whether it is done by a single

Figure 15.16   Illustration of ways in which a priori models of 3-D cities can be used for surveillance applications.

sensor that moves with the target or by multiple cameras that hand off the tracking process to each other as the target moves from one field of view to the other. In addition, the knowledge of the geometry of the scene can be used to optimally and adaptively adjust the pan, tilt, zoom, and other parameters of the cameras during the colocation and tracking processes. Finally, a priori knowledge of the 3-D textured models of a scene can be readily used to register multiple tracking cameras with respect to each other. Textured 3-D models of the scene can also be used in conjunction with spatio-temporal super resolution techniques from signal processing to enhance the ability to locate targets with subpixel accuracy and more frequent temporal updates, along similar lines shown in [28].

15.7 Summary and Conclusions

We have proposed a fast, automated system for acquiring 3-D models of urban environments for virtual walk-throughs, drive-throughs and fly-throughs. The basic approach has been to use ground-based and aerial laser scanners and cameras in order to construct ground and aerial models. The DSM resulting from airborne modeling is used in conjunction with MCL technique, not only to localize the ground-based acquisition vehicle accurately, but also to register the ground and aerial models. Future work involves developing fast, automated approaches to texture mapping airborne models using oblique aerial imagery, as currently this is the only computer-intensive step in our end-to-end processing. Even though this step can be sped up significantly using a semiautomatic approach, it is highly desirable to keep the human out of the loop during the model construction process. We also plan to explore the use of 3-D models in surveillance and tracking applications.

Acknowledgments

The authors would like to acknowledge Rusty Sammon, Siddarth Jain, John Secord, John Flynn, Steve Jian, John Lo, Dr. Hassan Foroosh, Ali Lakhia, and all other Berkeley students and staff who have participated in this project over the past six years. This work was sponsored by Army Research Office contract DAAD 19–00–1-0352, under a MURI program. This chapter was based on "Constructing 3-D City Models by Merging Ground-Based and Airborne Views," by Christian Frueh and Avideh Zakhor, which appeared in *IEEE Transactions on Computer Graphics and Applications*, Vol. 23, No. 6, November–December 2003, pp. 52–61. © 2003 IEEE.

References

[1] Kim, Z., A. Huertas, and R. Nevatia, "Automatic Description of Buildings with Complex Rooftops from Multiple Images," *Proc. IEEE Conf. on Computer Vision and Pattern Recognition,* Kauai, HI, 2001, pp. 272–279.

[2] Brenner, C., N. Haala, and D. Fritsch, "Towards Fully Automated 3D City Model Generation," *Workshop on Automatic Extraction of Man-Made Objects from Aerial and Space Images III,* 2001.

[3]   Nevatia, R., and A. Huertas, "Knowledge Based Building Detection and Description: 1997–1998," *Proc. of the DARPA Image Understanding Workshop,* Monterey, CA, 1998.

[4]   Heller, A., et al., "An Integrated Feasibility Demonstration for Automatic Population of Geospatial Databases," *Proc. of the DARPA Image Understanding Workshop,* Monterey, CA, 1998.

[5]   Firshchein, O., and T. Strat, (eds.), *RADIUS: Image Understanding for Imagery Intelligence,* San Mateo, CA: Morgan Kaufmann, 1997.

[6]   Gruen, A., E. Baltsavias, and O. Henricsson, (eds.), *Automatic Extraction of Man-Made Objects from Aerial and Space Images,* Boston, MA: Birkhauser, 1997.

[7]   Akbarzadeh, A., et al., "Towards Urban 3D Reconstruction from Video," *Proc. 3DPVT'06 (Int. Symp. on 3D Data, Processing, Visualization and Transmission),* 2006.

[8]   Pollefeys, M., "3D from Image Sequences: Calibration, Motion and Shape Recovery," in *Mathematical Models of Computer Vision: The Handbook,* N. Paragios, Y. Chen, and O. Faugeras, (eds.), New York: Springer, 2005.

[9]   Pollefeys, M., et al., "Visual Modeling with a Hand-Held Camera," *International Journal of Computer Vision,* Vol. 59, No. 3, 2004, pp. 207–232.

[10]   Nister, D., "Automatic Passive Recovery of 3D from Images and Video," *Proc. of the Second International Symposium on 3D Data Processing, Visualization and Transmission, 3DPTV,* September 2004, pp. 438–445.

[11]   Dick, A., et al., "Combining Single View Recognition and Multiple View Stereo for Architectural Scenes," *Intl. Conference on Computer Vision,* Vancouver, Canada, 2001, pp. 268–274.

[12]   Sequeira, V., and J. Goncalves, "3D Reality Modeling: Photo-Realistic 3D Models of Real World Scenes," *Proc. First International Symposium on 3D Data Processing Visualization and Transmission,* 2002, pp. 776–783.

[13]   Stamos, I., and P. Allen, "Geometry and Texture Recovery of Scenes of Large Scale," *Computer Vision and Image Understanding (CVIU),* Vol. 88, No. 2, November 2002, pp. 94–118.

[14]   Früh, C., and A. Zakhor, "Fast 3D Model Generation in Urban Environments," *IEEE Conf. on Multisensor Fusion and Integration for Intelligent Systems,* Baden-Baden, Germany, 2001, pp. 165–170.

[15]   Früh, C., and A. Zakhor, "3D Model Generation of Cities Using Aerial Photographs and Ground Level Laser Scans," *Proc. IEEE Conf. on Computer Vision and Pattern Recognition,* Vol. 2, Kauai, HI, 2001, pp. II-31-8.

[16]   Früh, C., and A. Zakhor, "Data Processing Algorithms for Generating Textured 3D Building Facade Meshes from Laser Scans and Camera Images," *3D Processing, Visualization, and Transmission 2002,* Padua, Italy, 2002, pp. 834–847.

[17]   Besl, P., and N. McKay, "A Method for Registration of 3-D Shapes," *IEEE Trans. on PAMI,* Vol. 14, No. 2, 1992.

[18]   Burgard, W., et al., "Estimating the Absolute Position of a Mobile Robot Using Position Probability Grids," *Proc. 13th Natl. Conf. on Artificial Intelligence*, 1996, pp. 896–901.

[19]   Chan, R., W. Jepson, and S. Friedman, "Urban Simulation: An Innovative Tool for Interactive Planning and Consensus Building," *Proc. of 1998 American Planning Association National Conference,* Boston, MA, 1998, pp. 43–50.

[20]   Cox, I., "Blanche—An Experiment in Guidance and Navigation of an Autonomous Robot Vehicle," *IEEE Trans. on Robotics and Automation,* Vol. 7, 1991, pp. 193–204.

[21]   Debevec, P., C. Taylor, and J. Malik, "Modeling and Rendering Architecture from Photographs," *Proc. of 23rd ACM Ann. Conf. on Computer Graphics and Interactive Techniques,* 1996, pp.11–20.

[22]   Dellaert, F., et al., "Structure from Motion Without Correspondence," *IEEE Conference on Computer Vision and Pattern Recognition*, 2000.

[23]   Fox, D., W. Burgard, and S. Thrun, "Markov Localization for Mobile Robots in Dynamic Environments," *Journal of Artificial Intelligence Research*, Vol. 11, 1999, pp. 391–427.

[24]   Thrun, S., et al., "Robust Monte Carlo Localization for Mobile Robots," *Artificial Intelligence*, Vol. 128, Nos. 1–2, 2001.

[25]   Früh, C., "Automated 3D Model Generation for Urban Environments," Ph.D. thesis, University of Karlruhe, Germany, 2002.

[26]   Turk, G., and M. Levoy, "Zippered Polygon Meshes from Range Images," *Proc. 21st Acm Ann. Conf. on Computer Graphics and Interactive Techniques*, Orlando, FL, 1994, pp. 311–318.

[27]   Gutmann, J., and K. Konolige, "Incremental Mapping of Large Cyclic Environments," *International Symposium on Computational Intelligence in Robotics and Automation (CIRA'99)*, Monterey, CA, 1999.

[28]   Gutmann, J., and C. Schlegel, "Amos: Comparison of Scan Matching Approaches for Self-Localization in Indoor Environments," *Proc. of the 1st Euromicro Workshop on Advanced Mobile Robots*, 1996.

CHAPTER 16

Multimodal Biometric Systems: Applications and Usage Scenarios

Michael J. Thieme

16.1 Introduction

To fully assess the utility of multimodal biometric techniques in real-world systems, it is useful to consider usage scenarios and applications in which multimodal techniques are implemented. This chapter presents results from an evaluation conducted by International Biometric Group (IBG) and funded by the National Institute of Justice under Grant 2005-IJ-CX-K059.

The evaluation compared the accuracy of various multimodal fusion and normalization techniques based on data generated through commercial fingerprint, face recognition, and iris recognition matchers. This evaluation attempted to situate results in the context of typical biometric applications and usage scenarios.

16.2 Multimodality and Multiple-Biometric Systems

Multimodal systems are subsets of multibiometric systems. The following terms describe different types of multibiometric systems:

- *Multimodal:* fusion of comparison scores associated with samples from more than one modality.
- *Multialgorithmic:* use of multiple matching algorithms to process the same biometric sample. Sufficiently divergent algorithms can be fused to produce a superior result.
- *Multi-instance:* use of two or more instances within a given modality. Many fingerprint systems are multi-instance, using 2 or 10 fingerprints per user for recognition.

Multimodality is the focus of this evaluation. The following multimodal fusion approaches can be applied to enhance system accuracy:

- *Decision-level fusion:* technique in which *accept* and *reject* decisions from multiple algorithms are utilized as inputs to a Boolean *and* or *or* decision. The results of the Boolean determine whether the event is a match or nonmatch. Decision-level fusion is a trivial case of max-score (*or*) or min-score (*and*) score-level fusion.
- *Feature-level fusion:* technique implemented within a biometric matcher in which feature or vector-level data is combined to comprise a new meta-modality that is processed through a matcher.

- *Score-level fusion:* technique in which scores from multiple matchers associated with different modalities are combined to generate a single, universal score used to render accept/reject decisions. Score-level fusion allows any matching algorithm to be fused with any other, so long as both algorithms output matching scores of some form. Score-level fusion does not require knowledge of the score generation mechanics. Normalization processes are implicit in most score-level fusion systems. Score-level fusion must consider the correlation between input sources of match scores, as any increased accuracy is derived from leveraging divergent modalities [1].

The normalization and fusion methods discussed below relate only to score-level fusion. A high-level score-level fusion system schematic is shown in Figure 16.1.

16.3   Multimodal Techniques Overview

16.3.1   Normalization Techniques

The fusion phase of score-level multimodal processing assumes that all comparison scores are functionally equivalent. However, matching algorithms' output scores

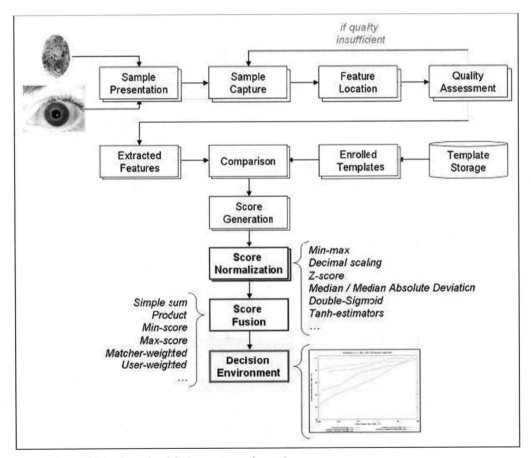

Figure 16.1   Score-level fusion system schematic.

are heterogeneous: comparison scores may indicate similarity or dissimilarity, may refer to different scales (e.g., linear or logarithmic), and may follow different statistical distributions.

Normalization is the process through which different algorithms' scores are transformed into a common domain by changing the location and scale parameters of the score distributions. Normalization is central to multimodal system design. If normalized scores are not indicative of the same level of match confidence, subsequent score fusion rests on an unstable foundation. Effective normalization requires robust and efficient estimates of score distributions' location and scale parameters. Robustness indicates insensitivity to the presence of outliers. Efficiency refers to the proximity of the obtained estimate to the optimal estimate when the distribution of the data is known.

Because score normalization is dependent on the creation of an accurate statistical model of input data, an a priori normalization method is inherently inaccurate. A biometric system must be initially tuned for the data it will process using statistical models describing the output ranges for each matcher. As the system progresses through its life cycle, these models must be updated.

16.3.1.1   Min-Max and Decimal Scaling

The simplest normalization technique is the min-max normalization. min-max normalization is best suited for cases where the maximum and minimum observed comparison scores produced by a matcher are known. In these cases, one can shift the minimum and maximum comparison scores to 0 and 1, respectively. Decimal scaling is similar to min-max normalization. It is calculated by scaling observed comparison scores to the minimum and maximum comparison scores that the matcher can generate. The advantage to decimal scaling is that it can be estimated from vendor documentation. While not ideal, this allows normalization to occur on datasets for which no reliable statistical information exists. It also can be used in conjunction with other statistical methods to determine a hard upper or lower bound for input scores. If comparison scores are not bounded, one can estimate the minimum and maximum values for a set of comparison scores and then apply min-max normalization. When the minimum and maximum values are estimated from the given set of matching scores, this method is highly sensitive to outliers in the data used for estimates.

16.3.1.2   Z-Score

The most commonly used score normalization technique, based on its ease of calculation, is z-score. Z-score is calculated using the arithmetic mean and standard deviation of the match scores resulting from training data. This scheme is expected to perform well if prior knowledge about a matcher's average score and score variations is available. Without this data, mean and standard deviation need to be estimated from a given set of matching scores. However, because mean and standard deviation are sensitive to outliers, z-score is not a robust normalization technique. That said, it often performs better than min-max due to its ability to process outliers in low-quality data. Z-score normalization does not guarantee a common numerical range for normalized scores from different matchers. Z-score assumes that input scores approximate a Gaussian distribution, in which scores cluster around an average value and frequency decreases away from the mean. If input scores are not Gaussian distributed, z-score normalization does not retain the same input distribution at the output. For an arbitrary distribution, mean and standard deviation are reasonable estimates of location and scale, respectively, but are not optimal.

16.3.1.3   Tanh

Tanh is a robust normalization technique that maps scores to the (0, 1) range. Tanh is a subset of sigmoid curves. The values for mean(S) and std(S) are determined as in the z-score method. As a robust technique, sigmoid normalization is less dependent on outliers in the input data. The formula brings outliers closer into the middle of the scale, preventing a small amount of outlying data from skewing the overall scale. For a biometric application in with noisy training sets, this robustness is useful, but scaling must be chosen carefully. Tanh curves allow for greater separation between genuine and impostor populations.

16.3.2   Fusion Techniques

Score fusion is the second step in the normalization-fusion process, in which normalized match scores are combined to generate a single numerical result. Information on matcher quality, image quality, subject identity, and the relation of component scores to one another may be used to calculate this value. This information is used to weight scores in relationship to one another, generally based on a confidence factor for each score.

In basic fusion methods, only the component scores are used. The advantage to such an approach is simplicity, as the combination of scores is not dependent on system-dependent variables such as performance.

16.3.2.1   Minimum-Score

In minimum score fusion, the lowest comparison score of all the normalized scores acquired from the matchers is designated as the fusion score.

16.3.2.2   Maximum-Score

In maximum score fusion, the highest comparison score of all the normalized scores acquired from the matchers is designated as the fusion score.

16.3.2.3   Simple Sum and Average (Mean/Median) Score

Simple sum entails adding the normalized scores generated by each matcher to yield a combined fusion score. It is computationally efficient and effective, assuming that it is preceded by effective normalization. The drawback to simple sum is that it is skewed toward users with more comparison scores than other users. This can be addressed by computing average scores for each user. For instance, a user with only fingerprint scores can be compared to a user with two fingerprint scores and an iris score.

Advanced score fusion methods are potentially more accurate than simple fusion methods but are more dependent on proper calibration and knowledge of error rates, image quality, or dataset training. For a large-scale system, investment in calibration, testing, and machine learning can improve error rates.

In general, advanced combinatorial methods seek to weight scores based on a confidence measure. This measure may be tied to the matcher, the data source, the image quality, or all three.

16.3.2.4   Matcher Weighting

Because matcher accuracy may vary, it may be desirable to designate a higher weight to more accurate matchers. Matcher accuracy can be measured in terms of equal error rate (EER). In practice, the equal error rate for a matcher will depend on the quality of the source data. Data source one may have low-quality iris images paired with high-quality fingerprints, while data source two has strong iris images with weak fingerprint samples. It is therefore useful to determine weights for each combination of data source and matcher, rather than simply for each matcher.

16.3.2.5   User Weighting

User weighting is a fusion technique that provides a custom weight for each combination of user and matcher. The technique focuses on users as opposed to enrolled samples. For instance, a thumb and a finger enrolled in a single matcher would require the same weight, regardless of data source or image quality. User weighting does not imply that decisions are based only on specific users, as doing so can incorrectly increase the confidence level in a biometric match. Instead, persons of interest should be enrolled in a watchlist (a subsection of the greater database), and the watchlist should be searched with a lower threshold than other segments. This ensures that the system operator will receive notice of a potential match for a person of interest, without a misleadingly high score.

16.3.3   Biometric Gain Against Impostors

Biometric gain against impostors (BGI) is a normalization and fusion technique that considers a score against both the impostor and genuine distributions. Mathematically, it is the ratio of the probability that a person is an impostor after consideration of biometric comparisons to the probability that he or she was an impostor before the comparison. The equations given in ISO/IEC JTC1/SC37 N1506 are as follows:

Score = PSi|I / P Si|G, where:
PSi|G = Value of PDFG at score Si
PSi|I = Value of PDFI at score Si

Because IBG evaluated two types of BGI in the evaluation, we refer to BGI described above as *modified BGI*. IBG also evaluated an alternative probability distribution function referred to as *TAR,FAR*. This technique uses genuine distributions and could thus be viewed as biometric gain toward genuine subjects. The formula for *TAR,FAR* is as follows:

$$Score = TAR\ Si\ /\ FAR\ Si$$

16.4   Matcher Evaluation: Sample Collection and Processing

16.4.1   Overview

IBG's evaluation utilized a database of 13,458 fingerprint, 12,401 iris, and 19,656 face images. These images had been collected from 1,224 test subjects in a controlled test environment prior to the beginning of the research. IBG enrolled and searched these images through (3) fingerprint and (2) face recognition matchers to generate comparison scores. Iris recognition comparison scores had already been generated through a single matcher prior to the evaluation.

For each intermodal matcher combination—for example, fingerprint system 1 and face recognition system 2–matcher-specific comparison scores were normalized (placed on a common scale) and fused (combined though a statistical method) to generate a fused score. The following techniques were applied:

- *Normalization:* min-max, z-score, tanh;
- *Fusion:* simple sum, matcher weighting, user weighting;
- *Probabilistic Confidence Methods:* biometric gain against impostors (BGI).

IBG plotted accuracy in terms of false match and false nonmatch rates. Single-modality results were compared against various multimodal results to assess the degree to which multimodal techniques and matcher combinations impact accuracy.

16.4.2   Data Collection

The manner in which the test image database is collected and test population composition are determinants of biometric system performance. The evaluation incorporated controlled data collection processes and a demographically diverse test population. Biometric data was acquired through the following devices:

- Flat fingerprint images were acquired through a Cross Match Verifier 300 single-fingerprint reader. Images were saved as 500×500, 8-bit grayscale bitmaps.
- Face images were acquired through a DFK 31F03 equipped with a Pentax C3516 lens. Images were saved as $1,024 \times 768$, 24-bit color jpegs.
- Iris images were acquired through a COTS imager (manufacturer name redacted as competition-sensitive). Images were saved as 640×480, 8-bit grayscale bitmaps.

Test operators guided test subjects in their interaction with each acquisition device. Acquisition processes differed from modality to modality. Four fingerprint positions were utilized: left and right thumbs and index fingers. For each position, two separate samples were acquired to enable genuine comparisons. Fingerprint images were acquired through a manual capture application. The decision to capture a given image was based on observed presentation quality. If necessary, test operators reacquired images if rotation, pressure, position on scanner, or fingerprint area captured were deemed insufficient.

Face image acquisition entailed capture of multiple frontal facial images from a digital camera attached to a PC. The test subject was seated against a gray background approximately 10 ft from the camera and instructed to look at the camera with a neutral expression. Test operators visually inspected acquired facial images to validate quality. Images were reacquired if facial aspect, expression, or position in the frame fell outside prescribed parameters. Test subjects began by facing forward, rotated the head left and right, and then rotated the head up and down during the capture process. The first 6 frames of the video sequence, as well as every 25th frame thereafter, were enrolled and searched. Images 1 to 6 were frontal, while images 7 to 12 were captured at different orientations.

Iris image acquisition entailed capture and automated quality assessment of multiple iris images. Iris images were acquired through an automated application that incorporated automated quality assessment in the capture sequence. This iris capture is differentiated from fingerprint and face capture. If an iris enrollment sequence failed to collect sufficient-quality images, then one or more samples would be declared a failure to enroll.

16.4.3   Comparison Score Generation

Fingerprint and face scores were generated at various intervals during the period of performance beginning in October 2005. Fingerprint and face recognition vendors supplied no-cost licenses for their matching technology. Iris comparison scores had been generated prior to the NIJ-funded evaluation during IBG's independent testing of iris recognition technology [2]. The following section describes how each matcher was implemented. Matcher-specific issues are important in that they are representative of issues that must be dealt with in operational systems.

16.4.4   Fingerprint System 1

All index and middle finger samples were utilized for fingerprint system 1 (FP1). Test subjects with one visit (and thus two samples per instance) had two commutative genuine comparisons per finger position. Test subjects with two visits (four samples per instance) had twelve commutative comparisons (six sample pairs in two directions). FP1 was implemented in identification mode, as the matcher was not capable of performing verification. To generate as many pairwise comparison scores as possible, each image was enrolled, comprising a gallery set. Each image was then searched against the entire gallery. Comparisons generated either scores (ranging from a minimum of 1 to a maximum of 100,000[1]) or null values. Scores resulting from comparisons of identical samples were removed from the database.

Ideally, all comparisons in each 1:N search would have generated a comparison score. In this fashion, one would have full score matrices for all fingerprint pairs. However, functions internal to FP1 prevented full comparison score generation. FP1 supports a configurable comparison score threshold below which no results are returned. This threshold prevents large quantities of very weak comparison scores being returned to the operator. If this threshold is set to return the lowest possible comparison scores, the system behaves erratically (e.g., returning more results than are in the gallery or returning phantom user IDs). The FP1 threshold was therefore set to 100 to eliminate irregular results. Further, FP1 limits the number of scores returned from each search to 50 results, based on the assumption that matches ranked below 50 are very unlikely to be genuine matches.

The images presented to FP1 numbered 13,458, of which 207 could not be processed for enrollment or searching. Intraposition comparisons were expected to generate ~43m comparison scores (27,466 genuine). Approximately 5,109 of 27,466 expected genuine scores were not returned. The following possible causes for missing scores were examined:

- *Cross-bin comparisons:* FP1 assigns fingerprints to one of five bins based on level I sample characteristics (e.g., loop, whorl, and arch). Matching logic begins with intrabin comparisons, expanding to cross-bin comparisons, if an insufficient number of intrabin hits are found. Of the 22,357 genuine comparisons that were returned for nonidentical images, 546 had the subject and probe images for the same finger placed in different bins. Further analysis may show that the majority of the missing 5,109 comparisons have been placed in different bins by the level I classifier.

- *Limited number of total hits per search:* FP1 returned a maximum of 50 results per 1:N search. This was determined to be the cause of the missing genuine scores, as all genuine probe/gallery comparisons that were missing a genuine result had a full complement of 50 results returned for the probe image.

- *Limited number of hits per gallery subject per search:* FP1 may return only a certain number of hits per person per instance. That is, if a search result already contains a certain number of subject A's index finger enrollments, further comparisons may not be returned. The presence of full complements of genuine results for many subjects eliminated this as a possibility.

- *Lack of sub-100 comparison scores:* In order to investigate matcher behavior when very low comparison scores were returned, a small number of samples were run as probes with a cutoff threshold of 1 to determine the number of scores between 1 and 100. For these subjects, 40% of the returned scores

---

1. Score values as represented in the report have been modified from their actual values to maintain matcher anonymity. The high-level scale, quantization, and directionality have been maintained.

were between 1 and 100. Attempts were then made to search at a threshold of 1. For some samples, this caused the system to hang indefinitely. Those that returned had a different result set than when searched at a threshold of 1, with a small overlap. Notably, the majority of scores were 0, and relatively high nonzero scores were not returned.

It was determined that in all cases where an expected genuine score was not returned for a probe and gallery image from the same subject, exactly 50 scores had been returned for that probe image. In some cases, all 50 of these results were impostors, while in others several genuine scores were included in the result set.

16.4.5   Fingerprint System 2

Fingerprint system 2 (FP2) was accessed via an SDK that provided separate applications for image conversion, image quality analyzer, template generator, and verification. Scores from each 1:1 comparison were logged with the exception of cross-position and identical-template comparisons. Due to time constraints, not all impostor matches were completed, although all genuine matches did complete. In all, 74,225,085 impostor comparison scores and 42,220 genuine comparison scores were computed.

16.4.6   Fingerprint System 3

Fingerprint system 3 (FP3) was provided as a sample application through which bulk comparisons of previously generated templates were executed. Of the 20,204 images available for comparison, 20,178 (99.87%) were able to generate templates. Due to time constraints, not all impostor comparisons could be completed, though all impostor comparisons were completed. Approximately 228,626,283 impostor comparisons and 42,216 genuine comparisons were executed.

16.4.7   Face Recognition System 1

Face recognition system 1 (FR1) was accessed through three command-line applications, which extract templates, store them, and search a probe template against a gallery of templates. Templates are stored on disk and read into memory for each query. Approximately 312 images of the 19,656 failed to enroll, resulting in a gallery of 19,344 templates. To execute searches against the full gallery of 19,344 images, 20 separate searches were executed, each against a separate 1,000-template gallery. FR1 generated almost 316 million results. Approximately 150 of 1,224 subjects failed to acquire on all 12 samples, resulting in a high FRR. Overall, 2,997 of 19,656 samples failed to acquire and 312 of 19,656 failed to enroll.

16.4.8   Face Recognition System 2

Face recognition system 2 (FR2) consisted of an application server, a database server, and a centralized search process. Search results whose match score exceeded a minimum threshold of three were returned. Of 19,656 images presented to the system, 170 failed to both enroll and acquire. Of 379,704,196 possible comparison scores, 5,044,477 results were returned. 293,411 of 336,216 possible genuine results were returned by the system. For facial images below 5° of rotation, 67,864 of 69,834 possible genuine results were returned. All images returned at least one genuine result when used as a probe against the full gallery.

16.4.9 Iris Recognition System 1

The following discussion of iris recognition system 1 (IR1) comparison score generation through IR1 refers to the process of score generation in ITIRT. IR1 was provided in the form of an SDK capable of executing rapid 1:1 comparisons (roughly 800 per second). IR1 was implemented across multiple machines, each processing a subset of the overall dataset. The IR1 gallery was comprised of 12,401 templates generated through enrollment transactions. The IR1 probe dataset was comprised of these same 12,401 templates. The iris capture devices used rejected lower-quality enrollments, resulting in failures to enroll. Enrollment transactions totalling 3,364 were conducted, with each transaction producing up to 4 images corresponding to left and right eyes under left and right illumination, for a maximum 13,456 possible templates. Templates were cross-compared in a 1:1 fashion; results from comparisons of identical templates, as well as AB:BA results, were discarded. A total of 18,133 genuine comparison scores and 61,383,769 impostor comparison scores were available for analysis.

16.5   Data Subselection

The purpose of the evaluation was to determine the error-reducing ability of different combinations of normalization and fusion techniques. Many combinations of data produced error rates that were too low for differentiation. In order to analyze the impact of biometric fusion on difficult test subjects (i.e., those with low genuine and/or high impostor results), test subjects whose samples generated disproportionate errors were selected as the focus of the evaluation.

For the three algorithms that contained substantial quantities of null results due to the 1:N matching process, results from non-null impostor scores were used for analysis, as these samples are capable of producing false acceptances. For each test subject-matcher combination, a single genuine score was selected at random from all available genuine scores.

16.5.1 Handling Null Scores

One of the challenges of implementing an operational multibiometric fusion system is that scores may not be available for all users and all systems. Fusion techniques generally assume the presence of a full input set, such that null inputs are not accounted for. Null values may occur as a result of a failure to acquire, failure to enroll, matcher error, or result not returned by a 1:N match. Null values also need to be treated differently than zero values, because many normalization methods convert zero values to nonzero values.

When performing matching on a 1:N multimodal system, each matcher queried will return a list of potential matches that exceed a rank or score. It is likely that results from systems A and B will be disjointed, such that some enrollees returned by system A will not be returned by system B and vice versa. This may be because individuals were unable to enroll in all systems or because enrolled individuals generated subthreshold or subrank results.

FR1, FR2, and FP1 contained internal thresholding and/or limits on result-set size that could cause null results for certain combinations. If no genuine result was returned due to an FTA, FTE, or noninclusion in a 1:N result set, the score was treated as a null. Certain fusion techniques use null values when tabulating results. Null values were set at 0.5 for iris recognition and 0 for all other matchers.

Null or unreturned impostor scores are uninteresting, as they result in true rejections due to matcher error, failure to acquire, or failure to enroll. A null or unreturned genuine score, on the other hand, may contribute to a false rejection. In order to focus on impostor subjects who generated high impostor comparison scores in error, impostor subjects were selected by the following methodology.

Operational systems will not necessarily know why a null score has been generated. Therefore a method of handling null scores needs to be developed that makes no assumptions about the null's root cause. The following logic was applied:

- *Simple sum fusion:* Null values are treated as zero.
- *Matcher weighted fusion:* Null values are treated as zero.
- *User weighted fusion:* Null values are treated as zero. EER for algorithms with null input is set to 1.
- *Modified BGI:* Null values are ignored, nonnull result is squared. If both results are null, 0 is returned.

16.5.2   Primary and Secondary Scores

The following multimodal system combinations were evaluated.

- FR1 and IR1;
- FR2 and IR1;
- FP1 and FR1;
- FP1 and FR2;
- FP1 and IR1;
- FP2 and FR1;
- FP2 and FR2;
- FP2 and IR1;
- FP3 and FR1;
- FP3 and FR2;
- FP3 and IR1.

In each case, one algorithm was considered the primary algorithm, while the other was considered the secondary algorithm. Subject pairs were selected for the primary algorithm's results first. Scores for those same subject pairs were then used to select scores from the secondary algorithm's results.

The fingerprint algorithm was considered the primary algorithm when combined with face or iris. Face was considered the primary algorithm when combined with iris. This approach is based on the premise that, in typical law enforcement systems, fingerprint will be the most available modality, followed by face, then iris. Different selection logic would result in different accuracy results, although it is fair to assume that the relative strength of fusion techniques, as well as the improvement of fusion relative to single modalities, will remain proportional.

Scores were selected as follows:

- For each of the 1,224 probe subjects, a list of gallery subjects was generated indicating those subjects for whom comparison scores against the probe subject existed. Five of these gallery subjects were chosen at random.
- For each of these subject pairs, one score was selected at random from the returned scores for the first algorithm. The number of scores returned between any two subjects depended on modality, and ranged from 0 to 144.
- All scores were then retrieved between the selected gallery and probe subjects for the second algorithm, and one score was selected at random from these.
- If no scores existed between gallery and probe for the second algorithm, the score was treated as a null (0.5 for IR System 1, and 0 for all other systems, representing the lowest possible score on each scale). This null marking was used by some fusion techniques to discount the value of that score.

This selection technique resulted in a focus on higher-scored impostor results, while ensuring that sets of probe and gallery data were internally consistent with the same subject. Algorithms produced a different set of impostor scores each time they were used as a secondary algorithm. For instance, FR1 had a different set of impostor pairs when it was fused with FP1 than when it was fused with FP2. In the graphs following, curves for FR1 differ slightly for each fingerprint algorithm it is fused with.

The false accept rates and false reject rates presented should be considered as pertaining to the more ambiguous cases (i.e., impostor results most likely to generate false acceptances and genuine results most likely to generate false rejections), rather than as definitive measures of overall matcher accuracy. As such, the points of interest are the extent to which each normalization and fusion technique reduces error rates for ambiguous subjects. Across a full user population, matchers will be more accurate than results suggest.

16.6 Results: Comparison of Fusion Techniques

In this section, a common set of techniques to show the strongest fusion-normalization approaches for each matcher combination.

The following results are shown in each figure:

- Single-modality performance (two matchers);
- Z-score (matcher weighting);

- Z-score (user weighting);
- Z-score (simple sum);
- TAR-FAR fusion;
- Modified BGI fusion.

Z-score was selected as the common normalization technique for matcher weighting, user weighting, and simple sum because, in general, it generated lower error rates.

Figures 16.2 through 16.12 show that in most cases, BGI variants deliver the lower error rates. Also, as one moves from figure to figure, the variations in overall accuracy—as well as the variations in performance clustering—become clear.

16.7   Analysis: Matcher Weighting and User Weighting

Matcher weighting and user weighting had varying results, depending upon combinations of specific algorithms, as summarized in Table 16.1.

User weighting proved to be an unpredictable method of fusion. Because a null score could represent either a score that was purposefully not returned or an inability to produce a score, results could be skewed in either direction, depending on which case was more prevalent.

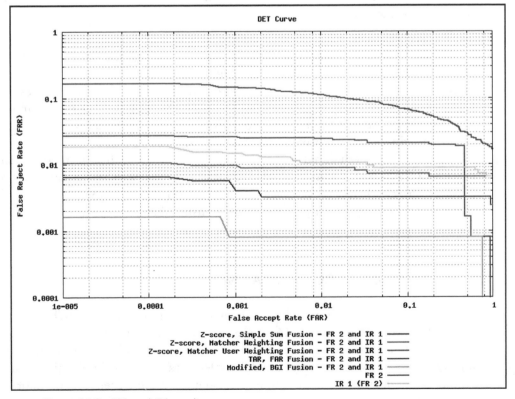

Figure 16.2   FR1 and IR1 results.

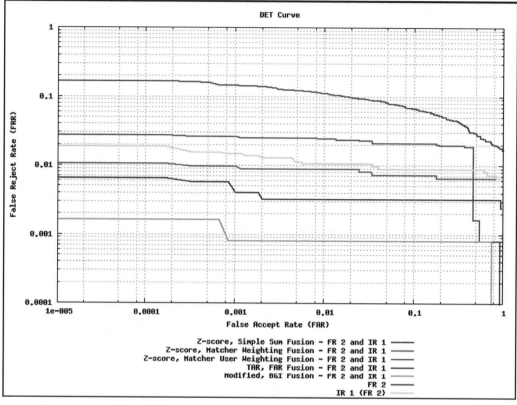

Figure 16.3 FR2 and IR1.

By not treating impostor nulls as zeros, user weighting fails to properly discount subjects who are, with good certainty, not a match. If one algorithm has many high impostor scores, while the other fails to return impostor scores for being too low, user weighting will mimic the high algorithm. On the other hand, if one algorithm has many null genuine results, user weighting will favor the algorithm with results, improving the overall fusion.

In a fielded, configured multimodal system wherein fusion module can determine the reason for missing scores, a form of user weighting may be more accurate. Preferably, the system would allow for both 1:1 and 1:N queries to each matcher, allowing a second pass that would ensure that each subject presented a full set of matching scores against the probe images.

16.8   Analysis: Modified BGI and *TAR,FAR*

Results suggest that multimodal techniques based on analysis of matching error distribution (BGI techniques) are more accurate than simple techniques that combine comparison scores. Therefore BGI techniques should be a starting point for organizations evaluating multimodal implementations. The implication of this finding is that deployers will need to build analytic tools into their systems in order to analyze data associated with impostor and genuine comparisons. In operation,

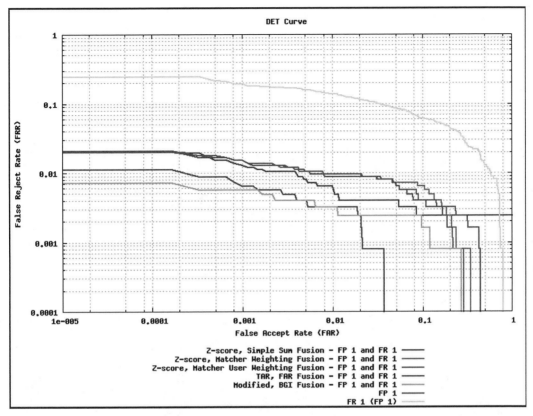

Figure 16.4   FP1 and FR1.

when matches are found or rejected, the score data that underlies these decisions—perhaps not visible to the operator—will be fed back into decision logic that governs multimodal responses. This sort of function will need to be implemented external to matcher systems.

The strength of BGI is that it can be determined directly from the probability density function (PDF) for each distribution, rather than a statistical approximation such as in z-score normalization. BGI does not require prior knowledge of the distribution of impostor and genuine subjects in the population. The PDF of the genuine distribution at any score can be determined as the false reject rate when the threshold is set to that score. The PDF of the impostor distribution can be determined as the false accept rate when the threshold is set to that score. The gain against impostors is higher as the probability that the score represents an impostor increases.

TAR,FAR generally results in a much more rapid decrease in false rejections as the number of false acceptances rises. For very low false accept rates, it is outperformed by modified BGI in terms of false rejections. It is generally capable of reaching a lower false rejection rate than modified BGI as false accept rate increases.

Both methods of fusion are highly dependent on the accuracy of the underlying distribution. In *TAR,FAR*, for instance, scores that are higher than the highest recorded impostor score are normalized to an infinite value, which propagates

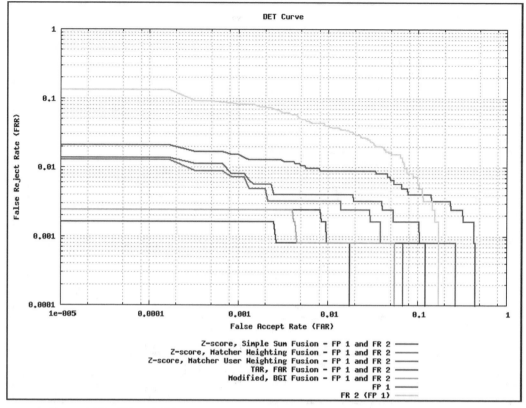

Figure 16.5 FP1 and FR2.

through to an infinite fused score. Modified BGI assigns a value of 0 to these scores, and an infinite value to any genuine match worse than the worst genuine calibration score. These values likewise propagate to a fused score of 0 or infinity, respectively. In both cases, an actual impostor whose score was higher than the highest calibration score for a single matcher would be marked as a genuine match no matter the result of other matchers. The existence of a perfectly minimal or maximal score precludes removing the erroneous match by simply raising or lowering the system threshold.

To avoid this, the PDF for each distribution can be extrapolated outward to produce miniscule but nonzero error rates at extreme operating points. The difficulty is in determining an accurate model for extrapolation—much of the accuracy and convenience from BGI methods comes from the ability to leverage the actual calculated data. In almost all cases, one or both of these methods is more accurate than any of the other nine normalization/fusion combinations at all points on the DET curve. In some cases, a form of BGI performed perfect disambiguation on the dataset, resulting in a range of thresholds that provided no false acceptances or rejections.

It should be noted that the set of data used to construct the impostor and genuine distributions is the same data that is utilizing the fusion. There is a possibility of overtraining the data in this situation. Further testing with separate training and testing datasets is required for full evaluation.

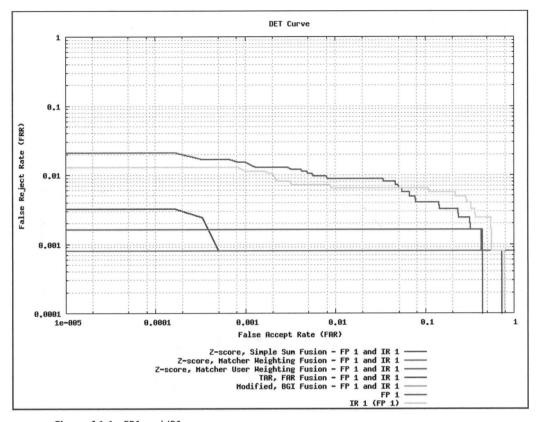

Figure 16.6   FP1 and IR1.

16.9   Results in Application-Specific Contexts

Evaluation results demonstrate the potential for multimodal systems in real-world applications. Certain combinations of fingerprint and face, face and iris, and fingerprint and iris can each provide more robust accuracy than fingerprint, face, or iris alone. A relatively weak biometric system (i.e., one with high false accept or reject rates) can be incorporated into a multimodal system in a fashion that improves overall error rates. However, certain combinations of matchers do not show performance improvement when implemented in a multimodal fashion. Further, the performance improvement associated with fusion ranges from substantia to marginal. When deciding on which matcher technology to procure and implement, deployers should consider whether a given technology lends itself to score-based multimodal fusion. However, to determine the utility of multimodal techniques in biometric applications, several factors must be taken into consideration.

16.9.1   Biometric Verification and Identification

Biometric systems can perform verification processes, identification processes, or both. The following definitions of biometric verification applications and identification applications are taken from an international standard on biometric

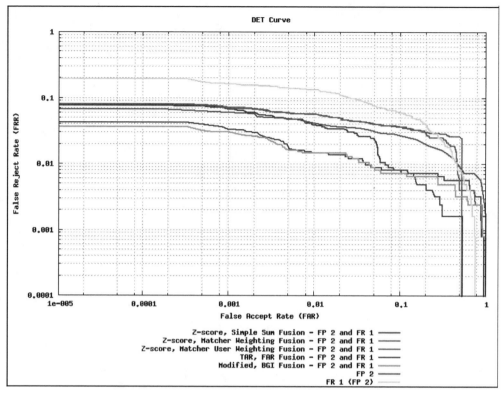

Figure 16.7 FP2 and FR1.

performance testing and reporting [3]. Verification applications are those in which the user makes a positive claim to an identity, features derived from the submitted sample biometric measure are compared to the enrolled template for the claimed identity, and an accept or reject decision regarding the identity claim is returned. Identification applications are those in which a search of the enrolled database is performed, and a candidate list of 0, 1, or more identifiers is returned.

For identification systems, a potential advantage of multimodal systems relative to single-modality systems is that multimodal systems may be capable of greater scalability without requiring more processing power. Filtering techniques that classify multimodal data according to characteristics in both modalities can substantially reduce the number of records that need to be searched in a given identification event. By searching fewer records, a justice agency can either gain shorter response time or increased accuracy (the latter by performing more intensive searches on a small dataset). In many cases, face images are available in justice applications in addition to fingerprints. Since face images are acquired in a controlled environment with relatively consistent pose and lighting, accuracy would typically be better than was observed in the evaluation.

Evaluations of multimodal techniques generally begin with the assumption that both genuine and impostor comparison scores are available for each matcher. This is clearly consistent with verification concepts of operations in which a comparison score is generated in almost cases. However, as was described above in the discussion of null scores, operational identification systems are typically con-

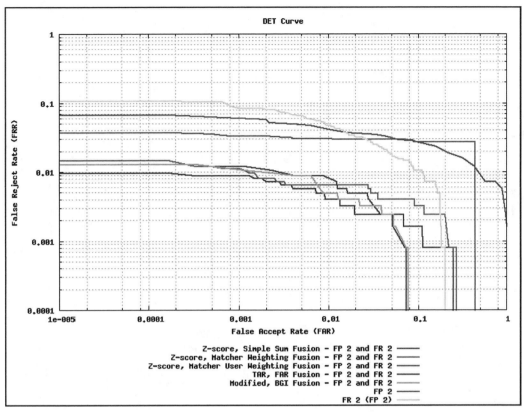

Figure 16.8   FP2 and FR2.

figured to return only a handful of matches from large-scale searches. In a search that returns candidate lists, one may have 20 to 50 potential matches, each with a score, for each modality. Therefore there may be very few subjects for whom two scores are available for multimodal processing. Research in which all users have scores for all available samples is not consistent with the operations of identification systems.

16.9.2   Legacy System Issues

Organizations looking to incorporate multimodal functionality will either have an existing system into which multimodal functionality would be integrated, or they will be procuring a new identification system that can incorporate multimodality as a fundamental design aspect. Unless a given agency has implemented a biometric platform that enables migration to additional biometric technologies, adding a new modality to an operational system is normally prohibitively difficult. Not only does operator workflow change, but the logic through which images are submitted to matchers changes as well. Evaluation results suggest that results vary so widely that multimodal functions and logic need to be highly configurable. Therefore incorporation of multimodal functionality is best undertaken as part of a major system redesign.

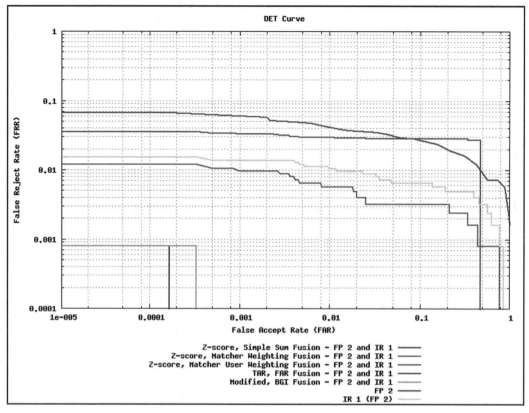

Figure 16.9 FP2 and IR1.

Furthermore, newly deployed biometric systems may be populated in part from legacy images, such as through digitized driver's license photographs collected over a period of years. In order to populate a multimodal database, samples from a second modality (e.g., fingerprints) may be collected from individuals who interact with the system. In this case, the system will have a partially populated database in which multimodal logic can only be applied to a subset of users.

16.9.3   Effort Associated with Biometric Sample Acquisition

Biometric data collection can be time-consuming and burdensome in certain applications. For example, convenience-driven applications such as the use of biometrics at the point of sale require that biometric data be collected quickly and with little effort. If a multimodal system requires two separate presentations—for example, a separate presentation of fingerprint and iris data—then the improvement in accuracy (or the improvement in enrollment percentage) must be sufficiently high to offset data collection effort.

16.9.4   Response Time Requirements

In identification systems, up to several million templates may be searched to locate one or more strong matches. Response times can range from fractions of seconds

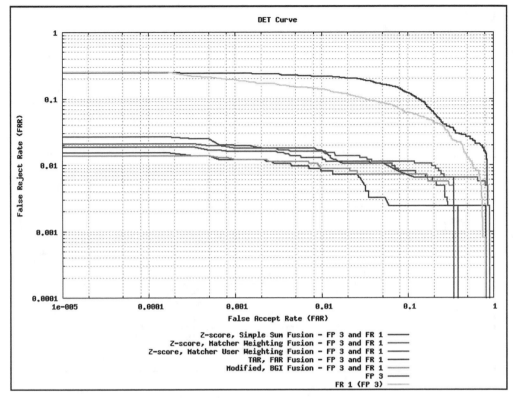

Figure 16.10   FP3 and FR1.

to several minutes, depending on processing power and computational load. In applications in which response time is critical, such as searching watchlists based on live-user data, systems must be configured to return results as quickly as possible. Therefore it may not be feasible to wait for the results from multiple searches, as the slowest matcher may be the determinant of overall response time.

16.9.5   Quantization

Certain matchers heavily quantize their scores such that weak matches are given very low similarity scores and strong matches are given very high similarity scores. This type of matcher does not lend itself to utilization in a multimodal system, as the score logic is essentially binary. The scenario in which several moderately high scores across multiple modalities reveal a match is unlikely to occur when using matchers whose scores are quantized.

16.9.6   Multipass Logic

In identification applications (e.g., AFIS-based fingerprint searches), an agency will likely implement multipass logic whereby only a percentage of transactions are routed to a secondary matcher, for example, those with marginal scores or with an abnormally small amount of sample data. In this case, the concept of a primary

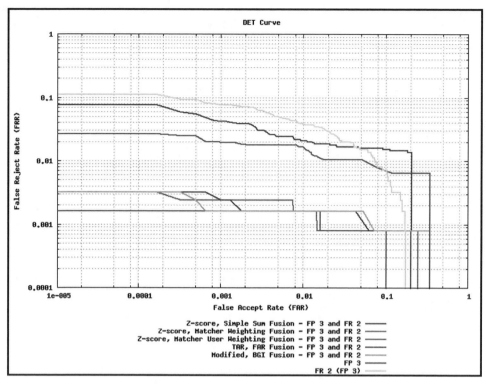

Figure 16.11   FP3 and FR2.

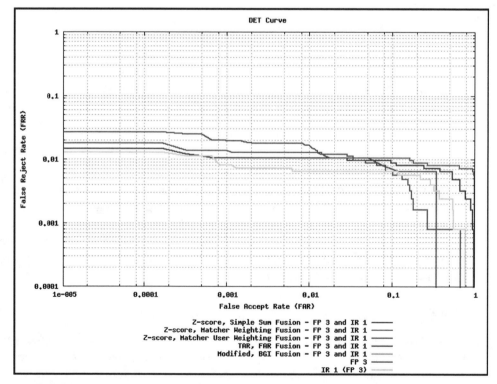

Figure 16.12   FP3 and IR1.

Table 16.1   Matcher Weighting and User Weighting Results

Matcher 1	Matcher 2	Comparison to Simple Sum
FP 1	FR 1	More accurate for non-z-score methods
FP 2	FR 1	Generally less accurate
FP 3	FR 1	More accurate
FP 1	FR 2	Little-to-no error for all methods
FP 2	FR 2	Less accurate
FP 3	FR 2	Less accurate for non-z-score methods
FR 1	IR 1	More accurate
FR 2	IR 1	Less accurate
FP 1	IR 1	Less accurate for non-z-score methods
FP 2	IR 1	Generally less accurate; tanh matcher weighting was the most accurate non-BGI method
FP 3	IR 1	Varying; tanh matcher weighting was the most accurate non-BGI method

and secondary modality would apply. It is fair to assume that multimodal utilization will be an exception case for many implementations.

References

[1]   ISO/IEC JTC 1/SC 37 N 1776, Draft Technical Report 24722, "Multimodal and Other Multibiometric Fusion," 2006.

[2]   http://www.biometricgroup.com/reports/public/reports/ITIRT_report.htm.

[3]   ISO/IEC 19795–1:2006 Information technology, "Biometric performance testing and reporting: principles and framework," 2006.

CHAPTER 17

SATware: Middleware for Sentient Spaces

Bijit Hore, Hojjat Jafarpour, Ramesh Jain, Shengyue Ji, Daniel Massaguer, Sharad Mehrotra, Nalini Venkatasubramanian, and Utz Westermann

In this chapter, we present SATware: a stream acquisition and transformation (SAT) middleware we are developing to analyze, query, and transform multimodal sensor data streams to facilitate flexible development of sentient environments. A sentient space possesses capabilities to perceive and analyze a situation based on data acquired from disparate sources. We present a multimodal stream processing model in SATware and its elementary architectural building blocks. These include a distributed run-time system that permits injection, execution, and interconnection of stream processing operators, a declarative language for composition of such operators, an operator deployment module that optimizes deployment of stream processing operators, the concept of virtual sensors that encapsulates stream processing topologies to create semantic sensing abstractions, and an infrastructure directory for storing the availability of resources. We describe how this basic architecture provides a suitable foundation for addressing the challenges in building middleware for customizable sentient spaces.

SATware is implemented in the context of Responsphere—a pervasive computing, communication, and sensing infrastructure deployed at the University of California, Irvine, that serves as a unique testbed for research on situation monitoring and awareness in emergency response applications. SATware provides a powerful application development environment in which users (i.e., application builders) can focus on the specifics of the application without having to deal with the technical peculiarities of accessing a large number of diverse sensors via different protocols.

17.1 Introduction

Advances in computing, communication, and sensing technologies has made it possible to create large-scale physical spaces with diverse embedded sensors, ubiquitous connectivity, and computation resources that together can enable pervasive functionality and applications. Today, such applications are built in a somewhat ad hoc manner directly on top of multimodal sensor streams, and application writers are exposed to the technical challenges of accessing a large number of different sensor types via different protocols over a network. We envision a different type of pervasive computing infrastructure, which we refer to as *sentient spaces*, that provides powerful and flexible abstractions for building pervasive applications. In a sentient space, observers possess capabilities to perceive and analyze a situation based on available multimodal data from disparate sources. Sentient spaces provide a much higher level of semantic abstraction compared to sensor middleware prevalent today. The key concept underlying sentient spaces is *events*. Events are essentially important occurrences (or state changes in a pervasive space) that are semantically meaningful for the application. For instance, entry and exit of people from a building might be a meaningful event in the context of a surveillance application. Instead of viewing data as sensor streams, sentient spaces provide the abstraction of the (semantically meaningful) event streams that form the building block for pervasive applications. Decoupling (and abstracting) events from the underlying sensors provides numerous advantages. First and foremost, it provides a level of separation that enables application writers to focus on the application logic instead of having to directly manipulate sensor streams. Another advantage is that it provides a framework for the underlying middleware and run time to optimize application execution based on variety of application and resource constraints. For instance, an event (such as location of a person of interest in an instrumented space) might be detected through tracking built using video cameras. It could also be detected using coarse level triangulation using WiFi access points. Which specific sensing technique is utilized can be decided by the run time based on available resources and the applications' accuracy needs.

In this chapter, we describe the architecture of SATware, a multimodal sensor data stream querying, analysis, and transformation middleware that aims at realizing a sentient system. SATware provides applications with a semantically richer level of abstraction of the physical world compared to raw sensor streams, providing a flexible and powerful application development environment. It supports mechanisms for application builders to specify events of interest to the application, mechanisms to map such events to basic media events detectable directly over sensor streams, a powerful language to compose event streams, and a run time for detection and transformation of events.

SATware is being developed in the context of the Responsphere infrastructure at the University of California, Irvine, campus, which is a unique publicly accessible testbed for interdisciplinary research in situation monitoring and awareness in the context of emergency response applications. Currently, Responsphere includes more than 200 sensors of 10 different types deployed over the geographical space that covers about half of the UCI campus. The sensors range from network pan-tilt-zoom fixed video cameras, microphones, wireless motes with accelerometers, temperature sensors, and motion sensors to RFID tag readers and networked people counters. It also includes mobile, WiFi-connected sensors such as an autonomous vehicle and a wearable EvacPack[1] carrying sensing devices like GPS devices, gas chromatographs, cameras, and microphones.

Responsphere is currently used to instrument and observe activities such as emergency drills that are conducted in the University of California, Irvine, campus. The data collected by sensors (e.g., video, audio, and so forth) is used by sociologists to observe and analyze human behavior as well as ascertain the effectiveness of new technologies, protocols, and strategies in crisis situations

---

1. EvacPack is composed of a backpack with a small computer box with WiFi connectivity, a pair of visualization goggles, a wireless mouse, a wearable keyboard, and a set of sensors including gas sensors, Webcam, microphone, and compass.

deployed within emergency drills. Furthermore, continuous data feeds from Responsphere sensors are used to build pervasive applications such as surveillance and infrastructure monitoring. Although Responsphere provides the main motivation of our work, it must be noted that SATware is a generic framework suiting any application scenario in which highly diverse data streams need to be processed. As a sentient system, SATware provides information about the situation in an environment to users from different application perspectives.

The chapter is organized as follows: Using our needs in Responsphere as the main source of motivation, we highlight elementary notions and issues that multimodal sensor data stream processing middleware needs to address in Section 17.2. In Section 17.3, we briefly analyze related work with regard to which extent it addresses (or does not address) the raised issues. In Section 17.4, we describe the basic architectural building blocks of SATware and the underlying processing model. Section 17.9 concludes this chapter with a summary and an outlook to ongoing and future developments of SATware.

17.2   Multimodal Sensor Data Processing: Notions and Issues

The processing of the sensor data produced by the large sensing infrastructure of Responsphere is a challenging task. The heterogeneity of the infrastructure and its elements provides a very powerful sensing infrastructure that enables a wide variety of applications. However, it makes the task of building applications using the infrastructure more complex. Responsphere produces a set of heterogeneous data streams that range from raw video data to temperature readings. The sensor hardware, software, and their access protocols are also heterogeneous; and so are their mobility capabilities, power constraints, and processing, storage, and networking capabilities. This heterogeneity, together with the wide range of applications that can be implemented in such an infrastructure and the fact that Responsphere encompasses a large number of sensors installed throughout the UC Irvine campus, poses the following challenges:

- *Multimodality streams:* The sensors in Responsphere produce a large diversity of data streams that need to be processed in a unified manner. Existing approaches to data stream processing focus mostly on a specific kind of stream: e.g., scalar streams as produced by motes (e.g., [1]) or event streams as produced by RFID tag readers (e.g., [2]). Only recently, the processing of raw media streams as produced by cameras and microphones have been considered as well (e.g., [3–5]).
- *Abstraction:* In order to provide an abstraction to ease sensor data stream processing in Responsphere and to allow automatic optimization of concurrent processing tasks, declarative languages for analyzing, querying, and transforming data streams are essential. While sensor programming environments prevalent today provide an SQL-like language to speed up applications, they have focused primarily on data acquisition from sensor

streams [1]. Also, data stream processing systems and the languages they support deal with homogeneous events (e.g., [6, 7]).

The different sensors in Responsphere produce streams of data at different semantic levels, ranging from low-level raw media data produced by video cameras or microphones up to rather high-level events produced by RFID tag readers. However, there is still a significant semantic gap between the information usually of interest for users (e.g., "Where is the evacuation warden?") and the data produced by sensors ("RFID reader 57 read tag 0815"). A sensor stream processing middleware for sentient spaces must provide a way of bridging this gap, permitting the assembly of high-level views onto lower-level sensor data streams.

- *Scalability:* In Responsphere, a large number of sensors potentially producing streams of considerable data rates requiring considerable computational power for processing (such as video streams) pose the difficult challenge of efficiently utilizing both limited available network bandwidth and limited available computing capabilities. Many of the sensor data processing infrastructures proposed in the literature, however, are based on centralized architectures thwarting scalability to Responsphere dimensions [8–11].
- *Extensibility:* With the large number and variety of different sensors, developers of multimodal sensor data analysis tasks on top of Responsphere face the challenge of discovering which sensor types are available, what sensor streams they produce, where sensors are located physically and which area they cover, under which addresses they can be reached on the network, what the network topology is, what computing nodes are available for stream processing, and so forth. A sensor network directory service offering access to this information is required.
- *Mobility:* The availability of mobile sensors in Responsphere constitutes a further challenge. While many sensor data processing infrastructures are capable of temporally synchronizing different streams for multimodal processing using different variants of time window-based joins (e.g., [1, 8, 9]), mobile sensors also require location-based means of stream synchronization. For instance, one may want to join the camera streams of two autonomous vehicles only when they are observing the same area.
- *Power-awareness:* Responsphere encompasses sensors that are battery-powered and sensors that are constantly connected to the power grid. Battery-powered sensors can also be further divided into rechargeable sensors and not (easily) rechargeable sensors. This heterogeneity implies that different power optimization approaches have to be considered when accessing each device and balancing the computation.
- *Privacy:* As Responsphere covers public spaces and work environments, privacy of observed individuals is of high importance. To avoid misuse and unauthorized access to information about individuals, it must be possible to enforce privacy policies already at the level of the sensor data processing infrastructure.

17.3   Related Work

There has been extensive research in different areas of data management in sensor networks and pervasive environment that is relevant to SATware. Each of these systems has focused on a subset of the requirements described in Section 17.2. In this section we provide a brief overview of some of these systems.

17.3.1   Data Stream Processing

There has been a considerable work on data stream processing systems including Aurora [12], TelegraphCQ [7], and STREAM system [5]. These systems provide many features such as data model, continuous query semantic, query language, and scalable and adaptive continuous query processing. However, most of these systems have concentrated on data streams and do not consider multimedia streams.

17.3.2   Sensor Networks

Sensor networks generate high volumes of data stream that need to be analyzed. Much work has been done on different aspects of data acquisition and analysis in sensor networks that includes a broad spectrum of issues from energy optimization to query accuracy [1, 6]. However, most of the work in this field has focused on single type of sensors such as motes [13].

17.3.3   Multimedia Stream Processing

Recently there have been systems for processing multimedia streams including IBM Smart Surveillance System (S3) [10], MedSMan [8], IrisNet[3], and Smart Camera Network [4]. These systems provide features such as media capture, automatic feature extraction, declaration and query language, temporal stream operators, and querying algorithms. In this section we describe IBM S3 system and MedsMan in detail. The reader is referred to the cited references for a detailed description of the other systems;

17.3.3.1   IBM S3 System

The IBM Smart Surveillance System is a middleware that provides video-based behavior analysis. It has two main components: Smart Surveillance Engine (SSE) and Middleware for Large Scale Surveillance (MILS).

The main functionality of the SSE is to provide software-based surveillance event detection. It uses video analysis techniques such as object detection, object tracking, and object classification. The MILS middleware provides data management services for S3. It converts surveillance video into a relational database table. It also provides querying capabilities on the surveillance video. S3 also provides a privacy-preserving surveillance feature that can hide personal information in different levels.

Despite all the rich functionality that it provides, the S3 system is based on a centralized database that may result in scalability issues in systems with large number of video sources.

17.3.3.2 MedSMan

MedSMan is a live multimedia management system. It consists of three main components: stream model, environment model, and event model. The stream model in MedsMan consists of two basic parts: *media streams* and *feature streams*. A media stream consists of tuples with sequence number, a time interval, and media content corresponding to that interval. A feature stream consists of a sequence number, a timestamp, and a feature value that has occurred on that time. MedSMan defines stream operators that can convert media streams to feature streams and vice versa. In MedSMan an event has start and end times. MedSMan allows a hierarchical event structure where an event can consist of a set of events.

Based on its three main components, MedSMan defines a query language, which is an extension of CQL [9], to handle new concepts such as feature streams. Using this language, a user can define different queries on top of video streams, and MedSMan evaluates them in real time. In addition to video streams, MedSMan supports audio streams; however, its architecture is not scalable, and it should be extended to accommodate systems with a large number of multimedia sensors.

17.4 SATware Architecture

This section describes the basic building blocks of SATware. These building blocks (Figure 17.1) are organized as a stack of layers where each layer provides an extra level of abstraction of the sensing infrastructure. User applications write queries over the pervasive space being sensed (e.g., using a high-level language such as CQL). These queries are translated into a graph of operators where each operator performs a function on a certain stream of data and produces another stream of data. Translating queries into operator graphs provides general applicability to the different stream data types, and a natural mapping to a distributed execution. The main SATware layers are SATLite, SATDeployer, and SATRuntime.

SATLite provides a language and a language processor for describing graphs of operators. After the operator topology has been expressed in SATLite, each operator is assigned a machine where the operator will be executed. The mapping of operator graphs to machines and the deployment of such operators and establishment of their connections is done by the SATDeployer. SATDeployer uses the methods provided by the SATRuntime layer in order to deploy operators in the network. The SATRuntime layer is distributed along machines in a network (including sensors) and provides a run-time environment where operators can be injected and executed. Through the infrastructure directory, SATRuntime also provides an image of the available sensors, operators, and resources. This information is used by SATDeployer to optimize the operator deployment. The following sections describe in more detail SATware's stream processing model, SATRuntime, SATLite, and SATDeployer. The last section introduces the concept

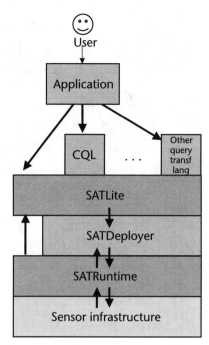

Figure 17.1 SATware architecture.

of privacy in such a pervasive space and its implications for SATware's architecture.

17.4.1 Coffee Room Scenario

Throughout this section, we use a sample application to illustrate different concepts and functionalities of SATware. In this application, a kitchen in a shared office space is instrumented with sensors that are used to monitor the resulting smart kitchen, including the status of various appliances in order to implement various policies of using a shared facility. In particular, video sensors monitor a coffee machine to determine policy violators. Examples of such policies include: "Warn a person who leaves the coffee pot empty with the burner on more than three times," "Charge people for number of cups of coffee they drink," "Determine people who leave the kitchen area dirty," and so forth. As part of an experimental testbed, we have created such a smart kitchen in our workplace. The kitchen is instrumented with a number of cameras and an RFID reader. Based on the reading from the sensors, we have implemented different operators to detect events such as change of coffee level in pot, change of pot position, burner key status switched, and person who entered the kitchen. These operators extract events from video captured by the camera or from identities detected by the RFID reader. Figure 17.2 depicts a snapshot from the camera. Such events are used to enforce the policies of our shared kitchen facility.

Note that SATware can easily be used for processing streams and detecting events in systems with large number of heterogeneous sensors. We use the simple coffee room scenario, since it allows us to illustrate the different concepts and

Figure 17.2   Snapshot of the coffee pot.

mechanisms of SATware without getting lost in the details of the examples itself. The concepts and mechanisms of SATware illustrated within the coffee room scenario are not particular to that scenario; they are generic concepts and mechanisms that are also applicable to more complex scenarios.

17.4.2   Stream Processing Model

This section presents the multimodal stream processing model that enables processing of a broad diversity of sensor data streams in a scalable and flexible manner. It is divided into two parts. First we present the stream data model adopted in SATware. To be able to process multimodal streams originating from different sensors, the system has to be able to accept a wide range of stream types (e.g., video, temperature values, RFID tags). In addition, the system needs to be able to handle event streams generated by the sensors and the system itself.

We then present our data stream processing model that provides both flexibility and scalability. Flexibility is an important requirement since SATware is designed to be a generic data stream processing middleware. This implies that SATware needs to support the seamless addition of a wide range of sensors and applications. A key point to support flexibility is component reusability. This reusability is achieved by writing applications as a set of simple operators wired together. For example, for detecting fires in a building, one can write one operator to get the sensed temperature at the sensor, another to determine if that temperature is unusually high, and another to determine the extent of the overheated region given the location of sensors that had sensed an unusual high temperature. Since each operator is implemented separately, they all can be reused by other applications. For example, getting the temperature sensed at the sensor can be reused for customizing temperature of spaces by controlling the air conditioning.

Writing applications as a set of simple operators wired together also makes the system more scalable.

The large number of sensors producing streams of data and the sharing of the same hardware infrastructure by several applications makes it indispensable to optimize the use of the infrastructure resources. The placement of each operator plays an important role in scalability of the system in terms of bandwidth, computing, storage, and energy. Placing operators as close as possible to sensors results in efficient filtering of information and prevents unnecessary propagation of information in the system. For instance, consider an operator that, given an image of a coffee pot, outputs the level of coffee in the pot. By placing this operator close to the camera (i.e., in the same subnet or even in the camera itself), we can prevent propagation of video data in the system and just forward the desired event.

In SATware, we compose a query using a set of operators. The first step is to determine the set of operators that are required for performing a query. Then the system decides how these operators should collaborate with each other in order to evaluate the query. This is done by chaining operators such that the output of an operator is used as the input of another operator based on the corresponding query. This results in an operator graph that represents relation among deployed operators in the system. Note that for some operators the input comes from sensors. As an example, consider a query in our sample system to detect if a user leaves the coffee pot empty with the burner on more than three times as the operator graph in Figure 17.3. In this example, we have:

1. An operator (*O1*) that gets video frames from the video camera;
2. An operator (*O2*) that gets ID tags from the RFID reader;
3. Three operators (*O3*, *O4*, and *O5*) that detect the status of the burner, the coffee level, and the coffee pot position;
4. Three filters (*O6*, *O7*, and *O8*) that detect the events *burner switched on*, *coffee-pot level changed to empty*, and *coffee pot placed on the burner*;
5. An operator (*O9*) that detects the *burning* event if the coffee pot is on the burner, empty, and the burner is on;
6. An operator (*O10*) that, given a person ID and the *burning* event, detects if a person has let the coffee burn more than three times;
7. An operator (*O11*) that provides a GUI for the user.

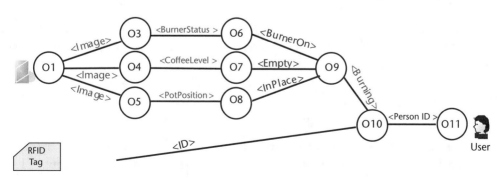

Figure 17.3   The coffee-pot application expressed as a graph of operators.

17.4.2.1 Data Model

In SATware, data streams go through a variety of transformations. To be able to define SATware's processing model, we first need to define the data model for the streams. A stream is an infinite discrete flow of packets, and we define it as an infinite list of packets:

$$Stream := list(packet);$$

A packet is a tuple of the form;

$$packet := <t, c>,$$

where t is a timestamp and c is the packet content. The timestamp indicates the time that the packet content is related to; the packet content is the data the packet carries. The data can be either data of a simple type (i.e., integer, float, character, byte), a list or tuple of a simple type, or an event. Formally, the packet content is defined as:

$$c \in domain(DT)$$
$$DT := integer \mid float \mid character \mid byte \mid void \mid boolean \mid event$$
$$DT := list(DT) \mid tuple(DT)$$

And types can be named:

$$DT := name <= DT.$$

Lists and tuples are a succession of elements. A list is a succession of elements of the same type, and a tuple can have elements of different types:

$$list(type) := t\ instanceoftype => (e1 \in domain(t), \ldots, en \in domain(t));$$
$$tuple(type) := <e1 \in domain(type), \ldots, en \in domain(type)>$$

Events are more expressive than raw data and are either an interpretation of those or of other events. Given the inherent inaccuracy on hardware (sensing circuits, clock synchronization, and so forth) and on software (sampling rate, approximations, and stochastic event detections), events always carry with themselves a confidence level. Formally, an event is a tuple with the following form:

$$event := <event\_id, confidence, DT>.$$

Apart from *event_id* that identifies an event and confidence level, events can also carry some extra data, such as references to streams that provide evidence of the event.

17.4.2.2 Processing Model

Streams are modified by operators. The set of operators is defined as a subset of functions such that every function has arguments of type *DT*, inputs of type *stream*, and an output of type *stream*:

$$OP = \{f_{DTx...xDT} \mid f_{DTx...xDT}: Streamx...xStream -> Stream\}$$

Based on its parameters, each operator performs some task on tuples of the input streams and generates one tuple and sends it as a tuple of its output stream. The well-defined format of parameters and input and output streams for operators makes them independent of specific queries, and they can be plugged and used anywhere in the system. As an example, consider an operator $op1_{http://coffee-cam.org}(\varnothing))$ that connects to an Internet video camera at http://coffee-cam.org and fetches images. Another operator like $op2_{100,60,30}(op1_{http://coffee-cam.org}(\varnothing))$ would be an operator that, given the output from $op1$, detects the amount of coffee in the coffee pot (*100*, *60*, and *30* are calibrating arguments for the image processing algorithm and specify the height (in pixels) of the coffee when the pot is full, half-full, and empty).

17.4.3 Virtual Sensors

In SATware we can encapsulate a query into a single operator: a virtual sensor. This is the similar idea to encapsulation in object-oriented programming and aims to hide complexity of operators and simplify the design of applications. It also speeds up application development in SATware and reduces probability of having faulty applications. Applications can also be modularized and tested independently. Also, reusability is increased since (1) operator topologies (and not only operators) can be now reused, and (2) virtual sensors can be replaced without changing the rest of the application. An example of a virtual sensor is:

$$WhoLeftCoffeeBurning = O10(CoffeeBurning, Person\_Id)$$

Using the virtual sensor concept, SATware provides a dedicated notion of views for sensor data streams. Virtual sensors provide an analog to the data hiding principle in programming languages. This way, virtual sensors provide a controlled avenue toward producing more semantic views closer to the thinking and problem domain of the users of sensor data stream processing applications.

17.4.4 Operator Graph

As seen in Figure 17.3, an application query can be viewed as a directed graph G such that G is a weakly connected graph formed by a set of vertexes V and a set of arcs A. We usually classify the nodes into source nodes, intermediate nodes, and destination nodes. Source nodes get data from sensors—they transform reality (or even a simulated or a predicted reality) into a stream of digitalized sensed values. These usually run at the sensor itself or on a PC in the same subnet. Intermediate nodes transform data streams. Among the intermediate nodes, we identify two special cases: transformers and synchronizers. Transformers are nodes that have only one incoming arc, and synchronizers are nodes with more than one incoming arc. Synchronizers synchronize/join streams depending on values such as time or location. Destination nodes are not connected to other SATware nodes but produce output intelligible for a human user or other applications (e.g., it outputs into an XML file or a database).

17.4.5   Physical Deployment

Once an application query has been designed as a graph of operators, it has to be instantiated in the SATware system. Instantiating an application query means to determine which operator executes in which machine. In order to be able to instantiate it, we need to know which code implements each operator, what is the cost of executing each operator, and what is the current state of our resources. Every operator has some code that implements it as expressed by the *instance* function:

$$instance: OP -> Code$$

Cost functions are defined for every operation. These functions express the cost of the operation in such terms as bandwidth, memory, CPU, and so forth:

$$cost: OP -> R$$

With this, we now define the *mapping* function as:

$$mapping: G \; x \; SATware\_state -> SATware\_state$$

where *SATware_state* is a snapshot of SATware that contains the current operator deployment (I_G), the available sensors (*Sensors*), and the state of the network and machines available to SATware (*Resources*):

$$SATware\_S = <I\_G, Sensors, Resources>$$

Thus, the mapping function decides the optimal (or good enough) way to deploy operators, given the cost of each operator and the current state of SATware in terms of deployed operators and availability of sensors and resources.

17.5   SATRuntime: The Stream Processing Run Time

The previous section described SATware's stream processing model. The model is based on decomposing every query as a graph of simple operators. This processing model allows for the reuse of operators and subgraphs of operators (virtual sensors), as well as efficient decision-making in regards to where each operator executes. Operators with high output rates can be placed in the same subnetwork as the operators that they are linked to, operators that need a lot of storage can be placed on nodes that have a large amount of free disk space, and so forth. Thus, a distributed run time is needed that is aware of the state of both the network and the nodes' resources. Further, it must be able to inject, move, and remove the operators as well as link them.

SATware's run time (SATRuntime) is a reflective distributed run time that meets the above requirements. It has a central directory that contains the characteristics and state of each SATware's resources (sensors, machines, and network) and a repository of operators and virtual sensors. Each operator is implemented as a mobile agent that can migrate to any of SATware's nodes. Mobile agents are autonomous software entities with the capability of dynamically changing their execution environments in a network-aware fashion [14, 15]. This adaptability allows operators to be deployed and dynamically redeployed as needed, to work in isolation when the network is temporarily down, and to migrate operators closer to where data is generated to optimize network resources. Moreover, the fact that agents can be thought of as reactive autonomous entities that know how to perform certain actions (instead of just objects that encapsulate data) simplifies the application design.

17.5.1   Architecture

Figure 17.4 depicts the system architecture. The system nodes are of three types: (1) sensor nodes, (2) processing nodes, and (3) the directory server. Sensor nodes correspond to the heterogeneous set of sensors in the pervasive infrastructure (e.g., Responsphere). These sensors range from RFID readers to video cameras and, thus, the programming platform in each of them is different. Due to this heterogeneity, SATware supports two ways of obtaining the sensed data. The first way is to create an agent gateway that connects to the sensor. For example, consider a network camera that runs a given operating system with a Web server.

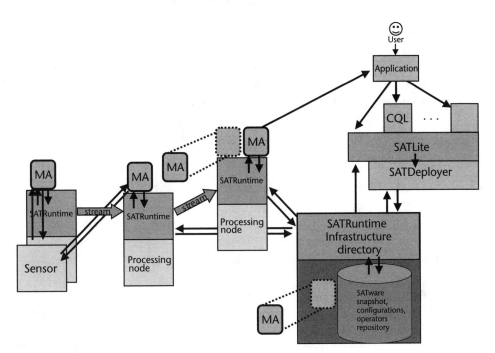

Figure 17.4   SATware-distributed architecture.

A SATware agent can be programmed such that it opens an http connection with the camera Webserver and requests the video being captured. The second way to access the cameras is to extend the SATRuntime in the sensor node. For example, a compatible run time (e.g., written in C or Java) is installed in the camera and a SATware agent is injected. The first allows for faster prototyping while the second one brings the flexibility, adaptability, and robustness of mobile agents all the way to the sensor.

The second type of nodes in SATware are the processing nodes. SATRuntime is installed on each one of the processing nodes, allowing agents to execute in each of them. SATRuntime provides mobility support and message-passing to SATware agents. When deploying a graph of operators, SATRuntime nodes (not the agents) are explicitly connected according to the topology. This way, agents are programmed in a topology-agnostic manner. Agents receive streams of data and output a stream of data, without needing to know the origin or destination of such streams. This simplifies the agent design and implementation, increases reusability, and reduces implementation time, errors, and agent size.

The third component of SATware is the directory server. Mainly, the directory server is an interface to the central directory. The directory server stores a repository of operators and the state of SATware. The repository of operators contains the mobile agent code that, when an agent moves into a SATRuntime node, is downloaded into the run time if it had not been downloaded there before. The state of SATware contained in the directory server includes which sensors are available and how to access them, which processing nodes are available and how much resources they are offering, the network topology and its state, and the current agent deployment. Interfaced via Web services, the directory server provides services such as insert, query, and update types (e.g., stream types, sensor types, machine types, router types, operator types) and instances (e.g., sensors, machines, routers, operators). It also provides Web services for deploying operators and graphs of operators. SATRuntime and other components, such as the query processor, interact with the directory server so both application and processing nodes constraints (e.g., bandwidth, storage) can be met and optimized. In addition to assisting on deploying applications, the directory server provides a permanent storage area where SATware agents can save permanent data (e.g., the encrypted automatons described in Section 17.8).

17.6   SATLite Stream Query and Transformation Language

SATLite stream query and transformation language in SATware provides an interface for describing and instantiating operator graphs. Stream querying and transforming are realized by SATLite language together with SATware agents. There are two types of SATLite users: a user that types commands in SATLite language to deploy, move, or terminate agent operators on the network, and an administrative agent that generates SATLite statements to control the running of other agents on different machines. The former usage facilitates the implementation of a *system dashboard*; the latter could support query plan generator for higher level continuous query languages such as CQL [5] and TelegraphCQ [7].

We assume that all the available operations in the system are known to SATLite. After receiving a query and based on the knowledge about the available operations in the system, SATLite transforms the query into an execution plan in order to deploy it on nodes in the system. Similar to conventional query processing systems, SATLite first parses the query and checks the syntax of it. When the correctness of the query is confirmed, SATLite takes three major steps to transform a query into an execution plan.

In the first step, it detects the required operator that should be used for executing the query. Then each of these operators is instantiated and their parameters are set based on the query. Since an operator may be used for different queries in different locations in the system, to distinguish between different instances of the same operator SATLite assigns unique identifiers for each created instance in this step. Note that the operators in SATware are implemented as mobile agents.

After instantiation of the operators, SATLite sets up the data flow among these operators through using channels. A channel is a data path that has one source and one or more destinations. Channels serve as input and output variables for agent functions. After setting up the data flow among operators, the query is represented as a directed graph. For most of the queries, this directed graph is a tree where its leaves are sensors and its root is the node that has issued the query.

The final step is to check the compatibility of the data flows that set up among operators. Stream compatibility check in SATLite is similar to type checking in a programming language. Before such checking is issued, agent functions (not instances) need to be registered into the directory service with their input/output stream type.

As an example, consider the coffee room scenario (Figure 17.3). Our purpose is to detect whether someone left the coffee pot empty and burning on the coffee machine more than three times. Let us assume that functions that provide certain transformation and synchronization functionality already exist.

For example, function *ReadCam* reads a video stream from a specified camera; function *DetectBurner* detects whether the burner of the coffee machine is switched on; function *And* does the logical *and* operation on multiple boolean inputs. The following statement illustrates how operators are created:

$O1 = CreateOperator(ReadCam, camera\_url);$
$O3 = CreateOperator(DetectBurner).$

The following shows set-up statements for several data flows:

$Image = O1();$
$BurnerStatus = O3(Image);$
$BurnerOn = O6(BurnerStatus);$
$O11(PersonID).$

Notice that $O1$ and $O2$ are sources in the topology graph. There is no stream output from $O11$.

17.7   SATDeployer: Operator Deployment in SATware

In this section we introduce SATDeployer, the operator deployment component in SATware responsible for deploying stream processing operators. SATDeployer translates a query plan (topology) described in the SATLite language into a deployment plan. Besides the query plan, SATDeployer needs information about the current operator deployment, the available sensors, and the state of the network and machines available to SATware. Different objectives can be defined for operator placement in a distributed stream processing system. Minimizing communication cost, latency, or operator computation cost are some of the objectives that can be considered in operator deployment for executing queries.

The main step of query deployment in SATDeployer is to map the query plan graph to the nodes in the network. The result of this mapping is an operator deployment graph. Each node of the deployment graph corresponds to a machine in the SATware network and depicts the operations that should be performed in that machine. The deployment graph is formed in such a way that the objective of the operator placement is satisfied. After forming the deployment graph, SATDeployer, using the API provided by SATRuntime, sends the operators to the corresponding nodes in the SATware network and connects them to each other according to the deployment graph. Then the stream processing to answer the query starts.

Consider a query to detect people who do not pay attention if the pot is burning. The query should be as follows:

SELECT PersonID from KitchenEvents where PotBurning = true

A possible instantiation of this query would use streams from two sensors, an RFID reader and a camera. From the RFID reader we obtain the identity of who is in the kitchen. With the camera, we detect whether the coffee pot is on the burner, the burner is on, and the coffee pot is empty. When there is someone in the kitchen and the pot is left empty on the burner and the burner is on, we want to get the identity of the person who was in the kitchen. The query plan generated by SATLite for this query is depicted in Figure 17.3.

After receiving the query plan from SATLite, it is the responsibility of SATDeployer to decide which operator should be placed on which machine. Consider the SATware network in Figure 17.5. One possible deployment plan is to perform $O3,\ldots,O9$ on the processing node *1,O2* on processing node *2*, and *O10* on the processing node *3*. However, depending on the network situation and other existing operations, SATDeployer may decide to place $O3,\ldots,O9$ on the processing node *3*.

Much research has been conducted on the operator placement problem in recent years [16–19]. Most of the existing work in operator placement concentrates on minimizing communication load, including bandwidth consumption and latency, and it ignores the processing load caused by performing operators in the nodes. However, in SATware applications there are operations, such as event extraction from video stream, that have a high processing load that cannot be ignored.

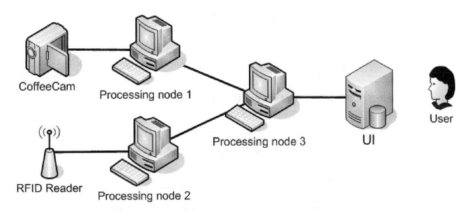

Figure 17.5 SATware resources for the coffee-pot application.

The other important feature that SATware considers is the reuse of the existing operators in the network. Many incoming queries might share some operators. If operator placement for each query is done independently, it may result in replicated execution of shared operators among queries causing higher processing load on the nodes. Since SATDeployer has the knowledge of the deployment graph of the existing queries, by reusing existing operators it can considerably reduce the processing load in SATware.

While processing streams, the SATware network may change, and some nodes may leave the system or fail. Also, because of adding new queries or removing existing queries, the communication and computation load on the nodes and links in the SATware network may change dramatically. In the current version of SATware operators are static. After placing an operator on a node, it stays there for the rest of its query life. However, in the next version of SATware, the operator deployment component will consider changes in the SATware network and adapt the deployment of operators to balance the load on links and nodes. This adaptive operator replacement increases scalability of SATware and makes it resilient to link and node failures.

17.8 Privacy

In this section we illustrate how the SATware system can be configured to implement human-centric applications in a privacy-preserving manner. For example, in context of a surveillance application, the goal could be the detection of events that violate certain policies or rules of the space. In general, policy-violating events are composed of one or more sequences of basic events. We can denote such a composite event by a finite-state automaton where the transitions between its states happen on some event. An application will have to maintain partial information about the state (automatons) of the pervasive environment in order to detect the occurrence of policy-violating composite events. Therefore, the question that arises is the following: "Is it possible to design an event-detection system in a manner that reveals an individual's identity (if at all) only when he violates a policy applicable to him and not at any time before that?" In other

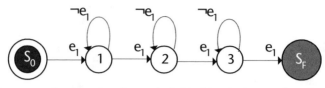

Figure 17.6   Automaton for detecting that an individual has had more than three cups of coffee.

words, the goal of the system would be to implement the set of automatons (corresponding to the set of policies applicable to each individual) in a privacy-preserving manner.

To illustrate, let us go back to our coffee-room scenario. Assume that individuals present in the pervasive space have roles assigned to them such that a set of policies apply to each role, for example, students can have at most three free cups of coffee in a day, and professors will need to pay for each cup. Figure 17.6 shows an automaton for detecting the event when an individual has had more than three cups of coffee. $S0$ and SF denote the start and the end states of the automaton, respectively. The event $e_1 = <t, < Coffee\text{-}drink, 0.9, ('coffee\text{-}room', 'coffee\text{-}pot') >>$ denotes the act of leaving the coffee room with a cup of coffee at time t. The self-loops labeled e_1 denote all other events and act as filtering edges.

We assume the adversary to be an inquisitive but noninterfering (passive) entity that can observe all stored data and intercept all communication between two components of the system. Such an adversary is also assumed to be knowledgeable of the policies and the corresponding automatons that implement these policies. In real life, such an adversary could be an insider, like a system administrator. In the future, we might want to enable various users of the system to execute various kinds of continuous/aggregate queries without violating the privacy of individuals. The problem of determining which queries can be safely executed in a privacy-preserving manner is difficult and will be a topic of our future investigation.

To keep things simple, we will restrict the notion of privacy simply to that of anonymity and the application semantics to that of implementing deterministic finite-state automatons (representing composite events). We say the system ensures *k-anonymity* of individuals if no observable pattern (of automaton access/manipulation plus stored data/meta-data) leads to identifying an individual to within a set of at most k individuals. The problem now is that of designing a system that can store, retrieve, and advance automatons in a k-anonymous manner, that is, where the adversary can at best attribute any event e in the stream to a set of k or more individuals even after observing a very large number of events after e.

To enable privacy-preservation, the system requires one or more trusted components with the capability to encrypt and decrypt strings with a secret key. We assume all trusted components have a common secret key that they can use for carrying out secure communication between themselves. All activity inside the trusted perimeter is assumed to be invisible to the adversary. We have formalized the problem of k-anonymity for such an application in [20], but the model assumes a more centralized system. In that architecture, the sensors that are deployed in the system are considered to be the trusted components, each having a limited amount of computation and storage capability. As a first step toward supporting privacy-preserving applications in the current setting, we propose a straightforward adaptation of the model proposed in [20] to the distributed architecture of SATware as follows.

In [20] the basic events are assumed to be generated in a secure manner by the sensors, which are the trusted components. In the SATware architecture, such a sensor might be a virtual sensor (i.e., a subgraph of trusted operators). The simple approach will be to assume that all nodes deployed in the system are secure, but this will lead to a very expensive solution in practice. Instead, our goal is to minimize the number of trusted operators that need to be deployed in the system and specify the protocols for communication between the trusted and untrusted components in the system. At the current moment we implement a simple solution that processes events involving humans differently from events without human-involvement. All sensors, like camera, RFID readers, and voice-recorders that are programmed to capture individual-centric information are assumed to be tamper-proof. Each such sensitive-information gathering sensor is only connected to trusted nodes (note that some cameras that explicitly focus on areas where no human presence is expected need not be connected to trusted nodes). A trusted node can transmit identifying information only to another trusted node. All other information transmitted from a trusted node to the server or to untrusted nodes (like video stream, RFID data, and so forth) have to be either scrubbed of all identifying information or sufficiently anonymized by employing various data transformations.

A generalization/clustering based scheme to anonymize event streams for a surveillance application are discussed in detail in [20], which is easily applicable in the SATware system as well. But, unlike in that case, the trusted nodes in the SATware system will also require implementing operators that can scrub data streams of identifying information in such a manner that other (untrusted) nodes are able to perform their tasks on the resulting stream. For example, in context of the coffee room surveillance application, a video stream containing an individual as well as the coffee pot might be processed at a trusted node that first scrubs (hides) the human figure from the video frames (Figure 17.7) before passing on the stream to another node that is responsible for detecting whether the coffee pot is empty or full.

Figure 17.7 Human identity anonymized on a video frame.

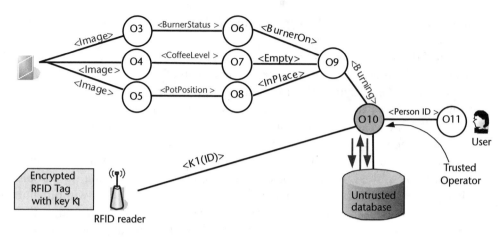

Figure 17.8 Privacy-preserving coffee-pot application.

To detect individual-centric events, which requires the identity of the individual to be known, the system might resolve it through another channel—for example, with the help of a secure operator that maps RFID tags or face images to identities. Consider the setup of the coffee room scenario where the camera is focused only on the coffee pot such that it does not capture video of individuals. Figure 17.8 depicts a privacy-preserving version of the query to detect who leaves the coffee pot burning depicted in Figure 17.3. In this case, RFID tags are printed encrypted with key *K1*. *K1* is a symmetric key known by the trusted operator *O10* and a trusted RFID printer (public-private key pairs could be used if the printer was not trusted). This way, $O10(\_, O2(\emptyset))$ is a trusted virtual sensor and is the equivalent of the trusted sensor in [20].

17.9   Conclusions and Future Work

Sentient spaces augment the users senses and cognizance in both spatial and temporal dimensions. Such spaces are permeated with multimodal sensors that capture the environment and the activities within. In contrast to most sensor middleware prevalent today, sentient spaces users are not provided with streams of raw data but rather with streams of events. Events provide a higher abstraction than raw data and are semantically meaningful. The higher level of abstraction provided by sentient spaces empowers users with the capabilities to perceive and analyze situations without needing to know (and deal with) the details of an underlying heterogeneous sensor network.

To be able to provide such abstractions from the sensor network, sentient spaces need to overcome a series of challenges. These challenges are rooted in the heterogeneity of data streams, sensor hardware, software, and protocols. The dimensions of the sentient space and the wide range of applications of such space are also important factors. Privacy preservation becomes as well a requirement since human and human activities are being sensed in a sentient space.

This chapter first discusses each of the challenges involved in building a sentient space. Given that no current system addresses all of design criteria, we propose SATware: a middleware for sentient spaces. SATware's basic blocks include a distributed run-time system (SATRuntime) that permits injection, execution, and interconnection of data stream processing operators, a declarative language (SATLite) for composition of such operators, an operator deployment module (SATDeployer), and an infrastructure directory for storing the availability of resources.

Currently, SATware is being implemented in the Responsphere pervasive space. Inside Responsphere, we have instrumented a shared kitchen space with a set of sensors including cameras and an RFID reader. Our first applications and proof-of-concept demos have been developed in that kitchen. As part of our ongoing work, we are building larger applications that involve the rest of the instrumented space. Future work involves the research and implementation of optimization algorithms for operator deployment, as well as supporting different network protocols between SATRuntimes to cope with different types of data. Last, privacy is an important issue that has to be taken into account when building pervasive spaces. Here we provide a technical solution for preserving privacy when detecting human-centric events. However, our solution for now deals with a subset of all possible problems. Investigating further privacy-preserving–related issues (such as determining which queries can be safely executed) is also part of our future research.

Acknowledgments

This research has been supported by the National Science Foundation under award numbers 0331707, 0331690, and 0403433.

References

[1] Madden, S., et al., "TinyDB: An Acquisitional Query Processing System for Sensor Networks," *ACM Trans. Database Syst.*, Vol. 30, No. 1, 2005, pp. 122–173.

[2] Hellerstein, J., W. Hong, and S. Madden, "The Sensor Spectrum: Technology, Trends, and Requirements," *ACM SIGMOD Record*, Special issue on sensor network technology and sensor data management, Vol. 32, No. 4, December 2003, pp. 22–27.

[3] Gibbons, P., et al., "Internet-Scale Resource-Intensive Sensor Network Service," http://www.intel-iris.net/research.html, 2006.

[4] Goldsmith, A., et al., "Distributed Vision-Based Reasoning for Smart Camera Networks," wsnl.stanford.edu/index.php, 2006.

[5] Arasi, A., S. Babu, and J. Widom, "The CQL Continuous Query Language: Semantic Foundations and Query Execution," *The VLDB Journal: The international Journal on Very Large Databases*, Vol. 15, No. 2, 2006, pp. 121–142.

[6] Babcock, B., et al., "Models and Issues in Data Stream Systems," *Symposium on Principles of Database Systems,* Madison, WI, May 2002, pp. 1–16.

[7] Chandrasekaran, S., et al., "TelegraphCQ: Continuous Dataflow Processing," *Proc. of the 2003 ACM SIGMOD International Conference on Management of Data,* 2003.

[8]   Liu, B., A. Gupta, and R. Jain, "MedSMan: A Streaming Data Management System over Live Multimedia," *Proc. of the 13th ACM Intl. Conference on Multimedia (MULTI-MEDIA '05),* Singapore, November 2005.

[9]   Arasu, A., S. Babu, and J. Widom, *An Abstract Semantics and Concrete Language for Continuous Queries over Streams and Relations,* Technical Report 2002–57, Stanford University, Stanford, CA, 2002.

[10]   Brown, L., et al., "IBM Smart Surveillance System (S3)," http://www.research.ibm.com/peoplevision/index.html, 2006.

[11]   Carney, D., et al., "Monitoring Streams—A New Class of Data Management Applications," *Proc. of the 28th Intl. Conference on Very Large Data Bases (VLDB 2002),* Hong Kong, China, August 2002.

[12]   Abadi, D., et al., "Aurora: A New Model and Architecture for Data Stream Management," *The VLDB Journal,* Vol. 12, No. 2, August 2003, pp. 120–139.

[13]   Crossbow Technology, Inc., *MPR/MIB User's Manual,* 2004.

[14]   Karnik, N., and A. Tripathi, "Design Issues in Mobile-Agent Programming Systems," *IEEE Concurrency,* Vol. 6, No. 3, July–September 1998, pp. 52–61.

[15]   Fuggetta, A., G. Picco, and G. Vigna, "Understanding Code Mobility," *IEEE Trans. on Software Engineering,* Vol. 24, No. 5, May 1998, pp. 352–361.

[16]   Abrams, Z., and J. Liu, "Greedy Is Good: On Service Tree Placement for In-Network Stream Processing," *Proc. IEEE 26th Int. Conf. on Distributed Computing,* 2006.

[17]   Gu, X., P. Yu, and K. Nahrstedt, "Optimal Component Composition for Scalable Stream Processing," *Proc. IEEE 25th int. Conf. on Distributed Computing Systems,* 2005.

[18]   Pietzuch, P., et al., "Network-Aware Operator Placement for Stream-Processing Systems," *Proc. IEEE 22nd Int. Conf. on Data Engineering,* 2006.

[19]   Amini, L., et al., "Adaptive Control of Xtremescale Stream Processing Systems," *Proc. IEEE 26th Int. Conf. on Distributed Computing,* 2006.

[20]   Hore, B., et al., "Privacy-Preserving Event Detection for Media Spaces" *ICS-UCI* Technical Report, School of Information & Computer Science University of California at Irvine, April 2007, http://www.ics.uci.edu/~bhore/papers/pped-extended-version.pdf.

List of Contributors

Chapter 1: Multimodal Surveillance: An Introduction
Zhigang Zhu (City College, City University of New York)
Thomas S. Huang (University of Illinois at Urbana-Champaign)

Chapter 2: The ARL Multimodal Sensor: A Research Tool for Target Signature Collection, Algorithm Validation, and Emplacement Studies
Jeff Houser and Lei Zong (U.S. Army Research Laboratory)

Chapter 3: Design and Deployment of Visible-Thermal Biometric Surveillance Systems
Diego A. Socolinsky (Equinox Corporation)

Chapter 4: LDV Sensing and Processing for Remote Hearing in a Multimodal Surveillance System
Zhigang Zhu, Weihong Li, Edgardo Molina, and George Wolberg (CUNY City College and Graduate Center)

Chapter 5: Sensor and Data Systems, Audio-Assisted Cameras, and Acoustic Doppler Sensors
Paris Smaragdis, Bhiksha Raj, and Kaustubh Kalgaonkar (Mitsubishi Electric Research Laboratories)

Chapter 6: Audiovisual Speech Recognition
Stephen M. Chu (IBM T. J. Watson Research Center)
Thomas S. Huang (University of Illinois at Urbana-Champaign)

Chapter 7: Multimodal Tracking for Smart Videoconferencing and Video Surveillance
Dmitry N. Zotkin, Vikas C. Raykar, Ramani Duraiswami, and Larry S. Davis (University of Maryland)

Chapter 8: Multimodal Biometrics Involving the Human Ear
Christopher Middendorff, Kevin W. Bowyer, and Ping Yan (University of Notre Dame)

Chapter 9: Fusion of Face and Palmprint for Personal Identification Based on Ordinal Features
Rufeng Chu, Shengcai Liao, Yufei Han, Zhenan Sun, Stan Z. Li, and Tieniu Tan (Chinese Academy of Sciences)

Chapter 10: Human Identification Using Gait and Face
Amit Kale (Siemens Corporate Technologies, Bangalore, India)
Amit K. Roy-Chowdhury (University of California Riverside)
Rama Chellappa (University of Maryland)

Chapter 11: Sensor Fusion and Environmental Modeling for Multimodal Sentient Computing
Christopher Town (University of Cambridge Computer Laboratory)

Chapter 12: An End-to-End eChronicling System for Mobile Human Surveillance
Gopal Pingali, Ying-Li Tian, Shahram Ebadollahi, Jason Pelecanos, Mark Podlaseck, and Harry Stavropoulos (IBM T. J. Watson Research Center)

Chapter 13: Systems Issues in Distributed Multimodal Surveillance
Li Yu (ObjectVideo, Inc.)
Terrance E. Boult (University of Colorado at Colorado Springs)

Chapter 14: Multimodal Workbench for Automatic Surveillance Applications
Dragos Datcu, Zhenke Yang, and L. J. M. Rothkrantz (Delft University of Technology)

Chapter 15: Automatic 3-D Modeling of Cities with Multimodal Air and Ground Sensors
Avideh Zakhor and Christian Frueh (University of California at Berkeley)

Chapter 16: Multimodal Biometric Systems: Applications and Usage Scenarios
Michael J. Thieme (International Biometric Group)

Chapter 17: SATware: Middleware for Sentient Spaces
Bijit Hore, Hojjat Jafarpour, Ramesh Jain, Shengyue Ji, Daniel Massaguer, Sharad Mehrotra, Nalini Venkatasubramanian, and Utz Westermann (University of California at Irvine)

About the Editors

Zhigang Zhu received a Ph.D. in computer science in 1997 from Tsinghua University, Beijing, China. He is currently a professor in the Computer Science Department, the director of the Visual Computing Laboratory, and the codirector of the Center for Perceptual Robotics, Intelligent Sensors and Machines (PRISM), at the City College of New York (CCNY). Dr. Zhu is also on the faculty of Computer Science Department at the CUNY Graduate Center. Previously he has been an associate professor at Tsinghua University and a senior research fellow at the University of Massachusetts, Amherst. From 1997 to 1999 he was the director of the Information Processing and Application Division in the Computer Science Department at Tsinghua University.

Dr. Zhu's research interests include 3-D computer vision, human-computer interaction (HCI), augmented reality, video representations, multimodal sensor integration, and various applications in education, environment, robotics, surveillance, and transportation. Over the past 15 years, he has published more than 100 technical papers in the related fields. Currently, he is mainly working with various research issues and applications of video computing, computer vision, and multimodal sensing fusion, supported by NSF, AFRL, ARO, NYSIA, CUNY Research Foundation, and the Grove School of Engineering at CCNY.

Dr. Zhu was the recipient of the CUNY Certificate of Recognition in 2004, 2005, and 2006; of the Science and Technology Achievement Award from Ministry of Electronic Industry China in 1996; of Outstanding Young Teacher in Beijing award in 1997; and of the C. C. Lin Applied Mathematics Scholarship First-Prize Award at Tsinghua in 1997. His Ph.D. thesis, "On Environment Modeling for Visual Navigation," was selected in 1999 as a special award in the top 100 dissertations in China over the last three years, and a book based on his Ph.D. thesis was published by China Higher Education Press in 2001. Dr. Zhu is a senior member of the IEEE, a senior member of the ACM, and an associate editor of the *Machine Vision and Applications Journal.* He has been involved in organizing such meetings as the IEEE Virtual Reality Conference, the 2004 IEEE International Workshop on Image and Video Registration, and the 2007 IEEE Workshop on Multimodal Sentient Computing: Sensors, Algorithms, and Systems.

Thomas S. Huang received an Sc.D. from MIT in 1963. He is a William L. Everitt Distinguished Professor in the Department of Electrical and Computer Engineering and the Coordinated Science Lab (CSL), at the University of Illinois at Urbana-Champaign, and he is a full-time faculty member in the Beckman Institute and the head of the Image Formation and Processing Group. His professional interests include computer vision, image compression and enhancement, pattern recognition, and multimodal signal processing.

Professor Huang's current research interests lie in three closely related areas: multimodal human-computer interface; 3-D modeling, analysis, and synthesis (animation) of human face, hands, and body; and multimedia (images, video, audio, text) databases. Although these problems are application motivated, the main goal is to develop general concepts, methodologies, theories, and algorithms that would be widely applicable to multimodal and multimedia signal processing in general. Professor Huang's research support includes the NSF, DOD, UIUC Research Board, and a number of industrial firms.

Professor Huang's honors include: member, National Academy of Engineering; foreign member, Chinese Academy of Engineering; foreign member, Chinese Academy of Sciences; IEEE Jack S. Kilby Signal Processing Medal (2000) (corecipient with A. Netravali); International Association of Pattern Recognition, King-Sun Fu Prize (2002); Honda Lifetime Achievement Award (2000); Professor, Center for Advanced Study, UIUC; IEEE Third Millennium Medal (2000); Honda Lifetime Achievement Award (2000); and the IST and SPIE Image Scientist of the Year Award (2006). Professor Huang has also been a keynote speaker at numerous international conferences, such as the IEEE International Conference on Multimedia and Exposition in Toronto in 2006.

Index

Related Titles from Artech House

Concepts, Models, and Tools for Information Fusion, Éloi Bossé, Jean Roy, and Steve Wark

Counterdeception Principles and Applications for National Security, Michael Bennett and Edward Waltz

Information Operations Planning, Patrick D. Allen

Information Warfare and Organizational Decision-Making, Alexander Kott

Knowledge Management in the Intelligence Enterprise, Edward Waltz

Mathematical Techniques in Multisensor Data Fusion, Second Edition, David L. Hall and Sonya A. H. McMullen

Role-Based Access Control, Second Edition, David F. Ferraiolo, D. Richard Kuhn, Ramaswamy Chandramouli

Security in Wireless LANs and MANs, Thomas Hardjono and Lakshminath R. Dondeti

Statistical Multisource-Multitarget Information Fusion, Ronald P. S. Mahler

Wireless Sensor Networks, Nirupama Bulusu and Sanjay Jha, editors

For further information on these and other Artech House titles, including previously considered out-of-print books now available through our In-Print-Forever® (IPF®) program, contact:

Artech House
685 Canton Street
Norwood, MA 02062
Phone: 781-769-9750
Fax: 781-769-6334
e-mail: artech@artechhouse.com

Artech House
46 Gillingham Street
London SW1V 1AH UK
Phone: +44 (0)20 7596-8750
Fax: +44 (0)20 7630-0166
e-mail: artech-uk@artechhouse.com

Find us on the World Wide Web at: www.artechhouse.com